Thomas Twining

Technical Training

Thomas Twining

Technical Training

ISBN/EAN: 9783743687103

Printed in Europe, USA, Canada, Australia, Japan

Cover: Foto ©berggeist007 / pixelio.de

More available books at **www.hansebooks.com**

TECHNICAL TRAINING

BY

THOMAS TWINING

(ONE OF THE VICE-PRESIDENTS OF THE SOCIETY OF ARTS).

BEING

A SUGGESTIVE SKETCH OF A

NATIONAL SYSTEM OF INDUSTRIAL INSTRUCTION,

FOUNDED ON A

GENERAL DIFFUSION OF PRACTICAL SCIENCE

AMONG THE PEOPLE.

LONDON:
MACMILLAN AND CO.
1874.

LONDON: PRINTED BY WILLIAM CLOWES AND SONS, STAMFORD STREET
AND CHARING CROSS.

CONTENTS.

PAGE

INTRODUCTION xiii

List 1: Books and Papers of Reference xviii

CHAPTER I.

INDUSTRIAL INSTRUCTION SYSTEMATIZED. NECESSITY FOR A CENTRAL TECHNICAL UNIVERSITY.

SECTION 1.

THE EDUCATIONAL SYSTEM OF THE FUTURE 1

General educational prospect. Attention to be specially directed to the Technical Element.

SECTION 2.

PRIMARY EDUCATION 3

Prevailing ignorance of Nature's Laws. The kind of Scientific Knowledge required by the working population; its usefulness for general purposes as the SCIENCE OF DAILY LIFE, and for industrial purposes in particular, as a FOUNDATION FOR TECHNICAL STUDIES. Feasibility of introducing it in the Primary Schooling of the Children of the People. Recommendations of H.M. Commissioners on Scientific Instruction. Various progressive measures to be adopted.

SECTION 3.

APPRENTICESHIP 15

Present Defects pointed out, with a view to their being remedied by the proposed System. Protection or *Patronage*.

a 2

SECTION 4.
INSTRUCTION FOR APPRENTICES AND ARTISANS IN GENERAL 19

Review of frequent causes of failure in Science Teaching; necessity for its careful adaptation to special circumstances and purposes. CONSIDERATIONS as to gradational instruction, style of Teaching, selection and condensation v. curtailment, breadth of foundation, links between theory and practice, &c. Value of Apprenticeship.

SECTION 5.
EXAMINATIONS. CERTIFICATES. DIPLOMAS 33

Minor and Major Technical Examinations in Germany. How they might be adapted to our requirements, and what advantages would accrue. Tests for Manual Skill. Trade Certificates of 3 Degrees. Subject Certificates.

SECTION 6.
INDUSTRIES SUFFICIENTLY PROVIDED FOR 43

Review of existing Institutions of which it is desirable to obtain the co-operation in the proposed System. SPONTANEOUS CENTRALIZATION.

SECTION 7.
THREE CATEGORIES OF INDUSTRIES 47

A. Industries already provided for.
B. Normal Industrial Pursuits for which regular Trade Curricula, Examinations, and Certificates will be provided.
C. Unimportant or ill-defined occupations which will only have the benefit of "Subject" Examinations and Certificates.
Occupations included in Category B. COMMERCIAL TRADES.

SECTION 8.
INSTITUTIONS REQUIRED 49

INDUSTRIAL INSTITUTES (including Mechanics' Institutes, Working Men's Clubs, and all analogous Institutions where Science Classes are held); their helps and hindrances; their desirable attributes.

TECHNICAL COLLEGES, their leading functions.

PROVINCIAL and COLONIAL TECHNICAL UNIVERSITIES: their relations to the CENTRAL TECHNICAL UNIVERSITY. Importance of securing UNITY OF PURPOSE, CONCORDANT ACTION, and UNIFORMITY OF STANDARDS.

CHAPTER II.

ORGANIZATION AND OPERATIONS.

SECTION 1.

PRINCIPLES OF ORGANIZATION 59

Union of Efficiency and Economy. The selection of a site and all arrangements to be governed by the resources as well as the requirements of Industrial Students.

SECTION 2.

TRAINING FOR INDUSTRIAL PURSUITS 64

Working of the proposed System illustrated by the example of a Dyer. Instruction offered to Commercial Trades.

SECTION 3.

ART INDUSTRIES 72

Considerations on the use to be made of the valuable existing System of Art Schools. Experience of the Society of Arts as to deficiency of origination. Favourable results to be expected from a scientific training of the mind.

SECTION 4.

TRAINING FOR PROFESSORSHIPS 77

Teaching Staff of various grades and denominations required for Training Colleges, and for Primary and other Schools, for Industrial Institutes and Technical Colleges, for Provincial and Colonial Universities, and for the Central Technical University itself.

SECTION 5.
INDUSTRIAL AND COMMERCIAL ECONOMY 83

Influence to bo exercised for establishing industrial and commercial relations on sound and conciliatory principles.

SECTION 6.
FEMALE TRAINING 86

Training (a) of Females for Industrial Pursuits, (b) of Female Teachers.

Courses of Science and Pedagogy for the Instructresses of the Middle and Upper Classes.

Training of Female Superintendents, Matrons, &c.

SECTION 7.
PUBLICATIONS 91

Text Books for Grades 1 and 2 of GENERAL SCIENTIFIC FOUNDATION.

Sets of Handbooks and Manuals respectively forming Junior and Advanced TECHNICAL CYCLOPÆDIAS.

SECTION 8.
THE MUSEUM AND DEPARTMENTS CONNECTED THEREWITH 93

A. Visual Instruction for the Industrial Institute, the Technical College, and the Central University. Leading Principles 93
B. Rules for Guidance in collecting Materials 97
C. Rules for Guidance in Arrangement and Display 100
D. The Economic Collection 103
E. The Industrial Collection 104
F. The Commercial Collection 104
G. The Art Collections 105
H. The Music Department 108
I. Miscellaneous Departments. Emigration 109
J. Means for propagating Visual Instruction 112

CONTENTS. vii

CHAPTER III.

FACTS IN SUPPORT OF PROPOSALS.

SECTION 1.

ECONOMIC EXHIBITIONS AND MUSEUMS 115

Facts and conclusions resulting from the study, at home and abroad, of popular ignorance, its consequences and its remedies. Plan for Educational Exhibitions of the Requirements of Daily Life approved by the Society of Arts in 1852, and carried out at Paris in 1855, and subsequently elsewhere. Advantages of Permanent Collections. The ECONOMIC MUSEUM. Its destruction. Its lessons.

SECTION 2.

RANGE OF ECONOMIC KNOWLEDGE REQUIRED BY THE INDUSTRIAL POPULATION 122

Principles of selection and classification. Considerations on Dwellings' Reform. Detailed Review of the 9 Classes of the Twickenham Economic Museum, for the purpose of facilitating the formation of similar collections, and the methodical study of the Science of Daily Life.

SECTION 3.

FORM OF INSTRUCTION FOUND MOST SUCCESSFUL 138

Necessity for supporting Visual Instruction in Domestic and Sanitary Economy by its Rationale. Difficulty of conveying the latter to unscientific minds. Desirableness of the co-operation of the Clergy. Attempts to render methodical Science palatable to Working Men. Lectures delivered on the BINARY SYSTEM.

SECTION 4.

APTITUDE OF THE WORKING POPULATION TO RECEIVE APPROPRIATE SCIENTIFIC INSTRUCTION 146

The "Science made Easy" Course, its character and its reception affording undeniable proof of the above aptitude.

SECTION 5.

EXAMINATIONS FOR INDUSTRIAL STUDENTS 152

Open Examinations in general. Frequent drawbacks. Measures of Improvement. Experiment at the Lambeth Baths. The OPEN-HANDED or QUESTIONARY SYSTEM.

CHAPTER IV.

RANGES AND GRADES OF INSTRUCTION. PUBLICATIONS. EXAMINATIONS.

SECTION 1.

THE FIRST GRADE OF GENERAL SCIENTIFIC FOUNDATION. 171

Purpose of this Chapter. Suggestive Syllabus of a GRADE 1 STANDARD COURSE of GENERAL SCIENTIFIC FOUNDATION; *Elementary* Division; *Applied* or *Economic* Division.—Remarks concerning Classification, number of Lectures, difference between Lectures and Lessons, &c.

SECTION 2.

THE SECOND GRADE OF GENERAL SCIENTIFIC FOUNDATION 189

Difference between this and the Grade 1 Course, especially as regards Chemistry. Remarks as to number of Lectures or Lessons.

LIST 2 : Works provisionally available for Grade 2 Studies.

SECTION 3.

THE THREE UPPER GRADES OF SCIENTIFIC INSTRUCTION. 203

Advanced Style of Third Grade Teaching. Existing Books available by means of a SYLLABUS OF REFERENCE. Examples given in LIST 3.

LIST 4 : Works available for Grade 4 Studies.
LIST 5 : Works of a Fifth Grade character.

CONTENTS.

SECTION 4.

ATTAINMENTS NOT INCLUDED IN THE FOREGOING SECTIONS.. 217

Special Scientific Preparation. Mathematical Preparation. Artistic Abilities; indication of five practical stages. Artistic Taste.

LIST 6: Works for special Scientific Preparation.
LIST 7: Works for Mathematical Preparation.
LIST 8: Works for Artistic Studies.

SECTION 5.

SCIENTIFIC TEXT BOOKS 227

Official production of the Grade 1 and Grade 2 STANDARD COURSES; suggestive remarks concerning them. Science popularized. Science for Primary Schools.

SECTION 6.

TECHNICAL TEXT BOOKS AND ILLUSTRATIONS 239

Deficiencies of our Technical Literature. Desiderata:—A JUNIOR CYCLOPÆDIA divided into Handbooks, and an ADVANCED CYCLOPÆDIA divided into Manuals; suggestions concerning them. Scientific and Technical Illustrations.

LIST 9: Sources available for the Cyclopædias: General Reference; Scientific Information; Technology.

LIST 10: Illustrations: Diagrams, &c.; Apparatus and Chemicals.

SECTION 7.

THE EXAMINATIONS 255

Tests of Manual Ability. Tests of Knowledge; difficulties to be surmounted. Proposed adoption of the OPEN-HANDED SYSTEM; measures to that effect. Time required. Examinations carried on personally, or from a distance.

x   *CONTENTS.*

CHAPTER V.

ANALYSES OF TYPICAL OCCUPATIONS.

SECTION 1.

INTRODUCTORY CONSIDERATIONS 260

Preliminary Attainments. Studies for Subject Certificates. Studies for Trade Certificates. Purpose of the proposed Analyses. Order of Trades. Order of Preparatory and Technical Attainments under six headings. Sources of information.

SECTION 2.

SELECT EXAMPLES OF CHEMICAL TRADES 275

The Dyer; Explanatory Remarks on Summaries, p. 281. The Tanner and Currier. Considerations on Manufacturing Industries. The Brewer. Miscellaneous Chemical Industries.

SECTION 3.

TRADES CONNECTED WITH THE CONSTRUCTION, DECORATION, AND FURNISHING OF DWELLINGS 305

The Builder. The Mason. The Brick and Tile Maker. The Bricklayer, Plasterer, &c. Woodworkers in general. The Plumber, the Glazier, the Painter, the Decorator, and Allied Trades. Metal Workers in general. The Thermologist. The Photologist. The Upholsterer.

SECTION 4.

TRADES CONNECTED WITH FABRICS AND CLOTHING 346

Manufacturing Industries; Subject Certificates. Commercial Trades; confused nomenclature. The Tailor. The Hatter. The Shoemaker. Occupations appropriate for Females, including Dressmaking, Millinery and Ladies' outfitting.

CONTENTS. xi

SECTION 5.
TRADES AND OCCUPATIONS CONNECTED WITH FOOD .. 363
 A. Industries yielding Raw Products 364
 B. Preparing and Preserving Industries 364
 C. Manufacturing Industries 368
 D. Commercial Industries 369
 E. Pistrinary and Culinary Industries. Training in Cookery 372

SECTION 6.
INDUSTRIES CONNECTED WITH WARMING, LIGHTING, AND CLEANLINESS. INDUSTRIAL HYGIENE 378

SECTION 7.
INDUSTRIES CONNECTED WITH CART AND COACH BUILDING, SADDLERY, &c. 381

SECTION 8.
INDUSTRIES CONNECTED WITH THE MANUFACTURING AND WORKING OF MACHINERY 384

SECTION 9.
THE WATCH AND CLOCKMAKER 389

SECTION 10.
THE OPTICIAN, THE PHILOSOPHICAL INSTRUMENT MAKER, &c. ELECTRO-TELEGRAPHY 392

SECTION 11.
THE STATIONER. ALLIED INDUSTRIES 396

SECTION 12.
THE PRINTER 399

SECTION 13.
PROFESSIONS AND TRADES CONNECTED WITH MUSIC .. 401

SECTION 14.
ART INDUSTRIES. PHOTOGRAPHY AND THE POLYGRAPHIC ARTS 404

SECTION 15.
COMMERCIAL PURSUITS 411

CHAPTER VI.
RECAPITULATION. CONCLUSIONS.

SECTION 1.
PAGE
QUESTIONS RESULTING FROM THE FOREGOING CHAPTER.

A. MANNER AND PLACE 413
Considerations as to how and where the various Categories and Grades of Industrial Students may best obtain the instruction they require; respective functions of Self-Instruction, and of the Institute, the College, and the University.

B. TIME 427
Leisure-time-Students, and whole-time-Students. How the convenience and means of Artisans may be consulted to the utmost.
Table showing LOCALITY and DURATION of studies.

SECTION 2.
PROPOSED ORGANIZATION OF A CENTRAL TECHNICAL UNIVERSITY 432
Outlines of a Constitution; objects to be effected, and authorities to be constituted.

SECTION 3.
PROSPECTS. PROPOSALS 437
Favourable juncture for instituting a National System of Industrial Training. Initiatory steps. How the measures required may be effected by the concordant action of existing Institutions, and of new ones on Model Principles.

SECTION 4.
CONCLUSION 450
Purpose of the proposed System to assist and complement, not to supersede, existing Institutions. Distinct provinces of usefulness open for South Kensington and for the Central Technical University.

INTRODUCTION.

Not many years ago the present volume would have required another one for its justification. It is written on the assumpton,—1stly—that Scientific Knowledge when judiciously selected and appropriately taught, adds very materially to the chances of Industrial Success,—2ndly—that this maxim having been assiduously adhered to by some Continental Nations, has given a very marked impulse to their industrial prosperity, and 3rdly—that our own can only be secured by following a similar course.—Now, not many years ago, Science, instead of being considered as almost synonymous with Practical Wisdom, was regarded with general distrust, the rapid progress of other Nations was stoutly denied, and it was a mark of patriotism to maintain that our industrial supremacy was and ever would be unshaken. —In vain did reliable Authors faithfully describe the Technical Institutions established in foreign countries, and first predict the probable results, then after a time relate the actual ones. In vain did the walls of the

Society of Arts and other meeting places ring again and again with facts and exhortations. The facts were unpalatable, and the eloquence failed to move unwilling minds that preferred dozing under the drowsy influence of self-reliance.[1] But about the year 1867 the tide began to turn. The information as to the educational resources of foreign Industries, and the wants of our own, which was brought to light in the Reports of the Workmen deputed by the Society of Arts to visit the French Universal Exhibition, helped very materially to bring home the arguments of twenty years to an actual bearing on the public mind, and since then such a flood of evidence has kept pouring in from authoritative sources, that very few persons are now inclined to deny that in the industrial world; "Knowledge really is Power;" that through the Power of Knowledge foreign nations are actually making gigantic strides, and that corresponding measures on our part, on a scale far beyond anything yet attempted, can alone maintain our commercial supremacy.[2]

(1) People were too ignorant of Science even to feel their ignorance. This lamentable fact is forcibly brought into relief by Mr. Bartley's interesting Paper on Science Schools in the 'Society of Arts Journal,' of the 14th January, 1870, where he describes the apathy with which the offers of the Science and Art Department were met by local institutions.

(2) "The facilities for scientific education are far greater on the Continent than in England, and where such differences exist, England is sure to fall behind as regards those industries into which the scientific Element enters. In fact, I have long entertained the opinion that in virtue of the

Whilst however this important change has taken place in public opinion, comparatively little has been actually done. It is true that in the eloquent addresses of which our National Industries have been the never-failing theme, many brilliant ideas have been expressed, that the Books and Pamphlets published in advocacy of Industrial Instruction, contain highly acceptable information concerning what exists abroad, and valuable indications as to what ought to exist at home, and that several exceedingly useful institutions are in course of development; but on the other hand no endeavour has

better education provided by Continental Nations, England must one day—and that no distant one—find herself outstripped by those nations, both in the arts of peace and war."—TYNDALL. Quoted by SMILES, in his Speech at Huddersfield, 1867, which contains much valuable evidence on this subject.

Mr. M. Arnold, quoting M. Duruy, says that the young North German or the young Swiss of Zurich or Basle, is seizing by reason of his better instruction, a confidence and a command in business which the young man of no other nation can dispute with him.—School Inquiry Commission, vol. vii. p. 627.

We read in Dr. Lyon Playfair's Inaugural Address, delivered at Aberdeen in February last, "Britain, whose immense wealth and prosperity hang upon the thread of Applied Science, is far behind France, and infinitely behind Germany."

At a meeting of the Chemical Committee of the Society of Arts on March 20, 1874, Dr. Versmann describing the latest feats of Chemistry in the production of the coal-tar colours, mentioned that only one factory exists in England for the manufacture of Alizarine (artificial Madder), but more than a dozen in Germany and Switzerland.

Further evidence will be found in several of the Books and Papers enumerated at the close of this Introduction, and particularly in the following:—Scott Russell's 'Systematic Technical Education.'—' Reports of the Artisans selected by the Society of Arts to visit the Paris Exhibition of 1867.'—' Modern Industries by 12 British Workmen,' similarly deputed by the Club and Institute Union.—' Replies of the Chambers of Commerce;' and ' Reports of H. M. Consuls.'

apparently been made to select and fit together the most appropriate among these various materials, to supply suggestively whatever connecting links might be necessary for uniting them in a National Scheme of Industrial Progress, and to add whatever practical details might be required for making the various parts of so complex a machinery work smoothly and harmoniously. Such then is the laborious task which I have been induced to undertake in the present Volume. I have been led to it through having for nearly a quarter of a century devoted special attention to the intellectual and physical well-being of the Working Classes, and through having thus arrived at the possession of peculiarly favourable resources. I have also felt greatly encouraged by the flattering manner in which my advice was sought, towards the beginning of 1871, respecting the establishment of a " National University for Industrial and Technical Training." For several months I acted as Honorary Referee to a Committee of Gentlemen associated for that object, and under its sanction my views acquired a certain amount of publicity in the form of a "Preparatory Programme." The movement having been suspended, partly in consequence of the non-fulfilment of a hope that the East London Museum might have been available as a centre of the proposed operations, I have introduced

INTRODUCTION. xvii

in the present volume, various portions of the Programme in question, as well as passages taken from a small book printed chiefly for private circulation in 1870, under the title of "Science for the People." My chief source however, has been a renewed and more exhaustive retrospect of the experiences obtained during a protracted stay in various parts of the Continent, and a long subsequent training as Member of the Committee of the Labourers' Friend Society, and of the Council of the Society of Arts.

I have avoided discussing what is no longer disputed, but on the other hand, I have spared no pains in sifting, classifying, and elaborating into a homogeneous system, the various means by which Industrial Improvement founded on Science, may best be carried into actual effect. My plan is to show in the First Chapter the necessity for a Central Technical University; to sketch in the Second the leading features of its organization and functions; to state in the Third certain facts serving as groundwork to a system of Instruction detailed in the Fourth; to analyze in the Fifth the Educational Requirements of our various Industries; and to endeavour in the Sixth to arrive at legitimate conclusions as to the measures to be adopted.

[LIST 1.]
BOOKS AND PAPERS
RELATING TO THE EDUCATION AND INDUSTRIAL CONDITION OF THE WORKING CLASSES.(1)

Systematic Technical Education for the English People, by J. Scott Russell, F.R.S. (Bradbury, Evans, & Co.) 1869. Price 14/-

Papers relating to Proposals for establishing Colleges of Arts and Manufactures, for the better Instruction of the Industrial Classes, by J. A. Lloyd, F.R.S. London : printed for private circulation by Clowes & Sons, Stamford Street. 1851.
Lectures on the Exhibition of 1851, by Dr. Lyon Playfair.
Industrial Instruction on the Continent by Dr. Lyon Playfair. (Longmans.) 1852.
Report of the Committee appointed by the Council of the Society of Arts, to inquire into the subject of Industrial Instruction. (Longmans.) 1853.
Scientific and Art Education in relation to Progress in Manufactures, Education in England, Foreign Schools, &c. By Alfred Tylor, Deputy Chairman and Reporter of Class XXXI. of the London International Exhibition of 1862.
Speech delivered at the Annual Soirée of the Huddersfield Mechanics' Institute, by Samuel Smiles. Oct. 1867.
Reports of the Artisans selected by the Society of Arts to visit the Paris Exhibition 1867. (Bell & Daldy.)
Modern Industries; a series of Reports on Industry and Manufacture by 12 British Workmen who visited the Paris Exhibition of 1867 in connection with the Club and Institute Union. (Macmillan.) 1868.

(1) Most of the Publications here enumerated may be inspected at the Library of the Twickenham Economic Museum. (See note to p. 121.)

WORKS OF REFERENCE. xix

Reports of the School Inquiry Commission.(1)
Replies of the Chambers of Commerce, in England, to the Circular of the Vice-President of the Committee of Council on Education, on the subject of Technical Instruction. 1867.
Reports of H. M. Consuls on the state of Continental Technical Instruction. 3/-
Report of the Sub-Committee on Technical Education appointed by the Society of Arts.
Report of the Meeting held in support of the Technical Education Movement, at Manchester, Dec. 17, 1869, under the auspices of the Society of Arts.
Report of the Workmen's Technical Education Committee. Club and Institute Union, 150 Strand.
The First, Second, Third, and Fourth Reports of the Royal Commission on Scientific Instruction and the Advancement of Science.
On the Assistance afforded by the State to the Advancement of Science, a paper read by Lieut.-Col. Strange, F.R.S. at a Conference held at the Society of Arts.
On the Necessity for a Permanent Commission on State Scientific Questions, by Lieut.-Col. A. Strange, F.R.S.; a paper read at the Royal United Service Institution, May 15, 1871.
On the Necessity of having a Minister for Science, a Letter to the 'Times' of Feb. 6th, 1874, by Lieut.-Col. A. Strange, F.R.S. reproduced in 'Nature' of Feb. 12th, 1874.
The Educational Calendar and Scholastic Year Book. 1/4 (Kempster, 9 & 10 St. Bride's Avenue, Fleet Street.) Affords a summary insight in the various Educational Institutions and Organizations for Examination.
The Revised Code of the Committee of Council on Education. Published annually. -/2 (Longmans.)
Syllabus of Examination for Students in Normal Schools, including Copies of the Examination Questions.
Science Directory of the Science and Art Department. -/6 South Kensington Museum.(2)

(1) All Reports presented to the House of Commons may be had at Hansard's Office for the sale of Parliamentary Papers, 32 Abingdon Street, S.W. I have found convenient for procuring other official publications, as well as books in general, the agency of Mr. C. Goodman, Bookseller, &c., 407 Strand, W.C.
(2) All Publications of the Science and Art Department are procurable at the South Kensington Museum, and at the official Publishers, Messrs. Chapman & Hall, 193 Piccadilly, W.

Programme of the Science Examinations at South Kensington (see Science Directory).
The Questions of passed Examinations are yearly published; price -/3 each.
Prospectus of the Examinations for Sir Joseph Whitworth's Scholarships in Mechanical Science, conducted by the Science Department, South Kensington. -/3
Art Directory of the Science and Art Department. -/6
A List of Publications issued by the Science and Art Department. Free.
Prospectus of the Royal School of Mines, Jermyn Street.
Annual Report of the Civil Service Commissioners. Office, Cannon Row, Westminister, S.W.
Regulations of the various Civil Service Examinations. (Ibid.)
Prospectus of the Indian Civil Engineering College, Cooper's Hill.
Opening Address to the Students of the Indian Civil Engineering College, by the Duke of Argyll, Aug. 5th, 1871.
Prospectus of the Naval College, Greenwich.
The Calendar of the University of London. 4/- (Taylor & Francis. Red Lion Court, Fleet Street.)
Prospectus of the Society of Arts, John Street, Adelphi.
Yearly Programme of the Society of Arts Examinations, and Examination Papers of passed years.
Reports of the Society of Arts Examinations. See the Society's Journal.
Programme of the Oxford Middle-Class Examinations.
Programme of the Cambridge Middle-Class Examinations.
Calendar and Syllabus of Examinations of the Pharmaceutical Society (17 Bloomsbury Square, W.C.)
Report to the Owens College Extension Committee. 1868.
Prospectus of the College of Physical Science at Newcastle-on-Tyne, affiliated to the University of Durham.
Report of the Liverpool School of Science.
Proposed College of Science in Liverpool. (A Pamphlet obtainable at the School of Science.)
Report of the Bristol Diocesan Trade School Society.
An Educational Experiment at Keighley in Yorkshire. (Bradford, Byles, Printer.) 1872.
History of Adult Education by Dr. Hudson.
Hints to Mechanics on Self-education (Claxton.)
On Primary and Technical Education. Two Lectures, by Dr. Lyon Playfair. Edmonston & Douglas, Edinburgh. 1870.
Suggestive Hints on Secular Instruction, by Dean Dawes.

The Schools for the People, by G. C. T. Bartley. (Bell & Daldy.) 1871.
Transactions of the National Association for the Promotion of Social Science. Annual. Office, 1 Adam Street, Adelphi.
On Museums for Technical Instruction in the Industrial Arts; a Paper by Thomas Webster, read at the Society of Arts, 14 Jan. 1874.
Inaugural Lecture at the Opening of the Industrial and Technological Museum at Melbourne.
Technical Training in Belgium, compiled by Oliver le Neve Foster from materials supplied by T. Twining. Society of Arts Journal, Sept. 1st, 1865.
Brief Outline of the System of Public Instruction in the Netherlands, by Dr. J. Yeats. 1868. Society of Arts Journal, vol. xvi. p. 284.
Technical Education in Austria.—The Imperial and Royal Polytechnic Institution of Vienna.—Society of Arts Journal Sept. 20, 1867.

Letter to the Earl of Shaftesbury on "Means for improving the Efficiency of British Artisans," by T. Twining. 1851.
Notes on the Organization of an Industrial College for Artisans, by T. Twining. 1851.
Letters on the Condition of the Working Classes of Nassau, by T. Twining. London. 1853.
Letter on the Technical Training of Artisans, by T. Twining. Journal of Society of Arts, Jan. 13th, 1865.
Letters on Industrial and Scientific Education, by T. Twining. Journal of the Society of Arts for Dec. 20th and 27th, 1867, and Jan. 3rd and 10th, 1868.
Science for the People, by T. Twining. Goodman, 407 Strand. 1870.
Preparatory Programme of the National University for Industrial and Technical Training, 1871.

On the Education of Women, by Mrs. Wm. Grey. (Ridgway.) 1871. 1/-
Education of Women, a Letter to the "Times," by Mrs. Wm. Grey. Printed separately. -/1.

Prospectus of the National Union for the Education of Women. Office: 112, Brompton Road.

Journal of the Women's Educational Union. -/6 (Chapman & Hall.)

The following Series are sanctioned by the National Female Educational Union, and are published by Ridgway 169 Piccadilly, at -/6 each.
- No. 1. On the special requirements for improving the Education of Girls, by Mrs. Wm. Grey.
- No. 2. Are we to have Education for our Middle-class Girls? by Mary Gurney.
- No. 3. The Work of the National Union, by Emily Shirreff.
- No. 4. The Importance of the Training of the Teacher, by Joseph Payne.

Education of Girls and Employment of Women, by Dr. Hodgson.

Prospectus of the Ladies' Educational Association in connection with University College, Gower Street, London.

The Prospectus of the Queen's Institute for Females. Dublin.

In French.

Les Ouvriers Européens. Études sur les travaux, la vie domestique et la condition morale des populations ouvrières de l'Europe, par M. F. Le Play. Large Folio. Paris. 1855. (Printed at the Imperial Press.)

Les Ouvriers des deux Mondes. A series of octavo volumes published by the "Société internationale des Études Pratiques d'Économie Sociale," in continuation of the researches recorded by M. F. Le Play in his great work 'Les Ouvriers Européens.' (Guillaumin et Cie.) Paris.

Enquête sur l'Enseignement Professionnel, ou Recueil de Dépositions faites en 1863 et 1864 devant la Commission de l'Enseignement Professionnel sous la Présidence de son Exc. M. Béhic.

It contains in two Quarto Volumes a vast amount of valuable information concerning Institutions for Industrial Education in France, Germany, Switzerland, and England, with Synoptical Tables, Plans, &c., the whole preceded by a Report to the Emperor by M. E. Rouher. Printed at the Imperial Press. Paris, 1865.

Annales du Conservatoire des Arts et Métiers.
Bulletins de la Ligue de l'Enseignement, Association pour la Propagation et le perfectionnement de l'Éducation et de l'Instruction en Belgique. Secrétaire Général, M. Ch. Buls, 103 Rue du Marché aux Herbes. Bruxelles.
Projet d'Organisation de l'Enseignement populaire adopté par le Conseil Général de la Ligue de l'Enseignement dans sa Séance du 18 Juillet 1871. Bruxelles.
Prospectus et Dispositions Réglementaires de la Maison de Melle Lez - Gand, Belgique, Institution Littéraire, Scientifique, Commerciale et Industrielle, sous la direction des Joséphites.
Programme des Cours de la Maison de Melle Lez-Gand, Belgique.
Notice sur les Collections Scientifiques, et sur le Musée commercial-industriel de la Maison de Melle Lez-Gand, Belgique.
Bulletin de l'École Industrielle et Littéraire de Verviers, Belgique.

In German.

Pedagogic Annals of the Year 1872, for the Schoolmasters of Germany and Switzerland, elaborated and edited by the late August Lüben of Bremen, in conjunction with Bartholemai, Dittes, Gottschalk, Lehr, Oberlander, Richter, Schlegel, Schulze, and Zimmermann. 25th year. Published by Brandstetter, Leipzig, 1874.
This book, together with the advertisements appended, affords a comprehensive view of the Works used for, or treating of, Popular Instruction in Germany and Switzerland.
A Reading and Class Book for Industrial Advanced Schools, elaborated on behalf of the Central Committee of the Swiss Teachers' Association, by Friedrich Autenheimer, Principal of the Industrial School at Basle, with Woodcuts. (H. Amberger, Basle, 1870.)
This Book gives a good idea of the average attainments of the Swiss Industrial Classes.
The "Bildungs-Verein" (Intellectual Improvement Association). A weekly periodical, being the central organ of intellectual improvement in Germany, and at the same time the Official Journal of the "Society for the Diffusion of Popular Instruction," and of the Institutions connected with it. Edited by Dr. Franz Leibing, 39 Köthener-strasse, Berlin.

This paper affords a lively picture of the progress made by the Industrial Classes of Germany through attendance at Schools of Improvement organized in all parts of the country; a movement mainly due to the exertions of Dr. Leibing. The "Bildungs-Verein" may also be referred to for Technological and other Publications suited for the Libraries of Industrial Institutes.

Draft of Regulations for Improvement Associations (Industrial Institutes) with advice for their Organization.

CHAPTER I.
INDUSTRIAL INSTRUCTION SYSTEMATIZED. NECESSITY FOR A CENTRAL TECHNICAL UNIVERSITY.

SECTION 1.—THE EDUCATIONAL SYSTEM OF THE FUTURE.

THE general Educational Requirements of the Country can only be satisfactorily met by combining in mutual support, as harmonious parts of one great system, the various institutions best calculated to train each portion of the Community to the full enjoyment of its resources, and to the satisfactory fulfilment of its duties.[1]

Of course in a Country where so many educational institutions already exist, and claim respect for their good points or veneration for their past services, where the ground is held by deep-rooted customs, and where vested interests are ever ready to defend their own, the most desirable reforms might be endangered by urging them on too fast, and it would be worse than

---

(1) Respecting the absence of system, forethought, and Scientific Principle, which has hitherto prevailed in educational matters in this Country, it may suffice to quote John Scott Russell, who, in his 'Systematic Technical Education,' page 436, says: "We want ..... the chaos of existing education to be rendered orderly, organized, and systematic," and the Reports of the School Commissioners who pronounce our English Education not a "System" but a "Chaos."

futile to attempt any sudden transition to a perfect state of educational organization. But what can be done, is to devise in conformity with the suggestions of past experience and present knowledge, the main outlines of the EDUCATIONAL SYSTEM OF THE FUTURE. Many portions of this vast scheme may be carried out forthwith, and blend the vigorous activity of new institutions with the tried energy of existing ones. Other portions will be developed as circumstances may permit, and provided all parties and persuasions can be induced to work harmoniously together as far as concurrent views allow, bending their interests to the common weal, we shall rapidly realise a Systematic Educational Organization fraught with the welfare of future generations.[1]

Let our present purpose be to consider what part should be assigned to the TECHNICAL ELEMENT in this great movement, and how the progressive development of our Industrial Prosperity may be ensured by systematically uniting the best and most fruitful influences of Science and Art, inventive ingenuity and practical cleverness, and by effectually diffusing these elements of success throughout the education of the various classes on whose intelligent and united exertions that prosperity depends.[2] And first of all let us follow the career of the Artisan as it were from the cradle upwards, that we can best take note of all educational measures

[1] "The whole national Teaching must be one complete organized whole, or it will prove waste and failure."—'Systematic Technical Education,' page 436.

[2] "The time is not far distant when Science and manipulative skill must be joined together."—HUMBOLDT.

"The nation most quickly promoting the intellectual development of its industrial population, must advance as surely as the country neglecting it must inevitably retrograde."—LIEBIG.

tending directly or indirectly to his improvement. It is thus that we shall best arrive at a clear idea of a National System of Industrial Instruction, culminating in a Central Technical University.

SECTION 2.—PRIMARY EDUCATION.

Success in manhood is greatly dependent on the care bestowed in developing and tempering the mind at an early age, and the way to improvement in Industrial Instruction must be prepared by measures establishing the Primary Education of the People on sound principles, so as to present a foundation at once broad and secure for any future intellectual superstructure.

The present generation is through ignorance of Nature's Laws continually transgressing them, and paying the penalty of it. The rising generation must be saved from this, by being trained to understand the plain interpretation of those Laws, which Natural Science supplies for our benefit. It is of vital importance, individually and nationally, that even the son of the LABOURER should sufficiently understand himself and his surroundings, to know what is good for him; and as he is to be dependent on the produce of his toil, it is essential that his intelligence should be so leavened with the "Rationale of Common Things," as to enable him to do cleverly what he undertakes, suiting his actions to his purpose and his living to his means. A small amount of Science well applied, may make a considerable difference in the duration of a man's life, and a still greater difference in his happiness.

I will not fill pages with an enumeration of the blunders committed and evils incurred by Working Men and their Families in the ordinary routine of Daily Life, and this simply through the want of a good insight into the requirements of the human frame, into the nature of the resources and dangers which surround it, and into the manner of dealing with these respectively. The value of such a guiding insight for all classes, and especially for those who depend on their health and strength for their earnings, and on their own knowledge and intelligence for a good use of them, is too obvious to require lengthy arguments. Equally obvious is it that fallacious guidance would be worse than no guidance at all. Thoroughly sound as well as thoroughly practical must that knowledge be, which is to act as helmsman at all times and in all weathers, and nothing could make it such but a genuine and methodical Scientific Training.

It is not any single Science, though ever so completely mastered, that could serve the purpose; not even Chemistry itself, the Queen of Sciences. If we investigate the circumstances of a Working Man's condition, his inward temperament, and the little world of facts, influences and contingencies with which he is more immediately surrounded, we find here Chemistry required to help him in his daily struggle, there Physics equally indispensable; then, more frequently still, an amalgamation of the two is necessary, whilst everywhere Physiology must be at his elbow, and occasionally Natural History has a word to say.

At first one is startled at the mass of information apparently required; but on close examination one finds that only the fundamental facts and simplest principles of these various Sciences are absolutely necessary, or at all events that with these a large majority of the obstacles which beset the Working Man may be overcome, provided he possesses naturally, or has acquired through training, the faculty, and what is more the habit and the inclination, of thinking logically and quickly, of putting two and two together, and his shoulder to the wheel. One finds that the Physics *he* requires do not necessarily involve any difficult mathematical problems, that *his* modicum of Chemistry leaves untouched considerably more than one-half of the elementary bodies, and a vastly greater proportion of the compound ones, that the long list of organic serials and substitutionals, may be ignored, that for *him* a starch need not be " the oxygen ether, or the anhydride, of a polyglucosic alcohol of a high order," [1] nor the natural fats and oils " tribasic ethers of the triatomic alcohol glycerin,"[2] and that he can even learn what he most wants, without symbols, equivalents or algebraical notations.

It is on investigations and conclusions like these, tested and confirmed by practical trials of the most conclusive character, that is based the methodical selection of scientific Knowledge, *elementary* and *applied*, which will be frequently adverted to in the present Treatise under the title of " The Science of Daily Life." A full Syllabus will be given in Chapter IV., and in the mean time it may suffice to say that :—

FIRSTLY — the ELEMENTARY DIVISION embraces the most essential elements of Mechanical Physics, Chemical Physics, Inorganic and Organic Chemistry, a few outlines of Natural History, and a toler-

---

[1] See Fownes' ' Chemistry ' (Churchills), 10th edition, page 684.
[2] Miller's ' Organic Chemistry ' (Longmans), 4th edition, page 287.

ably comprehensive sketch of Human Anatomy and Physiology.

SECONDLY—the APPLIED DIVISION reviews in regular succession, the applications of these branches of Science to the various departments of Household and Health Economy, making manifest the principles which should prevail in the construction and furnishing of Dwellings, assigning their proper value to the various articles and appliances used for Clothing, Food, Warming, Lighting, and Cleanliness, and explaining as far as necessary their origin or manufacture, introducing the best means of Safety from Harm and Accident, and in short canvassing the most essential requirements of Public, Domestic, Personal, and Industrial Hygiene.

It will be seen by the detailed account given in Chapter III. of the circumstances which led to the adoption of this standard range, of the manner in which I have been induced to embody the Elementary part of it in a Course of Nine Familiar Lectures, and of the unexpected earnestness with which this methodical instruction has been accepted year after year by varied audiences, including some of the humblest classes of the Working Population in some of the least cultivated parts of the Metropolis, that the apprehensions raised by previous experiments, perhaps less carefully made, were groundless, and that the gulf supposed to exist between Science and the Popular Mind was in great part imaginary. Let Elementary Science be carefully selected, arranged in a methodical and progressive sequence, expressed in plain English, and attractively illustrated, and it will obtain ready appreciation even among the most illiterate ranks of the Sons of Toil, provided they be continually made to feel that the knowledge they are imbibing is fraught with practical and tangible benefits, and that there is something in this exposition of the resources opened by nature to the wise, that lifts their thoughts in pleasing emotion towards Him from whom all blessings proceed.

Though peculiarly favourable opportunities had for many years attracted my chief attention to the economic side of Industrial Life, I had never, in devising the manner in which a Working Man could best spend his wages, forgotten that he must first earn them. In fact nearly a quarter of a century ago, and before the formation of my Economic Museum had led me to analyze the requirements and resources of a Working Man's Home, I had occupied myself with those of the Workshop. It was therefore natural that having ascertained the scientific studies required for understanding the rationale of Home Comforts, I should revolve in my mind the possibility of making those same studies subserve the purpose of Technical Training; well knowing that the saving of Time and Teaching Power effected by double-acting instruction might for artisan students render practicable what would otherwise be utopian. I was delighted to find that the more I looked into this question, the more evident it became that the range of knowledge constituting the SCIENCE OF DAILY LIFE, was as good as any that could be devised for serving as a GENERAL SCIENTIFIC FOUNDATION to Technical Studies. It may be seen by the heads of subjects mentioned above, and still better by the detailed Syllabus in Chapter IV., that the Elementary Division of that knowledge forms a preparatory compendium of the branches of natural Science most essentially involved in

Technical Studies, whilst the Applied Division, affording as it does in a familiar way an insight into the most important manufacturing processes connected with Domestic Economy, teaches just that amount of general Technology which is necessary for preventing special Technical Studies from becoming narrow and cramped. These explanations will sufficiently account for my using the expressions, "Science of Daily Life," and "General Scientific Foundation," as nearly synonymous; the former in reference to the general improvement of the condition of the Working Classes, the latter in special reference to Technical Training.

The next question that presented itself to my mind was the following :—this knowledge so pregnant with elevating principles, so useful for a twofold purpose, and which in the form of Familiar Lectures has for several years been so attentively listened to in various parts of the Metropolis by audiences admitted with open doors, and including in many instances minds that were, as far as educational culture is concerned, in comparative childhood,[1]—would it not be feasible to propagate this knowledge far more extensively among the rising generation of the Sons of Toil, as a part of their School Education?[2] In my own mind this question has long been answered in the affirmative, and I feel no diffidence in saying so, seeing that the principle of infusing Science into the minds of Children is distinctly advocated in the Second Report of H.M. Com-

(1) See Chapter III. Sect. 4.
(2) Dr. Lyon Playfair gives in a note to p. 38 of his Lectures on Primary and Technical Education, a highly interesting Summary of the instruction provided by the *Messageries Impériales* of France, for the Children of the Men in their employ. It includes among other subjects, "Man's duty to God, his fellow-creatures, and himself; the Elements of applied Physics; Industrial Chemistry; Industrial Mechanics; the rudiments of Sanitary Science, and Gymnastics."

missioners "On Scientific Instruction and the Advancement of Science," which contains the following

"RECOMMENDATIONS.[1]

"I. We recommend, as regards the elder children in the Elementary Schools, that the teaching of such rudiments of Physical Science as we have previously indicated, should receive more substantial encouragement than is given in the Regulations of the New Code.

"II. We recommend, as regards the younger children, that Her Majesty's Inspectors should be directed to satisfy themselves that such elementary lessons are given as would prepare these children for the more advanced instruction which will follow.

"III. We recommend that the mode of instruction of Pupil Teachers; the conditions of admission to Training Colleges; the duration of the course of study in them; and the syllabus of subjects taught, should be so modified as to provide for the instruction of students in the elements of Physical Science."

The task of enabling the Training Colleges to provide suitable Teachers, is of paramount importance, and will be taken up further on as constituting one of the essential features of the proposed System; but for

[1] It is not generally known how much the Country is indebted to Lieut.-Colonel A. Strange, F.R.S., for originating at the Meeting of the British Association at Norwich, in August 1868, and advocating at the Society of Arts, in March 1870, the important movement of which the furtherance has been entrusted to the above-named Royal Commission consisting of the Duke of Devonshire, Marquess of Lansdowne, Sir John Lubbock, Sir James Phillips Kay-Shuttleworth, Bernhard Samuelson, Esq., William Sharpey, F.R.S., Thomas Henry Huxley, F.R.S., William Allen Miller, F.R.S., and George Gabriel Stokes, F.R.S.—Henry John Stephen Smith, M.A., has since been appointed to succeed Dr. Miller deceased. J. Norman Lockyer, F.R.S., is Secretary to the Commission.

the present let us confine our attention to the manner in which the scientific element might be introduced in the Primary Schools of the People.

Many of the visible and tangible features or properties of COMMON OBJECTS, which children at an early age are found capable of noticing and retaining, may advantageously be amalgamated with illustrations of simple scientific facts, and worked into a kind of visual entertainment or show, in small progressive parts, which will be all the better remembered for being mixed up with anecdote and fun. These properties and facts may further be impressed, whenever opportunities present themselves, or can be created, by exciting little curiosities which Science may be made to satisfy, and by raising little difficulties which Science may be made to overcome: in short, every available means should be pressed into the service for rendering the instruction visible and tangible, impressive and entertaining, and making Scientific Lessons a boon and a treat, never a bore. Evidence is supplied by many existing Schools and especially those conducted on the Pestalozzian system, as to the capability of children for learning things demonstrated visibly and tangibly, vastly better than those which are matters of thought or memory, and especially things of which they see the purpose or enjoy the zest, far better than those which they only know to be deserving of attention because they are told so.[1]

(1) Excellent indications for the early infusion of scientific notions in Children's minds, are afforded by the Dean of Hereford's (Rev. R. Dawes, M.A.) 'Suggestive Hints on Secular Instruction.' His admirable exertions at the King's Somborne School will be ever gratefully remembered by the friends of intellectual progress; and it is to be regretted that the same tact and discrimination have not been exercised by certain advocates of the introduction of Science Teaching in Elementary Schools, whose exaggerated proposals are only calculated to injure the cause.

After a time the properties and facts, which were before matters of opportunity, may be renewed in connected sequence, and separated into groups belonging to distinct Sciences, of which the names will now be introduced. When the ELEMENTARY KNOWLEDGE is sufficiently consolidated, the Teacher will pass on to the APPLIED KNOWLEDGE, still proceeding on the same thoughtful plan of creating a spontaneous absorption of intellectual matter into the brain, instead of the old method of forcible injection.[1] Thus by the expiration of the PRIMARY PERIOD, say by the conclusion of the 12th year of age, he may succeed in storing the minds of his pupils, if not with a very notable, yet with a very serviceable amount of genuine knowledge.[2]

I reckon that these Rudiments of the Science of Daily Life will be duly supplied to schools in the form of appropriate Text Books, but the judicious use of these as regards the distinctions to be made between Boys and Girls, and other particulars, must necessarily form part of the special training of Elementary Teachers alluded to in the above "Recommendations." On many points their judgment will be called into action, and it will be their duty to exercise considerable discrimination

[1] I was told not long ago by a person who in his youth had passed through both methods, that what had profited him intellectually more than all the rest of his schooling, was a series of oral lessons delivered in an interesting way by a Teacher whom infirm health obliged to call the Boys to his bed-side.

[2] I am perfectly aware of the harm done to the cause of Popular Education by proposals for too comprehensive and indiscriminate a selection, and too advanced a treatment of scientific subjects, but it will be shown in Chapter III. Section 3, that what is here described as a serviceable amount of knowledge, has been proved by actual experiment to be susceptible not only of being adapted to the perception of uncultivated minds, but also of being condensed within the compass of a Lecture a week for one short season; so that a competent Teacher would have comparatively easy work by spreading it over the whole period of Elementary Schooling.

and forethought, carefully considering what, in view of the probable condition and occupations of the children, will be most likely to render them essential service, and what can be made lively and interesting, and may have a chance of being understood and retained. What however they should prize and foster more than any amount of facts and precepts stored up in the memory, is a ready ability to bring them to bear on the right purpose at the right time. It is through common sense and presence of mind, that Scientific Knowledge becomes actually and effectively the SCIENCE OF DAILY LIFE; ever thoughtful, vigilant and active, ever ready to secure a legitimate benefit or to avert an injury; having a word of advice for every difficulty, and a word of comfort for every mishap; a trusty guide at all times, and a true friend in the hour of need.

I omit all details as to the manner in which the present routine of Primary Schools may best be accommodated to the introduction of the new subjects. This is a matter not free from difficulty, but I am consoled by a prospect of its being in excellent hands.[1] Already,

---

[1] For information respecting Improvements in the organization of Primary Schools for the children of the People, the comparison of results obtained at small district Schools, and at large amalgamated Schools, and also the results obtained by the Half-time System, and other contrivances for an improved application of Teaching Power, see G. C. T. Bartley's 'Schools for the People,' and various Papers on Primary Instruction by Edwin Chadwick, C.B., and other eminent Educationalists, in the Journal of the Society of Arts.—I would particularly draw attention to the number of that Journal for the 13th of August, 1869, containing the following Notice of a Parliamentary Motion by Dr. Lyon Playfair, of which the importance can only be appreciated by those who have looked into the practical working of the capitation system under the Revised Code. "That in any scheme for National Education, the Revised Code should not limit State aid in elementary schools to the subjects of reading, writing, and arithmetic, but should also offer inducements for the study of such subjects of elementary Science and Art, as bear upon the occupations of the people, and tend to the advancement of industry."

thanks to the exertions of enlightened educationalists, and the examples of judicious reform crowned with the most gratifying success, men's minds are expanding with a conviction that children's minds are also capable of expansion. In proportion as common sense takes the place of prejudice, and that kind of information comes into favour which may best help the Working Man to help himself, and honestly to make his way in the world, room will be provided for the said information by the omission of all that is of questionable utility.[1] Moreover the art of imparting knowledge and of school management in general is undergoing improvements proportionate to the extent to which its importance has been recognized, and it may be hoped that the time is not far distant when all classes of society will practically receive the benefit of the modern educational principles which were thus theoretically described in 1862:—" To instil religious and moral instruction in such a manner as to pervade all the actions, sentiments, and circumstances of life; to supersede as much as possible the mere drilling of the mind by labour without intrinsic purpose; to train the intellect in natural and healthy development by an easy concatenation of ideas, husbanding the powers of the youthful brain, carefully directing them according to capabilities and prospective requirements, and teaching to make an intelligent use of the knowledge acquired. Finally, to couple physical with mental hygiene, and to realise the motto, 'Mens sana in corpore sano.'"[2]

(1) For valuable hints in this direction, I may refer to the speeches at Educational Meetings held by the Society of Arts, and especially to those of the Rev. W. Rogers, whose authority in such matters is undeniable.

(2) See 'Handbook of Economic Literature,' page 7, where a note is appended to the following effect:—"It is perhaps worthy of remark

Though in most country districts it is very difficult to obtain the attendance at School of the son of the Labourer beyond 12 years of age, which may be considered as the conclusion of the ordinary Primary Period, actual examples have proved that it is not too much to assume that Boys intended to become Artisans, may have the benefit of a Secondary Period up to 14, and we may venture to hope that this very capable interval will allow of such a further improvement and consolidation of the Scientific Knowledge previously acquired, as to make it serve as a substantial Scientific Foundation for subsequent Technical Studies.[1] At all events we may reckon that by the time he is transferred from the School to the Workshop, he will have acquired a sufficient perception of the usefulness of Science, to love it heartily, and to be determined to turn to good account any opportunities for regular and earnest study, offered him in the course of his Apprenticeship.

that this system appears to be making more progress in the schools of the Working Classes than in those of higher pretensions. This may in some measure be explained by the consideration that as the poor have neither the power nor the desire to interfere with the instruction bountifully provided for their children, it has happily fallen into the hands of a few eminent men specially qualified by the study of educational science, as well as by enlightened views and Christian benevolence; whereas among the rich, the education of the children is influenced by the notions of the parents who pay for it, and these have hitherto been remarkably slow to patronize improvement in educational methods and appliances." The reason of this is tolerably obvious:—every mind has a good opinion of itself, and of the process by which it has become what it is.

[1] Foreign Educational systems are in many localities so far ahead of our own, that I scarcely like to use them as points of comparison; I may however refer for a variety of suggestive details to the School Regulations respectively adopted by the more advanced of the Swiss Cantons.

SECTION 3.—APPRENTICESHIP.

There can be no question as to the value for technical training, of a *good* system of Apprenticeship,[1] but at the same time no one denies the grievous defects of the present system, both as regards the provisions of the Law, and the customary manner of carrying them out, not to say of neglecting them altogether. Yet this subject does not seem to have received its fair share of the attention bestowed of late on educational matters, and there is still an opening for those who can favour us with practical suggestions on the following and other similar points :—

a. Antiquated and anomalous condition of many laws and enactments relating to Apprenticeship, framed in industrial and social times essentially differing from our own.

b. Grounds for dissatisfaction on the part of the Master.

c. More frequent causes for legitimate complaint on the part of the Apprentice and his friends.

I have reason to believe that the latter category of grievances are frequent items among the cases brought before the Justices of the Peace, and to these therefore

---

[1] " I have the strongest feeling against any attempt to substitute collegiate teaching for practical apprenticeship."—Prof. FLEEMING JENKIN at the Meeting of the British Association in August 1871.

we might look for much valuable information in this matter. It appears that the chief causes of dissatisfaction and dispute are :—*a*. That there is not any suitable criterion by which persons, and especially poor and illiterate persons desirous of apprenticing a lad, can safely judge of the abilities and trustworthiness of the Master to whom they think of confiding him. I wish to lay particular stress on this point, because the proposed system will be found to provide a remedy. *b*. That the Master seeks too much his own profit, employing the lad on errands or other desultory work, to the detriment of his technical progress, to say nothing of his physical comfort and moral improvement. *c*. That Masters who undertake to teach a Trade, frequently know or practise only certain branches of it. *d*. That Masters are apt to show a kind of jealousy of communicating to a clever Apprentice, the whole amount of their own knowledge and ability.[1]

(1) The following observations are borrowed from my 'Letters on the Condition of the Working Classes of Nassau,' printed for distribution in 1853, as a "Report to the Council of the Society of Arts."

As far as I have been able to ascertain, serious disagreements between Masters and Apprentices are less frequent in Germany than with us. This is partly referable to the fact that the incompetency of the Master, which is so frequent a cause of complaint in England, is in some measure obviated in Germany by the Examination which must be undergone before an Artisan can settle anywhere as a Master, and take charge of Apprentices. In all cases redress is facilitated by the practice of paying the stipulated sum *by instalments*, so that one-third or one-half of the amount stands over till the conclusion of the term. If an Apprentice has just cause of complaint, he is released by the local authorities from further obligations towards his Master, and his friends from further payment.

The question of the number of years that should be allowed for Apprenticeships in the various Trades, is one which cannot satisfactorily be discussed till many other points have been settled; but I may mention that in the Duchy of Nassau, the term for ordinary Trades, such as those of the Shoemaker, Tailor, Joiner, Baker, &c., is, or at least used to be, three years; except when the term of four years was agreed upon, without payment, the work of the Apprentice in the last year being expected to

It will be seen as we proceed, that the proposed System is calculated to obviate nearly all these causes of complaint, connecting the respective positions of the Master and his Apprentice, by every appropriate guarantee of reciprocal services and mutual satisfaction. Should the proposed measures prove ineffectual for restoring solidity to a dilapidated Institution, they will greatly facilitate ultimately dispensing with it, but I am hopeful of its restoration. Generally speaking, when a practice, of which the main principle is intrinsically good, has acquired a certain hold on the Population, one must endeavour to take advantage of this circumstance, without however neglecting any new resources that may become available.

Though I must leave for the following Chapters the educational measures to be connected with Apprenticeship, I may here mention two points of a somewhat separate character which claim urgent attention:—1stly. The necessity for a careful revision of the laws relating to Apprenticeship, some of which are absolutely at variance with modern ideas. 2ndly. The desirableness of securing to the Apprentice a moral protection and guidance similar to those afforded in other countries, by Institutions of *patronage* watching over his interests, lending him a helping hand, and anxious to keep him in the right path. Happily a most favourable prospect is opened both in a legislative and a paternal direction, through the laudable desire manifested by our Guilds and Corporations, fully to reassert their antique and honourable privilege of being the pioneers of Com-

form an equivalent. Even for the more difficult Trades, such as those of the Watchmaker, Mechanician, Lithographer, &c., the term of Apprenticeship did not exceed four years.

mercial Progress, and the enlightened and liberal Guardians of Technical Industry.[1]

(1) See the 'City Companies and their Early History,' obtainable at the Society of Arts, John Street, Adelphi, 1874.

The following extracts are from an excellent article on Technical Education in France, borrowed from the 'Pall Mall Gazette' by the 'Journal of the Society of Arts,' for October 3rd, 1873.

1stly. As regards the patronage of apprentices by the Corporations of the olden time; and the state of things resulting from their indiscriminate abolition:

"The old guilds, with their antiquated rules, and their rage for public banqueting, were not perhaps progressive bodies; but in their own rough way they kept an eye on apprentices, reproved and even punished the masters who were remiss · in instructing them, and maintained among the apprentices themselves a wholesome emulation by means of frequent examinations, badges, and money prizes. There were, in fact, trade degrees like those in a university,; and an apprentice, however rich he might be, could only become a master and set up shop after having obtained three certificates of proficiency. The first was bestowed after two years of apprenticeship; the second with a coloured badge, at the end of the fifth year; and the third with a silver badge, when the apprenticeship was concluded in a brilliant manner. . . . . . In the guild times care was taken that every tradesman who accepted apprentices should be thoroughly qualified to teach them. Nowadays, no qualification being needed, it is naturally the most incapable tradesmen who are keenest in trying to secure apprentices by low premiums. They take in a boy as pupil and treat him as a servant, send him out to carry parcels, make him sweep the shop, wash up plates and dishes, and let him pick up many more bad habits than good lessons."

2ndly. As to proposed Technical Schools:

"Apprentice schools, of which two are being founded, one at Havre and the other at La Vilette, in Paris, purpose to cope with this state of things by giving boys a trade education at a cost but little higher than that of the primary education in communal schools. The special aptitudes of boys will be taken to account, and the mechanics set to teach them will be the best that can be procured; the boys will also be admitted very young, so that their training may begin two or three years earlier than a usual apprenticeship. Originally, the committee of gentlemen who started this scheme intended to manage the schools privately by the aid of voluntary contributions; but the Havre and Paris town councils having taken the whole plan under their patronage, and voted funds for the support of the schools, it is probable that the first scheme will be much extended, and that a special school will eventually be set apart for each sort of handicraft."

As an encouraging example of an English Trade School, I may draw attention to the admirable one at Bristol, so deservedly eulogized by the late Canon Moseley, and more recently by Sir John Pakington.

SECTION 4.—INSTRUCTION FOR APPRENTICES AND ARTISANS IN GENERAL.(1)

It is of great industrial importance that the Apprentice should enjoy all reasonable facilities for attending Evening Classes or Lectures, and this equally applies to Working Men in general, who, for want of the opportunities now sought to be established, have hitherto remained under the mark in Scientific Knowledge, and who will at all times be benefited by Instruction of the right sort given in a suitable manner.

Looking around us from this Stand-point, what do we perceive?—As regards the Arts of Design, thanks to the systematic organization of Art Teaching originated through the valuable exertions of Mr. Henry Cole, and of which the Art Schools at South Kensington form the centre, their mode of diffusion leaves little to desire that may not easily be accomplished by the proposed movement.—As regards Science for the

(1) Since this Section was written, I have had the pleasure of seeing its arguments derive authoritative support from the Fourth Report of Her Majesty's Commissioners on Scientific Instruction, &c., the concluding paragraphs of which enunciate the following conclusions:—

"That the diffusion among the people of a general knowledge of science is in itself an object of great importance, and that, in particular, an acquaintance with the manner in which abstract science is brought to bear upon industrial occupations is of the greatest moment to the working classes of this country, not merely as tending directly to increase the skill of the artisan in his handicraft, but as the best means of awakening his intelligence, by forcing him to reflect upon the general laws which are exemplified by the processes with which he is familiar in his daily life.

That no real advancement of knowledge and none of the higher benefits from science as educational discipline are to be hoped for from merely general and occasional scientific instruction, whether it be derived from books or from lectures, but that such advancement and benefits will result only from systematic and sustained study."

People, that is to say, Science selected, arranged and delivered according to the requirements of Youths or Adults of the Working Classes, engaged in or under training for Industrial Pursuits, the case is different. Notwithstanding the activity displayed by the intelligent Scientific Agent from South Kensington, apathy is still slumbering in most parts of the Country, and in many where a fitful impulse has given birth to Science Classes, they have fallen to the ground for want of leading strings. In most instances, the Teaching has been too Technical in the sense of addressing Technical Phraseology to unaccustomed ears, and not sufficiently Technical in the sense of supplying the necessary link between theoretical knowledge and its practical application. In some cases the instruction, guided by a special Professor, has gone the way that suited him rather than the way that suited the wants of general or local industry. There is indeed a tendency among Professors, each to isolate more or less his own Science or branch of Science, upholding it as the one thing which, if thoroughly mastered, would carry all before it, and either ignoring other departments of knowledge, or at all events neglecting to make manifest the admirable results derivable from a proper interweaving and anastomizing of the twigs and branches of the Tree of Science.[1]

More serious is the fact that the non-professional promoters of Technical Instruction and leading patrons of Institutions for its furtherance, have not generally paid sufficient attention to the peculiar conditions which

---

[1] " However lamentable the fact, it is certain that men engaged in one branch of Science are very apt to underrate the importance of all others." —See Lieut.-Colonel STRANGE's Paper on the 'Relation of the State to Science.'

scientific instruction must fulfil, in order that without ceasing to be thoroughly genuine and substantial, it may adapt itself to the more or less divergent requirements of different categories of working-class lads or adults, as well as to their mental culture, their leisure and their means. The fact is that Scientific knowledge has till of late been almost ignored in the Education of the wealthier Classes, and even among those influential persons on whose judgment the direction and character given to popular studies largely depends, very few have taken the trouble to dive into the manifold applications of Science to Arts and Manufactures, or on the other hand to examine the ins and outs of Industrial Life, and to study the grouping of the various categories of youths and adults, according to their present capabilities and prospective requirements. Hence the incongruities which mar many well-meant efforts to make up for former national apathy and neglect: scientific courses of heavy calibre weighing down Schools of Design which Science of a buoyant kind might have raised to still higher success;[1] detached branches of Science offered to those who are ignorant of the necessary preparatory knowledge; three years embraced by a single Course, leaving to chance which part a student may begin with, and making it very uncertain whether he can possibly attend the whole; or intellectual food fit for University Students dealt out to Men and Lads of the Working Class, on the principle that what is intrinsically good, must suit every constitution. Hence also the exaggerated anxiety often manifested to give Working Men the benefit of listen-

(1) I could name a School of Art to which was hooked a Course of Thirty Lectures on merely the first Division of Chemistry, that of the Nonmetallic Elements. It is easy to divine the end of such beginnings.

ing to tip-top Professors.[1] Scientific eminence, and even eminence in lecturing, *ex cathedrâ*, to advanced students, does not necessarily imply ability to render instruction easy and interesting to beginners. It is difficult for a *savant* who through the toil of many years has arrived at a familiar acquaintance with the most recondite depths of science, to resume in thought the unfledged mind with which he began his career, to identify himself with the inexperience of his hearers, and to see the difficulties which they see. He is too apt to deal out instruction in ponderous masses, that neither fit the wants, nor the minds, nor the available leisure of industrial students, or to indulge in lofty theories, and to talk of things that are never seen; or he must at least explain Physics with Mathematics, and Chemistry with what looks to the uninitiated like Algebra. There are brilliant exceptions to this rule, but generally speaking, the Professors best suited for Science Teaching among the Million, are men of incipient fame and moderate pretensions, less remarkable for depth of learning, than for felicity of manner in communicating knowledge, for judgment in selecting what may be most conducive to the future advantage of their pupils, and for tact and good-nature in adapting it to their present capacities.

A clear distinction must be made between *bonâ fide* Science Teaching, and professional lecturing on detached scientific subjects, selected for their sensational character, or for the sake of some local or temporary *àpropos*. Much talent is often displayed in this way, and novelties are brought out with real benefit to the

[1] More than one comparatively unsuccessful scheme has of late years been founded on the idea that the success of Lectures to Working Men, would, like the success of those addressed to the upper classes, be proportionate to the eminent scientific position of the Lecturers.

few among an audience, who possess sufficient preparatory knowledge, while for the rest it is less a question of understanding and retaining useful information, than of being amused. It has been said, with a show of truth, that some audiences like to be mystified, and it is no wonder that such a propensity should be often gratified; but lecturing to those who seek instruction in earnest is a very different thing, and Professor Barff hit the right mark in saying in the introduction to his Cantor Lectures on the Silicates, delivered at the Society of Arts in the Spring of 1872: "My object is not to make you think that I know this subject well, but it is to endeavour to impart to you as much as I can of the knowledge I possess."

Maxims are sometimes laid down, and proposals made by persons not devoid of knowledge, but wanting in special acquaintance with the intricacies of Industrial Instruction, which take aback the educationalist who has studied this particular subject for many a long year, and simply silence him through hopelessness of any attempt at argument. On these I will be silent; but there are also certain points on which divergences of opinion highly detrimental to progress may easily arise, even among well-informed persons, from the mere want of a mutual understanding on the meaning to be attached to certain terms. Thus for instance it is perfectly plausible, and in a broad general sense perfectly true, to say that there is and can be but one Chemistry, one Physics, and so on. Of course the Facts and Laws of Nature are true to their Divine origin, and consequently as immutable as the Divine will; but independently of the variety of deductions and interpretations which human minds draw from those facts or give to those laws, it would be absurd in practically administering scientific instruction to ignore on the one hand the varied *circumstances* and *previous culture* of various classes of Students, and on the other hand, the divergent manner in which the several branches of Practical or Applied Science spring from a common Elementary Stem. It will be my object in the following Chapters, rather to let these things speak for themselves in the course of a methodical and dispassionate review of the educational aspects of Industrial Life, than to substantiate my views by argumentative proof; but for the sake of marking the direction in which we are about to traverse the subject before us, viz. that of " Instruction for Apprentices and Artisans in general," I will note at once a few CONSIDERATIONS which will, I trust, meet with little opposition.

a. It will be seen as we progress, that a vast difference exists in the educational requirements of the various levels of the Industrial Community. In the upper ones we see young men who have been well to do from the cradle, brought up without stint of time or expense, and for whose well-trained minds nothing more appropriate can be recommended in the way of scientific culture, than that supplied by, or under the auspices of, the Science and Art Department. As we descend in the industrial scale, that which is most desirable under the circumstances, takes more and more the place of that which is most desirable *per se*, and we must be content to stoop very considerably if we wish to extend a helping hand to those who are striving under difficulties to climb the first rounds of the intellectual ladder.

b. The rule of progressing from the simple to the complex, and from the known to the unknown, and of

not mixing in a present lesson things that will not be
explained till a future one, if they are to be explained
at all, seems so obviously conducive to satisfactory
progress, that it is surprising to see how often a different course is pursued. It is true that sometimes it
may suit a Teacher's purpose to excite at the first the
earnestness of his Pupils, by making them feel the
awkwardness of attempting Applied Science before
they are grounded in Elementary Science. Thus taking
for example Agriculture, it may be instructive to show
how a mere cursory survey of a Farm reveals at every
step the Tyro's ignorance of one or more of the numerous branches of Natural Science which are interwoven
to form a highly composite intellectual structure; and
the same applies to Physical Geography, Nautical Science, and other analogous intellectual Compounds. But
this mode of rousing the ignorant must be very discreetly employed, even with Students who have time
to spare, and as regards the Lads and Adults of the
Working Classes, I may safely say that with the occasional exception of some preliminary Lecture, full of
striking illustrations, intended to make manifest the
interest of a proposed Course of Studies, a very plain
and straightforward plan should be pursued, of beginning the Alphabet of Science with A, and making each
letter serve as a stepping-stone to the next. I am the
more induced to recommend earnestly that all scientific
instruction delivered under the proposed system should
be progressive, well supported in all its parts, and well
cemented together, as I have come across instances
of laxity and hollowness of teaching where it would
certainly not have been expected.[1]

(1) It is with surprise and admiration that one hears Students after two
or three Lessons in a College Laboratory, talk with ease of the precipitates

c. In Chapter IV. will be found a detailed explanation of five progressive Grades into which it is considered expedient to divide the various levels of scientific Instruction for industrial purposes, from the lowest to the highest. The three upper Grades require scarcely any essential deviation from the ordinary routine of Science Teaching, but Grades 1 and 2 are both designed for those Artisan-Students, and their name is legion, who labour under such deficiencies of culture, and want of means and opportunities for retrieving it, or other disabilities, as to make it a question of *easy* and *short* Science, or no Science at all.

d. By *easy* Science, I mean Science reduced to a simple, familiar, and as far as possible untechnical form of expression, without the sacrifice of any essential fact or principle, and without any laxity or want of method.[1] Sometimes a simple Hypothesis that conveys to the mind a good working idea, may be preferred to a Theory more recent and perhaps more correct, but less susceptible of easy mental manipulation. Thus for instance the idea of a *Ray* of Light may be followed out

---

formed with " Chloride of Barium," " Ferrocyanide of Potassium," &c.; but other feelings supervene when one discovers that they have been promoted at a headlong rate into a mere semblance of Qualitative Analysis, and that their knowledge is confined to the names and special uses of a few common tests, of the general nature and properties of which they are ignorant, as they are indeed of the most elementary chemical truths. Sometimes the Student scarcely knows the appearance of the solid salt of which he has learnt to use the solution.

(1) Professor Odling in the preface to his 'Outlines of Chemistry,' p. 7, allows the use of trivial or popular names for designating familiar articles, and Prof. Barff in his introductory lecture on Silicates, published in the 'Society of Arts Journal' for August 9th, 1872 (p. 766), decidedly advocates the avoidance of technical language, or even of chemical formulæ, where they would prove a barrier to the acquisition of useful knowledge by untutored Students.—See also the 'Advocacy of Simplified Science Teaching,' by T. Sopwith, F.R.S., in the 'Social Science Transactions' for 1870, page 336.

through the vicissitudes of Reflection and Refraction more easily than any representative of the theory of Undulations.[1] In Chemistry in particular we have the sanction of many eminent men, for exercising in the training of juvenile or illiterate Students, a considerable amount of discretion as to the provisional use of simple and clear hypotheses or definitions, and of forms of nomenclature somewhat familiar to the popular ear, in preference to the startling ones suggested by theories or discoveries of recent date.[2] As for the expediency of affording to beginners in Chemistry an insight into the more obvious characters and properties of Bodies, before loading their memory with symbols, equivalents, notations, valencies, and the like, this important question will be duly considered in a more suitable place than the present (Chapter IV., Section 2).[3]

e. By *short* Science, I mean that a careful selection of representative data must be *condensed* into brevity, without being rendered either obscure or dry. I em-

(1) Respecting the use of Hypotheses, see Ganot's 'Physics,' 4th Edition, pp. 579-612.

(2) Thus Miller, in his 'Elements of Chemistry' (paragraph 6, repeated at par. 547), agrees that "any substance which is produced by the action of an acid on a base, is termed a Salt;" though after having discussed the merits of this as well as of the Binary Hypothesis, he confesses that probably neither the old nor the new view is absolutely correct.—Not all *Savants* are so candid, and it is not quite without reason that the author of an article on "Force and Matter," in the 'Rectangular Review,' for January 1871, warns us against the insecurity that will prevail "as long as cosmic speculations are not based on the solid foundation of self-evident natural facts;" and "men will adopt some theory or hypothesis, and then endeavour to force reality into this straight jacket of preconceived opinions."

(3) Some judicious remarks on this subject will be found in Fownes' 'Manual of Chemistry,' tenth edition, p. 237. They conclude as follows:—"So much more difficult is it to gain a clear and distinct idea of any proposition of great generality from a simple enunciation, than to understand the bearing of the same law when illustrated by a single good and familiar instance."

phasize the word "condensed" because a much less troublesome expedient of *curtailment* is frequently resorted to; that is to say, a slice of a Science is given to those who cannot afford time to learn thoroughly the whole. It is of course easy to rig up an apparent justification of this, or of almost any other convenient plan, by combining a few misapplied aphorisms; such as, "half a loaf is better than no bread," "whatever you learn, learn thoroughly," and the like; but these conversational arguments cannot stand for a moment the contact of the stern actualities of Industrial life. Let us take Chemistry, and consider what is demanded of this invaluable Science by the various Artisans who depend on it for their bread. One has chiefly to do with the Metals or their compounds; another with some department of Organic Chemistry, and so on. To tell either that as he has little time to spare, he must content himself with mastering the Non-metallic Elements, might be pronounced ridiculous if there were no one entitled to take offence. It would be equally futile to attempt teaching Organic, without Inorganic Chemistry; and even the acquisition of Inorganic Chemistry without some notion of the Organic Division, would be stinted and lame; so that in fact the only course open for adoption in dealing with Students of lowly culture and limited leisure, is somewhat to the following effect.

1stly. A simple and brief exposition in familiar language, gives to the mind of the Tyro as clear an idea as is possible under the circumstances, of the whole range of Inorganic and Organic Chemistry, bringing into prominence the most important facts and principles, like so many landmarks to be always remembered for safe guidance. This, which will be termed Grade 1, is no more than every one should know.

2ndly. Strengthened by this initiation, the Artisan whose pursuit actually involves Chemical Knowledge, attacks a Grade 2 Course, which, though still reduced to the simplest possible form of expression, is earnest and comprehensive enough for most practical purposes. Possibly he may want only certain parts of it, and in that case he will feel the full benefit of having acquired in Grade 1 a General Foundation capable of binding these parts together. Possibly he may require the whole, and may even have occasion to use his second Grade as a Foundation for Grade 3 Studies. If so, the more intricate theories and abstruse phraseology of this Grade, will (except as regards Etymology) be now no more an impediment to him than they would be to a young gentleman from Eton.[1]

3rdly. The necessity for a General Foundation to special studies, which is so obvious in Chemistry, applies more or less to Natural Science in general, of which the various parts are so intimately connected by mutual support, that they can only be satisfactorily enjoyed or utilized, when the mind has been first prepared by a panoramic view, like that which has been entitled the "Science of Daily Life."

4thly. Again, what applies to Elementary Science holds good with respect to Applied Science, and especially to that scientific insight into the resources and

[1] In alluding to the advantage which a classical education affords for understanding and remembering scientific words by their Etymological construction, I am far from a desire to magnify the disadvantage under which labour those for whom a classical education is simply an impossibility. I believe that much might be done for them by a few methodical lessons, and a good special Handbook of Etymology.

processes of Technical Industry, which mainly constitutes Technology. It will be one of the purposes of the proposed System, to improve the practical intelligence of Artisans by basing a thorough knowledge of what belongs specially to their respective trades, on a certain broad insight into the rationale of Trades in general. This insight will be afforded by the second or Applied Division of what it has been agreed to term General Scientific Foundation (see Section 2). In Grade 1 Studies it will necessarily be slight, but in Grade 2 it will be sufficient to prevent any of that narrow-minded isolation of ideas, which is so apt to be occasioned by the modern plan of reducing, as much as possible, each man's work to a mere fraction of a specialty.

It is in pursuance of the foregoing Considerations, that Grade 1 of the General Scientific Foundation, Elementary and Applied, briefly described in Section 2 of this Chapter,[1] will be rigorously prescribed to all Artisan Students, as a *minimum* of Preparatory knowledge. This may to many seem rather exacting; some will question the desirableness of including a greater breadth of Science than many Trades require; but independently of the economic and sanitary value of the prescribed range, it will unquestionably give a stamen to the Technical Knowledge superposed, which could not be afforded by a narrow base cut close to measure, and without a margin. Some again, will doubt the feasibility of this measure; but they must bear in mind that the kind of Science now before us, though describing Nature and interpreting her laws with perfect conscientiousness, is in outward form and character, very

---

(1) A detailed Syllabus is given in Chapter IV., Section 1.

different from the idea people generally have in their minds in speaking of Science. The *Savant* means the stern and austere Science he has learnt with labour and possesses with pride ; whilst the ignorant man means of course something beyond ordinary comprehension. Both will at first be amazed on hearing of a whole range of Scientific Knowledge proposed for being acquired by Artisans as a mere first step to ulterior knowledge ; but I trust that this surprise will cease when they see actual proof (Chap. III.) that by dint of careful and methodical elaboration, the essential features of the entire range can be reduced within the compass of a child's mind as to difficulty, and of a Working Man's single winter's study as to time ; whilst Visual Instruction in every available form, new contrivances for cheapening and multiplying sound oral instruction, and Books brought out on new principles, will render the young Artisan's Scientific Preparation what it ought to be—a pleasure rather than a toil, and a saving rather than an expense. Each of the foregoing means for rendering Instruction sound, easy, and agreeable, will be treated of in detail in its proper place, and a plan will also be explained by which most of the obstacles presented to Working Men by the present system of Examinations, may be removed.

Unprofitable as it would be to attempt piling up Applied Knowledge without first establishing a solid Elementary Foundation, it would be almost equally illusory as regards industrial requirements, to teach Elementary Science without its applications. It has been sometimes supposed that a mind stored with facts and principles, would of itself work out their applications at the first opportunity of practical employment; but the fact is that pure Science, such as is studied for the love of it, is quite inadequate to help a Man through the difficulties of a complex Trade or Manufacture, unless his mind be supplied with the various links which connect principle and practice. In other words, he must study the technical application, as well as the elementary foundation of the scientific knowledge which that Trade involves.[1] So difficult, however, is it to render technological teaching perfectly practical without actual practice, that even anticipations founded on carefully specialized studies, are sometimes falsified in the Factory or the Workshop. The experience of foreign Technical Institutions is said to show, that even for a superior class of Industrial Students who can afford the expense of a regular academic training, the latter does not satisfactorily supersede a practical Apprenticeship.[2] As regards our Artisans with limited means, it is for the present out of the question to propose either *superseding*, or *preceding* their Apprenticeship with anything like a complete Course of College Training, though well-conducted experiments in either direction may afford useful indications for future guidance, in case it should be found impossible to work the existing Apprenticeship System with satisfactory results.

(1) It will be seen in Chapter V., devoted to an Analysis of the requirements of a number of typical Trades, that in every Curriculum it is intended to follow up Preparatory, with Applied or Technical Studies.

(2) See Professor Jenkin's Address to the Mechanical Section of the British Association, August 1871, in which, referring to the engineering profession, he speaks of young foreigners taught in Colleges, who have afterwards " served their apprenticeship at the cost of their employers."

The Earl of Derby in his speech at the inaugural meeting of the " Society for the Promotion of Scientific Industry," held at Manchester on the 16th January, 1874, expressed himself as follows :—" It is the belief of the promoters of this new society that a great deal may be done for technical training without interfering with that training of the workshop which is, in one sense, the best of all."

SECTION 5.—EXAMINATIONS. CERTIFICATES. DIPLOMAS.

The indications afforded in the foregoing Section, will suffice to give an idea of the leading principles according to which the training of the Artisan, prepared in the School, is proposed to be pushed forward during his Apprenticeship, with the aid of such resources as his locality may afford. The extent to which those resources are susceptible of development, will become apparent in surveying the Institutions of various grades, which under the leadership of a CENTRAL TECHNICAL UNIVERSITY, are intended to conduct the clever and industrious Workman to the top of his profession; but we must first discuss certain highly important movements in the machinery of the proposed system, which will essentially serve to regulate the whole tenour of its influence; namely, the EXAMINATIONS by which Candidates belonging to any branch of Industry, may have their proficiency tested, and the corresponding CERTIFICATES by which their studious exertions will be appropriately recompensed.

Among the technical institutions or customs of the Continent, there is scarcely any one from which a more useful hint may be taken, than the Examinations in technical knowledge combined with manual ability, through which it is or has been the practice in certain countries, to sift the qualifications of matured Apprentices aspiring to become accepted Journeymen, and of experienced and clever Journeymen or Workmen, wishing to take rank as Masters in Trade, to set up shop on their own account, and to become the authorized instructors of another generation of Apprentices.[1] I will transcribe from my 'Letters on the Condition of the Working Classes of Nassau,' the following account of the method in which the Examinations in question used to be conducted at the time when I addressed those Letters to the Council of the Society of Arts, that is to say in 1853. By candidly pointing out certain defects of that method, I hope the better to obtain the approval of a plan for arriving at the proposed purpose more safely, and much more conformably to English notions of commercial freedom.

"At the expiration of his term, the Apprentice must furnish proof of the extent of his acquirements, by executing some appropriate piece of handiwork, in the presence of the official judges of the Trade, forming a kind of jury, which from its usefulness deserves some attention. Every three years the masters in each Trade residing in a district, or in a group of districts if the trade is a scarce one, assemble to elect or re-elect three representatives for the purpose of examining the Certificates, and of testing and recording the abilities of industrial Candidates. Such is the Board of Examiners, which we now find sitting in judgment on the merits of the young artisan anxious to emerge from his apprenticeship, and which we shall meet with again in a further stage of his career. If the Examiners are not satisfied with the young man's performance, he must find means of improving himself within half-a-year, against another trial; if on the contrary, they are well-

(1) In reference to Minor Examinations and Certificates, the terms "Workman" and "Journeyman" will be used as equivalents, according as the one or the other may seem more appropriate.

pleased, he obtains his Certificate of *Gesell*, or Journeyman, and sets out for his travels, or *Wanderschaft*."

As I do not see how the peculiar custom of *Wanderschaft* could be satisfactorily revived to any extent in this country as a regular part of an artisan's training,[1] I refer for an account of it to the Pamphlet just quoted, and proceed at once to take from the same source an account of the second or *major* Examination which the Journeyman must pass to become a Master. * * * " He is required to accomplish single-handed, for strict inspection by the Board of Examiners, some model piece of workmanship, sufficient to show, not merely a moderate amount of skill as when he was a candidate for a Journeymanship, but his thorough knowledge of the *arcana majora* of his calling. If he can follow up the display orally with theoretical evidence, he is entitled to be admitted forthwith to the honourable company of the Masters of the Trade."

Such is the system which, with local modifications, has largely prevailed in Germany, and has worked well, as may be seen by the following extract of a Letter which I received a few years ago from a very intelligent manufacturer of Oschersleben, in Prussia:—

"I enclose an abstract of the law in Prussia concerning the exercise of Trades and Handicrafts. * * * It works admirably, and the public are certain to be served by men who understand their Trade. In your country any person can commence a Trade, whilst in ours he must first show that he has the necessary capacity and knowledge for its successful exercise."

Yet in Nassau and elsewhere this system has been

[1] The word "*Journeyman*" sufficiently indicates the *former* prevalence of the practice in question.

abandoned, and I have been told that in many parts of Germany where it is still kept up, there is a growing inclination to drop it. The reasons however are tolerably obvious, and do not at all affect the good which might be derived from introducing Trades' Examinations in England in an altered form. The chief reason is that in the countries in question, these Examinations have been *compulsory*, and coupled with obnoxious restrictions to the free exercise of handicrafts. Nor could one always rely on the impartiality of the Trades' Juries, especially as regards the admission of Journeymen to Mastership. Suppose for instance a clever Journeyman Shoemaker, who has lately been working at one of the most fashionable establishments of Paris, and who now wishes to set up on his own account in his native German Town, with an ill-disguised intention of importing the goods of his late employer. Now it would be scarcely consonant with our knowledge of human nature, to imagine a conclave of three Master Shoemakers acknowledging the merits of this intruder, and the chances are they will find some good reason why he should stick to his last for a few years longer. Again it is obvious, that any arrangement like that above described, tends to isolate the members of the several trades in the several Towns or Districts, grouping them into close Corporations. Though all will probably be rather antipathetic to innovation and progress, yet some will be more enlightened and clever than others. Consequently they will measure technical abilities, whether mental or moral, by different standards of merit, and thus a Certificate of competency for Journeymanship or Mastership obtained at one place, will have a different significancy and real value from one obtained at another place.

Now two very simple measures would do away with all these inconveniences, and secure for our Industries the unquestionable benefit of a wholesome sifting, without any of the drawbacks resulting from its misapplication. The first is to make the System of Examinations a VOLUNTARY DISCIPLINE, the second is to render the Awards just, reliable, and uniform, and to ensure for the Certificates a standard indisputable value, by placing the examining and awarding power in the hands of a Central Technical University, and of Bodies united to it by the bonds of VOLUNTARY AFFILIATION. The precise meaning attached to these terms, will become apparent as we proceed.

A future Chapter specially devoted to Analyses of the chief types of Technical Industry, will afford abundant evidence of the unquestionable benefit which they may severally derive from that voluntary submission to the discipline of enlightened guidance, which a system of Technical Examinations, carefully organized on liberal and progressive principles, will be the best means of rendering practically effective, and which trustworthy Certificates and Diplomas, conferred with due solemnity, will be the best means of rendering palatable and popular.

I do not wish it to be inferred that these Certificates are equally wanted for all Industries. They are not a matter of urgent necessity for Art Workmen, who can mostly display proofs of their skill on demand; but they will on the contrary be particularly valuable among the Working Trades (especially those of a complex nature), connected with the construction and decoration of Dwellings, with the supply of Furniture, Clothing and Food, and generally with ministering to the direct wants of Daily Life. Here the palpable advantages resulting from a System of genuine Trade Examinations and Certificates, for the Workman who possesses good abilities, for the Master who adds the impress of his own, and for the Customer who pays for the cleverness of both, can scarcely be overrated. Of course the alacrity of the matured Apprentice to complete his attainments up to the standard required for a Minor or Journeyman's Certificate, will much depend on the one hand, on the facilities placed in his way, and on the other hand, on the inclination of the Employer of Skilled Labour to engage a young man with such a Certificate, in preference to one without it. Now seeing how much good-will our Master Tradesmen display in attending to their own interests, and trusting to the progressive diffusion of Science for enlightening them still more, I see no reason to doubt that the possession of such a Certificate would add a good percentage to the marketable value of its owner, and nothing more is required to make the Examinations in question grow popular and flourish. Then, as regards Studies for a Major or Master's Certificate, analogous facilities will induce similar exertions for obtaining a token of merit, to be stamped with its full value by public appreciation. It is true that in pursuance of the principle of allowing to Trade the utmost freedom compatible with public security against ignorance and deceit, the JOURNEYMAN might set up shop on his own account, but a blank escutcheon over his shop-door or window, would declare his small pretensions; whereas the MASTER TRADESMAN would glory in an appropriate BADGE, betokening his possession, not merely of the routine, but of the rationale of his Trade. To understand fully the vast difference this makes in a thousand matters bearing on Health and Comfort, one must have had occasion to study those details of Daily Life which I was led to investigate so minutely in forming my Economic Museum; but every intelligent Customer will *cæteris paribus* prefer the Tradesman of proved abilities to the one who may be an ignoramus, and thus, deserved patronage will remunerate the time spent in well-directed studies.

It is well known that in certain pursuits, ignorance is so dangerous as to render compulsory examinations a matter of necessity, but these are exceptions, and in the general run of the commercial world, it is not easy to prescribe limits to the liberty of the ignorant to deceive the more ignorant, without being accused of interfering with freedom of Trade; still I certainly think it might be made illegal for any person not possessing a proper Major Certificate, to receive an indentured Apprentice.[1]

Doubts have sometimes been expressed as to the possibility of putting Manual Dexterity to the test of a practical Examination, and I am well aware that we must not expect the majority of Industrial Employments to be so easily dealt with in this respect, as the Art Industries which have been submitted year after year to so useful a trial of strength by the Society of Arts, or the artistic Trade of Turnery, prizes for practical excellence in which have been repeatedly distributed by the Lord Mayor of London. Nevertheless, the feasibility of the general principle of obtaining, either through direct Tests or through Certificates from Employers, a reliable criterion of practical skill, may be considered as satisfactorily established by its adop-

[1] Something analogous to what is here proposed, already exists in the Regulations respecting the taking of Pupil Teachers by Elementary Schoolmasters.

tion in Sir Joseph Whitworth's munificent scheme of Competitive Scholarships, as well as in the Technical Examinations organized by Major Donnelly in concert with the Council of the Society of Arts. Moreover it will be found as we proceed that the proposed System includes measures calculated to dispose of every impediment in the way of Industrial Examinations, not only as regards Manual Tests, but also as regards the much more serious difficulties likely to arise from the want of intellectual culture among Candidates of the Minor Degree.(1)

It will be seen in the Analytical Review of Industrial and Commercial Pursuits (Chapter V.) that certain great Manufacturing Industries conducted on the factory plan of subdivided labour, with many *hands* directed by few *minds*, seldom demand a regular Technical Training for the former, but frequently claim for the latter the benefit of Major Examinations and Certificates, and not only this, but afford at the same time scope for the institution of a third and higher Degree of Examinations, at which may be adjudged SUPERIOR CERTIFICATES, or DIPLOMAS OF EXCELLENCE.(2) It will also be seen that certain Manual Trades which like that of the Dyer present in the

---

(1) An account will be given, in Chapter III., Section 5, of experiments successfully tried in London with a new form of Examinations, free from some of the inconveniences most likely to deter working-class Candidates, and in Chapter IV. Section 7 will be indicated the extent to which it is considered that these innovations may be adopted in the proposed System of Technical Training.

(2) These may somewhat correspond to the "Honours" which form the highest Stage or Class in several existing Scientific Examinations. The proposed Minor and Major Degrees will also in some measure correspond to the first (Elementary), and second (Advanced) Stages of those Examinations, except that the Minor Degree will have a lower scientific level than its existing analogues. It will somewhat make up for this by a greater breadth of knowledge.

common way the normal positions of Journeyman and Master, involve when carried out on a large scale, such a variety and depth of scientific attainments, as to be suitably managed only by such a Foreman or Director as would likewise deserve the distinction of a Diploma of Excellence. A similar conclusion will be arrived at on examining the claims, not only of the Builder who has to know more or less the Trades of all the men he employs, but also of the professed Decorator, who should be competent to direct operations involving no inconsiderable amount of chemical rationale, with a taste refined by the study of several branches of high Art. I could add if it were necessary several other instances of Handicrafts rising to the rank of Professions.

Thus we have Certificates of 3 Degrees for Industrial Trades and Occupations, viz. MINOR, MAJOR, and SUPERIOR; and we shall find as we proceed that these Degrees extensively apply to Commercial Trades.—As regards that branch of the proposed operations of the Central Technical University, yielding to none in importance, which will consist in training Science Professors for Normal Schools, Technical Colleges, and other establishments of which it is essential to raise the Teaching Power to uniform efficiency, appropriate Professorial Diplomas will mark a thorough possession of the attainments respectively involved.

The occupations which we have been thus far considering, whether manufacturing, commercial, or professorial, involve each of them a more or less comprehensive range of obligatory attainments to be determined by a Special CURRICULUM. For each Curriculum there will be a corresponding Examination, which from its embracing a range or set of Subjects, might be named a COMPOSITE EXAMINATION, though it will more conveniently be called a TRADE EXAMINATION when it embraces the requirements of a Trade or Industrial Pursuit.(1) But of course only well-defined Pursuits of a certain standing, can be accommodated with special Examinations and TRADE CERTIFICATES. Many are too varying and indefinite, or otherwise too unimportant to expect such a privilege; to say nothing of the mere bodily toil of Labourers, Factory Hands, and others more or less approximating to the condition of living machines.(2) The best thing to be done for an individual belonging to this miscellaneous Category is, *firstly*, to recommend him to acquire that insight into the Science of Daily Life which is particularly valuable when earnings are scanty, and when all the more ingenuity is required to make ends meet; *secondly*, to advise him to add to this indispensable General Scientific Foundation, the knowledge of one or more DETACHED SUBJECTS, selected for him with an eye to what is most likely to be of practical benefit in connection with his work and way of life. In reviewing by and by the various items of scientific and technical Instruction, we shall come across many departments of study suited for being thus chosen for their practical merits. It is proposed to encourage under proper conditions the study of Subjects thus recommended, by means of "SUBJECT EXAMINATIONS" and "SUBJECT CERTIFICATES."(3)

(1) Among the existing examples of Composite Examinations, I may indicate as particularly interesting the Minor and Major Examinations of the Pharmaceutical Society.

(2) As instances of the disconnected bits of Handicrafts, of which numbers float along in the industrial tide, may be named Spur-roll Makers, Bobbin Turners, Scale Board Cutters, Pill Box Makers, Dolls' Eye Makers, Dolls' Wig Makers, Spring Wire Cap-shape Makers, Bookbinder's Plough Knife Makers, Broom-stick Makers, &c.

(3) Information concerning Studies for "Subject Certificates" will be found in Chapter V., Section 1.

SECTION 6.—INDUSTRIES SUFFICIENTLY PROVIDED FOR.

To the *natural* difficulties lying in the way of Industrial Improvement, *unnatural* ones are too frequently added by over-sanguine advocates, who see none at all. The magniloquence with which they speak of sweeping operations more vast than even the field of Industry itself, prepares future disappointment in the more credulous portions of the public, and at once disgusts the more intelligent. What is worse, they give offence where favour should be most earnestly sought, by ignoring the steps already taken in the right direction, and by including in their monopolising grasp, many Industries which either already possess *special* Institutions, or legitimately claim to be thus provided for, on account of the importance or peculiarity of their requirements. I have no doubt that if a suitable central Nucleus of sufficient magnitude can be established in the right place, on thoroughly liberal and disinterested principles, a process of spontaneous crystallisation will gradually bring each of these kindred Institutions to its proper place. Some of them will work best as offsets of the Central University, others receiving its acceptable aid, will in the same measure acknowledge its guiding influence; all will, it is to be hoped, maintain with it a harmonious working connection. But too much caution can scarcely be exercised at the beginning, in order that Brother-workers may be brought to recognize each other as such. I hope that throughout these pages, my wish to see every patriotic effort encouraged and utilised, will be equally apparent with my desire to indicate plainly, and without bias or reserve, the principles which should unite those efforts in concordant usefulness.

The following are a few typical examples of Industrial Pursuits and corresponding Institutions, to which the foregoing considerations may apply.

MINING AND METALLURGY. The high-class studies required either for the training of Special Professors and Teachers for Mining Colleges, or for turning out first class Geological Surveyors and Engineers, or for supplying Men capable of taking charge of large Mining Establishments and Metallurgical Works, may best remain entrusted to the Government School of Mines.[1] There in fact they will have the benefit of the superior scientific culture which may be expected to be centred at South Kensington, in conformity with the first Report of the Royal Commission on Scientific Instruction and the Advancement of Science. This will be in perfect accordance with the principle of the distribution of Teaching Power for the greatest public advantage, which will be one of the fundamental rules of the proposed National System of Technical Training.

CIVIL ENGINEERING. The practical tendency which distinguishes our Indian Government, has lately given rise at Cooper's Hill to a special Engineering College, admirable in its kind, and presenting a highly suggestive model for general imitation.

(1) So great has been the demand for young men of good scientific abilities that those who have creditably passed through a Curriculum of the School of Mines have frequently been diverted from the satisfactory prospects thus opened to them in a direct line by still more promising positions offered in Manufacturing Establishments.

AGRICULTURE. The Agricultural College at Cirencester offers another well-organized example of Studies grouped with a special view to the requirements of a particular occupation, or class of occupations. Its success could but be confirmed by a voluntary connection with the Central Technical University, which it would in some measure represent in a branch of scientific knowledge requiring local facilities for its development.(1)

MARITIME PURSUITS. Among the pursuits appertaining to Maritime Industry, many of a decidedly scientific character will I trust be at once invited, others gradually admitted, to a participation in the advantages prepared for the Navy within the walls of that highly appropriate edifice, Greenwich Hospital, which I have long wished to see thus utilized. Other maritime Pursuits of a more commercial character will receive their best training at the Central Technical University, amidst collections illustrating the resources of the World.

PHARMACY. One of the most interesting examples of a special Technical Institution raised by independent exertions, and one for which a close alliance with the proposed University could not fail to secure a still more prominent place in the great national system of

(1) The following particulars published at the beginning of 1873, serve to remind us how much we have to do to keep pace with foreign progress.

Germany stands foremost among Continental nations in regard to the encouragement of agricultural science, and this is especially true of Prussia. An important School of Agriculture will be opened next April at Bittburg. This establishment, which is largely subsidized by Government, possesses an excellent body of professors. Besides this new school, agricultural institutes are annexed to the Universities of Kiel, Halle, Göttingen, and Königsberg, and to the Academies of Berlin, Rappelsdorf, and Eldena. The independent Academy of Breslau, and thirty agricultural schools, of which seventeen are elementary, complete the ample provision made by Prussia for this important study.

Technical Instruction, is the PHARMACEUTICAL SOCIETY, the judicious exertions of which for affording alike facilities for the Student, and guarantees to the Public, claim attentive consideration and deserve the highest praise.

The Trade or Profession of BUILDER is one which has been specially considered in the provision made for Technical Instruction in some Continental countries, and in favour of which it may not be amiss for the Central Technical University to promote the institution of a special affiliated College.

Whatever degree of autonomy may seem inherent to the branches of Industrial Instruction which have been thus selected by way of illustration, or to others which they may suggest, it is evident that their particular interests, and the great national ones which they are all intended to serve, cannot fail to be equally benefited by bonds of SPONTANEOUS CENTRALIZATION; and the facilities and inducements which will be afforded them by the proposed University for the accomplishment of this desideratum, will not be the least among its claims to universal support.

Considerations analogous to the foregoing, apply to Institutions of a broader character, constituted for, or capable of, undertaking the general furtherance of Technical Knowledge on a Scientific Basis, either by the direct training of the Industrial Population, or by the training of Scientific and Technical Instructors. Several already occupy in the United Kingdom and the Colonies, positions so conspicuous, that it would be natural to mention them; but it would be invidious to attempt arranging them according to relative importance, and difficult to assign limits to the selection. I will therefore only repeat that the Central Technical

University will recognize in the most disinterested spirit, the value of all other Institutions working in the right direction, whilst at the same time it will do all in its power for prevailing on them by legitimate inducements, to join it in carrying out systematically those improvements in Technical Training, which unity of purpose, and concordant action, can alone render great and durable.

SECTION 7.—THREE CATEGORIES OF INDUSTRIES.

If we now recapitulate the conclusions arrived at respecting the claims of various Industries to a share of the advantages derivable from a Central Technical University, we are led to the following Classification:—

CATEGORY A.—Industrial Pursuits, which from their peculiar importance either already have, or deserve to have special Institutions devoted to them. It will behove the Central Technical University to assist the action of these, not to interfere with it.

CATEGORY B.—Normal Industrial Pursuits, of frequent occurrence, and presenting suitable scope for Regular "Curricula of Studies," and Composite or "Trade Examinations," with corresponding "Trade Certificates."

CATEGORY C.—Pursuits of too ill-defined a character, too void of intellectual requirements, or of too unfrequent occurrence for being admitted to Category B, but which may be materially benefited by the study of well-selected "Detached Subjects," with a view to "Subject Certificates."

It may be well further to explain that Category B, which will be chiefly had in view, is intended particularly to comprize the ordinary standard "WORKING TRADES;" as for example, Chemical Trades, like those of the Dyer, the Tanner, the Sugar Refiner, the Brewer, the Distiller, and the Vinegar Manufacturer; to which may in some measure be associated the Pickler, the Confectioner, the Baker, and the Cook;—Mechanical Trades, like those of the Watchmaker, the Wheelright, the Carpenter, the Smith, the Tailor, the Shoemaker, the Miller and the Butcher;—Artistic Trades such as those of the Decorator, the Jeweller, the Silversmith, the Engraver, and the Pattern Designer. To these, many kindred Industries will naturally associate themselves, whilst here and there others of character mixed and undefined, will interpolate themselves with unavoidable awkwardness.

COMMERCIAL TRADES, involving any considerable amount of Technical Knowledge, will also be included in Category B, as for instance those of the Ironmonger, the Retailer of Glass and Ceramic Wares, the Draper and Haberdasher, the Oil and Colourman, the Grocer, and the Wine Merchant. Nor will those Commercial Pursuits be neglected which supply from the abundance of one part of the world the deficiencies of another, and of which the theory is mainly based on History and Statistics.[1]

Industrial Pursuits suited for Females will be included.

(1) For suggestions respecting Commercial Trades, see pp. 70 and 71; also Sect. 5 and Sect. 8 of Chapter II., and Sect. 15 of Chapter V.

SECTION 8.—INSTITUTIONS REQUIRED.

It is now time to consider, from the School upwards, the various Institutions for which educational work has been cut out in the foregoing Sections. It will be remembered that the young Artisan was first conducted through a Primary Schooling, during which the thin end of the wedge of wisdom was inserted by a painless process in his brain. Circumstances allowing, more definite ideas of the nature of Science, and of the value of its applications, were added in a Secondary Schooling up to 14, and we left him undergoing his Apprenticeship under an imaginary model of a Master, who favoured his attending in the Evening, instructive Lectures, or Science and Art Classes. Now Classes of this description, whether or not connected with South Kensington under the title of SCIENCE and ART SCHOOLS, or otherwise, may be devised and got up in many different ways, and held for instance either at the Parish Schoolroom, or at some Mission-room,[1] or at one of the Institutions formerly known as Mechanics' Institutes, and now more frequently denominated " Working Men's Clubs and Institutes." Much depends on the mode of organization of these Establishments, on which point I cannot do better than refer to the papers brought out from time to time under the sanction of the Working Men's Club and Institute Union, 150 Strand, London. There are however one or two points worth discussing

---

[1] I may mention as a suggestive example of scientific teaching in connection with religious instruction, the peculiarly gratifying reception met with by my Course entitled "Science made Easy," at some of the London Mission Halls.

E

as regards the special consideration in which we are now engaged; namely, the affording of permanent facilities for regular and earnest scientific or artistic studies, to Apprentices and Artisans preparing themselves for becoming Candidates at a Minor Examination in their respective Trades:—DRAWING CLASSES can be set up with Models and other necessaries at a very reasonable outlay, thanks to the liberal arrangements of the Art Department at South Kensington, of which the benefits are greatly enhanced by the care bestowed on the production of first-rate articles under enlightened direction.[1]—SCIENCE CLASSES likewise derive valuable assistance from the Subsidies accorded by the Science Department for the erection or adaptation of appropriate buildings,[2] and more particularly from the grants, amounting to as much as 50 per cent. towards the purchase of Apparatus, Diagrams, &c.[3] I trust that when the production of the new sets of Illustrations

---

(1) Messrs. Chapman and Hall, Piccadilly, and Brucciani, Russell Street, Covent Garden, the authorized Publishers for the South Kensington Art Department, are officially supplied with Designs and Models of considerable merit, which they reproduce in large quantities, and are able to sell at very moderate rates, to bonâ-fide Art Classes, of whose outlay as much as 75 per cent. is subscribed by the Department. I could name an Art Class which thus lately obtained for £5 or £6, a set of Models, &c., probably worth £17 or £18, according to the ordinary retail prices.

(2) 'Science Directory,' Nov. 1872, page 26. BUILDING GRANTS. Section LXIII.

"A grant in aid of a new building, or for the adaptation of an existing building, for a School of Science, may be made at a rate not exceeding 2s. 6d. per square foot of internal area, up to a maximum of £500 for any one school, provided that the School

" a. be built under the Public Libraries Act; or—

" b. be built in connection with a School of Art aided by a Department building grant. And provided that there is a population in the neighbourhood which requires a School of Science; that it is likely to be maintained in a state of efficiency; and that the site, plans, estimates, specifications, title, and trust deeds are satisfactory."

(3) 'Science Directory,' Nov. 1872 (pp. 26 and 27). GRANTS FOR APPARATUS AND SCHOOL FITTINGS, Sect. LXIV. "A grant towards School

required for the proposed System of Industrial Education shall have been satisfactorily organized, they may be included in similar grants, and I shall moreover point out in the next Chapter (Section 8, J), further means for reducing considerably the cost of a scientific outfit, without impairing its efficiency. Of course only well-managed Institutions of a substantial character can be relied on for properly developing and maintaining their educational *matériel*, and only such will be meant when in future pages I allude to the important functions to be assigned to INDUSTRIAL INSTITUTES under which title will however be included all Institutions or Establishments suited for undertaking equivalent functions.

One of the difficulties which the Organizers of Industrial Institutes must take into account, is the unwillingness of Adults to associate with Lads in learning things in which they feel that they ought to be considerably ahead of them. Let us suppose a Working Man with good natural abilities, who becomes a Member of an Institute where a studious spirit prevails. He soon discovers that his Reading, Writing, and Arithmetic, the pet R's of the Revised Code, have been taught him at the National School, less with a view to his intellectual development, than to the most lucrative pull at the capitation grant.[1] His Reading has been done by the eye and the mouth, and it is not easy for

---

fittings *of special construction for laboratories or lecture rooms*, and the purchase of apparatus, diagrams, &c., of 50 per cent. on the cost of them, is made to Science Schools and Classes taught by duly qualified teachers under the supervision of Committees approved by the Department. Sect. LXVII. Catalogues containing priced lists of apparatus, instruments, diagrams, books, &c., from various manufacturers have been prepared, and can be had on application."

[1] It is well known to those who are acquainted with the practical working of the present system of Government Subventions according to

him to check the vacant wanderings of his thoughts, concentrating his mind on the sense of what he reads, and impressing facts rather than words on his memory. His specimens of Writing are fair to look at, but to make his pen act as a substitute for his tongue, and convey the gist of what he has been reading, in somewhat like a readable style, approaching to correctness of grammar and orthography, is a very different matter. Accordingly before attending a Science Class, he is anxious to make good the shortcomings of his mental culture; but the very feeling that prompts this anxiety, and makes him so earnest a Student among fellow Men, causes him to shun juxtaposition with youngsters.[1]

Should the proposed improvements in Primary Education take place under the favouring influence of a Code re-revised in accordance with the " Recommendations " of the Science Commissioners, the want of preparatory culture will gradually cease to be an obstacle to the smooth working of Industrial Institutes, and then year after year the indispensable range of Elementary and Applied Science, will be diffused by a running fire of Lectures and Class Lessons to Working Men and Women seeking trustworthy guidance in Daily Life, and to young Artisans intending to raise on this

---

Results, how much can be done by management and tactics for increasing the amount of the Capitation Grant and how depressing is the educational influence of the parity of Grants, for the six different Standards. Then again the three subjects included in each Standard, are not separable. A boy possessing a peculiar talent for one of the three R's, cannot push that branch of his studies to a higher Standard than the other two, but on the contrary he is examined for all three in that Standard in which the Master thinks that his least advanced Subject is likely to pass.

(1) I could cite a case which occurred at Bilston in Staffordshire some 15 or 20 years ago, in which a Reading Class for Adults was kept up for several months with much spirit and success, till a well-intentioned Patron introduced a certain number of youths, when disgust ensued, and the Class was broken up.

secure foundation the fabric of their Technical Knowledge. The extent to which their Training can probably be supplied by Industrial Institutes, will be shown in Chapter VI., but I may at once point to the desirableness of cultivating there if possible, the easier branches of Mathematical Studies, and especially Mensuration. The facilities afforded by South Kensington for the formation of Drawing Classes have already been mentioned. Mechanical Drawing, of which the importance has been so emphatically advocated by Professor Fleeming Jenkin, will naturally be in most places the favourite branch,[1] but in many instances the local Industries will call for Freehand and Decorative Drawing, and the general rules of ARTISTIC TASTE in Outline and Colouring, must nowhere be neglected.

There are many further items in the instructional and recreational Bill of Fare of an Industrial Institute, of which the appropriateness is so obvious that the only question can be that of means and resources. I need scarcely allude to French or German, but will name Music, particularly vocal, as uniting such a variety of advantages when well managed, that it would be indeed a pity not to include it.

Let an Industrial Institute expand by the develop-

[1] See in the 'Journal of the Society of Arts' for August 11th, 1871, a discourse delivered by Professor Fleeming Jenkin at the Mechanical Science Section of the British Association, and in which the following passages occur:—"I cannot say too often, that the great want of the workman is a knowledge of mechanical drawing." . . . "The name of mechanical drawing is given to one and all those representations the object of which is to enable the thing drawn to be made by a workman. Artistic drawing aims at representing agreeably something already in existence, or which might exist, and for the sake of the representation; mechanical drawing aims at representing the object, not for the sake of the representation, but in order to facilitate the production of the thing represented. Now I say that it is this latter kind of drawing which is so vastly important to our artisans."

ment of its intrinsic energies, and let it deserve like that energetic specimen at Keighley in the West Riding of Yorkshire, to have the funds of some obsolete Endowment placed at its disposal by the enlightened discrimination of H. M.'s Endowment Commissioners,[1] or let some other liberal hand sufficiently raise its permanent resources, and it may become a TECHNICAL COLLEGE: that is to say, it may unite with the ordinary means of popular refinement, whatever else is involved up to a certain level in the *regular* Technical Training of Artisans, and become a recognized branch of the proposed National System.

There should be a Technical College in every large Town, say of 30,000 inhabitants and upwards, and populous Cities should have several, which will then take the name of DISTRICT TECHNICAL COLLEGES. Each Technical College, besides maintaining friendly intercourse with neighbours of its kind, and official relations with any Provincial Technical University within reach,[2] should if possible be directly affiliated to the Central Technical University, not only for its own advantage, but also for acting as intermediary in dispensing benefits to the Industrial Institutes of its *arrondissement* or district, of which Institutes it will be considered in some measure as the official guardian and representative.[3]

The Museum department forming an almost indispensable feature of every Technical College, should be

---

(1) See "An Educational Experiment in Yorkshire," reprinted from 'Blackwood's Magazine' for February, 1872. (Byles and Son, Bradford.)

(2) Allusion to this latter class of Institutions will be made presently.

(3) Technical Colleges should either have Industrial Institutes on model principles annexed to them, or include their educational functions. It will be seen as we proceed that similarly each Technical University, or at all events the Central one, should have a model Technical College attached.

carefully developed, and comprise, besides the essential illustrations of the Science of Daily Life, select contributions to the study of the Physical Sciences, both pure and applied, and it should specially include the Technology of the leading local Industries.[1]—Chemical Laboratories will be a necessary appendage of these Institutions, and well-appointed Ateliers would greatly add to their educational value.[2]

It may be found desirable to accommodate populations at a distance from the great industrial centres, by arranging that Examinations of a Minor Degree may be held by delegation at the Technical Colleges; but as a rule everything relating to the determining of Standards, the prescribing of Curricula, and the providing of Text Books, as well as of Certificates and Diplomas, and generally the administration of the proposed System, will rest with the Central Technical University, assisted by the Provincial and Colonial Institutions associated with it in this national undertaking.

Various names have been proposed for the central focus of Technical Instruction. The title of POLYTECHNICUM given to the admirable technical Institution at Zurich, loses perhaps somewhat of its value in this country, by its resemblance to that of the " Polytechnic Institution " of ghostly notoriety. The word

---

[1] For further indications see Chapter II., Section 8.
[2] H. M. Commissioners on Scientific Instruction, &c., in their Fourth Report speaking of towns of considerable population which at present have no museums, or only such as are worthless for purposes of popular education, but which seem well fitted for becoming centres of scientific instruction, proceed to say :—" If a Science School provided with laboratories and a typical museum existed in such a centre, it would exercise a most important influence on the scientific education of the district. The Museum would also be eminently attractive and humanizing as a place of popular resort."

PANTECHNICUM does not seem acceptable. NATIONAL TECHNICAL UNIVERSITY has many points in its favour. It is however very essential to avoid the apparent assumption of any superiority over the Provincial and Colonial Institutions of the kind, other than that arising from a *central* position, and the ability to dispense more benefits to them than they can offer in return. Consequently I think it best to drop the word National for the present, and to adopt either CENTRAL TECHNICAL COLLEGE or CENTRAL TECHNICAL UNIVERSITY. The latter appellation is in some respects preferable. It allows of giving similarly the name of Technical Universities to a certain number of first-class Technical Institutions in various parts of the United Kingdom and the Colonies, thus distinguishing them from the Technical Colleges described above. Using provisionally this nomenclature for convenience, I shall suppose the whole British Empire to be divided into suitable Industrial Provinces, having their respective Provincial or Colonial Technical Universities. These may either be adaptations of, or offsets from Institutions established for general educational purposes, or they may consist simply of special TECHNICAL FACULTIES, working under one roof with Faculties of other descriptions, or they may be constituted in other ways which it is unnecessary to enumerate. The essential condition is that they should be perfectly free to adopt principles of enlightened and energetic progress, and that acting on these they should, for their own and the public advantage, unite with the Central Technical University in such close bonds of amity and association, as to ensure their carrying on the joint task of Technical Instruction, Examinations and Awards, with one spirit and one standard.

Whatever equality of title and pretensions may prevail between the local Technical Universities and the one proposed to be established in the Metropolis, it will evidently be their interest to throw on it a large share of the responsible privilege of initiation, in order that they may claim a large share of those benefits which can only be derived from a certain accumulation of power at a convenient centre. A similar rule will apply to the relations between Provincial or Colonial Universities and the Colleges in their respective spheres of influence; and again between the several Colleges and the Industrial Institutes within their reach, and thus will be established that unquestionable desideratum UNITY OF ACTION through VOLUNTARY CENTRALIZATION.

Hitherto Industrial Instruction has been more discussed than practically pushed forward, and the few attempts at actual progress, have lacked the strength of concerted action. All we now see in the wide domain of our National Industry, is but a comparatively resultless guerilla warfare against ignorance, whilst what we want to see, are the regular tactics of a well-trained host, in numbers and equipment national, and national also in the unity of its strategy.

It would be difficult to overrate the value of the services that may be rendered in this direction by the proposed Central University. Developed on such a scale of magnitude and efficiency, and on such principles of systematic organization, as will entitle it to take rank with the best foreign Institutions of the kind, it will present to its Associates in the Provinces and the Colonies, a model which in their respective degrees they will be proud to emulate, and an exceedingly convenient central repository of progressive knowledge, maintaining a constant intercourse with each of them for the information of all, and always up to the level of the times. Its influence for inducing everywhere the improvement of existing Institutions, and for furthering the establishment of new ones of every calibre on the best principles of organization, will be so much the greater as it will be exercised on purely disinterested and patriotic motives, and backed by liberal assistance,—liberal, but not promiscuous; for every Institution thus benefited will be expected to show corresponding exertions, and to give itself the benefit of the most approved systems of instruction. Greatest however will be the boon conferred by supplying Professors and Teachers corresponding as nearly as possible to the varied nature of local wants and means, yet all specially trained in the art of rendering sound instruction, clear, easy, and entertaining, and of so adapting themselves to their audiences, as by degrees to raise their audiences to themselves. In return for the services rendered to them, the local Centres of Instruction will only be expected to diffuse freely around them the assistance freely received, to report all information likely to be of public benefit, to maintain as far as possible concerted action in all matters in which the general interests of the Industrial Community are involved, and especially to promote everywhere Unity of Principle in administering Industrial Instruction, and Unity of Standard in measuring its results.

CHAPTER II.
ORGANIZATION AND OPERATIONS.

SECTION 1.—PRINCIPLES OF ORGANIZATION.

IT is natural to ask ourselves:—Is this Central Emporium of Technical Knowledge to be provided by Government, and if not, by what other agency? It would doubtless be interesting to discuss at once this and other kindred questions; but it will be safer to let their solution flow spontaneously from a detailed and matter-of-fact examination of all the duties of the proposed Institution, beginning with certain points of organization and mode of action, which have hitherto been only faintly indicated as parts of a general outline.

It may of course be assumed that in organizing an Institution destined to be entrusted with vast influence over the industrial interests of the Empire, no outlay will be spared that may be necessary for securing Men as well qualified for their respective positions, as those composing the Staff of the best foreign Institutions; but in all matters in which Economy can be practised without detriment to real Efficiency, Economy must constantly be the order of the day. The best practical guarantee of a determination to uphold this principle, together with a clear and systematic management of accounts, will be the selection of an economical site. It is to be hoped that whilst it will be healthy and of convenient access, affording ample space for the development of the proposed edifice and appurtenances, on such a scale as to make it a worthy centre for the industrial energy of the most industrial nation in the world, it will at the same time be of so unpretentious a character, that students in fustian may not be considered as intruders, nor smoke and hammering denounced as nuisances. In a building of chaste rather than elaborate design, the most successful features of foreign Technical Institutions will be imitated without being copied. The first-rate instruction given in its Lecture Halls and numerous Class Rooms, will be illustrated with the aid of a Museum composed of Collections specially contrived for the purpose, and conspicuous for their comprehensiveness, yet still more for the principles of thoughtful selection and arrangement which will everywhere be made manifest. Laboratories and Workshops of the most approved construction, will be provided for the various specialities, with an unassuming completeness marking the actuality of their purpose; and if found feasible, the principle adopted at the Cornell University will be introduced, of affording facilities for alternating study with remunerative work.[1] In short, whilst instruction, illustration, and actual practice will be carried on in the most perfect style, care will be taken that in every respect the true interests of Technical and Commercial Industry, may be constantly kept in view, and the highest level of Industrial Knowledge and Ability, made

(1) There are many Trades that cannot be carried on in the workshops of the University, but some of these may perhaps be practised in their commercial reality in factories or workshops within reach. Moreover clever and deserving Students will be encouraged to undertake for a moderate remuneration the tutoring of their less advanced companions, as well as to direct various operations in the Laboratories and Workshops, and to take charge of any cheap Evening Classes that may be organized in favour of the studious youths or adults of the neighbourhood.

accessible to the greatest number at the most moderate expense.

Particular care will be taken so to organize the Examinations, that the various Certificates, Diplomas and Prizes may always reach the right hands, and recompense the right sort of attainments, and that every Candidate, no matter where his knowledge may have been acquired, may feel that he has at least a fair chance. Liberal Exhibitions and Scholarships will serve to attract from all parts of the Empire, young men of humble origin but of shrewd and practical genius, who will be regularly trained to the honourable mission of diffusing in all directions through the local centres of instruction, that knowledge which alone is wanting to complete the efficiency of our Artisans, and thus to perpetuate our industrial supremacy.

Every reasonable measure will be adopted for instilling honourable sentiments and securing decorous conduct; but all creeds will be respected, and the byelaws and regulations will be such as no earnest and sensible Student could object to. At the same time the encouragement of easy terms will be offered to OCCASIONAL STUDENTS; that is to say, to Students inscribed simply for attending certain Lectures or Classes without becoming Candidates at the Examinations, for using the Library and Museum beyond the hours of Public admittance, or for enjoying laboratory practice and the like.

Possibly it may be found practicable to induce the erection, in close proximity to the University, of large Buildings on the plan of Club Chambers, in which the contrivances best calculated to secure cheap comfort, would be carried out on hygienic principles. Arrangements on an extensive scale might also be entered into with respectable lodging-house keepers in the neighbourhood, for securing on very reasonable conditions the accommodation respectively required by the various classes of Students.[1]

However great the influence of the Industrial Element in this Country, comparatively few persons have made the requirements of Industrial Life the subject of their earnest study; and though Science is fast becoming fashionable to talk of, it will be some time before the bulk of the public knows what Science really is. Hence the assurance with which impossibilities have been announced, and the applause with which they have been echoed. Now, sensational bombast may sometimes render service in awakening the dormant energies of the Country, but when we require those energies to be expressed in Pounds Shillings and Pence, in other words when the public purse has to supply an amount which even reduced to its *minimum* will unavoidably run into hundreds of thousands, plain common-sense proposals, founded on patient impartial research, and regulated, not stinted, by thoughtful economy, may have the best chance of favour.

If there is one word which more than any other expresses the guiding spirit of the system now before us, it is this word Economy, taken in the sense of striving to achieve unqualified success by the cheapest means, with the least sacrifice of time and trouble. There will be Economy in teaching the young Artisan, as a part of his Primary Education, those Scientific Rudiments which now, if learnt at all, are learnt

---

[1] If the number of students congregated at the Central University would have to be in the direct ratio of the extent of its benefits the accommodation required would probably be something enormous, but it will be seen that such inconvenient concourse will be obviated by the systematic diffusion of knowledge at a distance.

in years that might be devoted to something more advanced. There will be double Economy in so devising that rudimentary instruction, as to make it serve for all who acquire it, as a GUIDE to an intelligent and thrifty management of the resources of DAILY LIFE, whilst it will also serve the Young Artisan as a GENERAL SCIENTIFIC FOUNDATION to his Technical Training.

There will be Economy in utilizing as far as practicable the present system of Apprenticeship, supporting rather than replacing it with new resources; Economy also in letting the Artisan acquire during his Apprenticeship at the nearest Industrial Institute, as large a portion as possible of the attainments required for his Minor Examination, and in the general plan of making instruction within convenient reach of the Student, enable him to dispense with seeking instruction at a distance.[1]

If the Artisan, having exhausted the instructional resources of his local Institute or College, must have recourse to those of a higher Institution, he will find there everything done within reason to economize his means, his time, and his brains. To show how this Economy can be effected without any notable detriment, how a simplified doctrine can be brought within the grasp of limited mental culture, and also how Teaching Power of the right sort may be provided by means, varying according to circumstances, but always sound, practical, popular and economical, will be the purpose of future pages. In the meantime a few features of the *modus operandi* may be indicated by a light sketch of a typical case, imaginary, yet probable.

(1) Particulars on this point will be found in Chap. VI. Sect. 1.

SECTION 2.—TRAINING FOR INDUSTRIAL PURSUITS.

The DYER affords one of the most convenient examples of a common and well-known scientific Trade. Accordingly we will assume that a Dyer's Apprentice is serving his time to the satisfaction of an intelligent Master, and attending Evening Classes at the Science and Art School of a neighbouring Institute.[1] When first he went there the old-fashioned plan prevailed of encouraging every one to take up whatever branch of studies he fancied he should like, irrespectively of its having any connection with his Trade, and he took to "Geometrical Drawing." Then hearing by chance that Dyeing had something to do with Chemistry, he attended a Course of Thirty Lessons on the Chemistry of the Non-metallic Elements, delivered by a Professor who maintained that it was far better to go thoroughly into one branch of the Science, than to obtain only a superficial acquaintance with the whole, and who made up for paucity of experimental illustrations, by constantly filling the Black Board with theoretical formulæ, and demonstrating the advantages of the new over the old system of notation. Our Apprentice's intellect not having received that degree of refinement

---

(1) It might be worth considering to what extent and in what manner the Saturday afternoon could be utilized for affording art and science instruction to the Working Classes, in lieu of the portion of the Sunday appropriated in continental countries to the visiting of Museums, the holding of Drawing Classes, &c.

which makes people admire most what they least understand, he felt little edified, and left off attending when he discovered that this high class exposition of pure Chemistry, even if it should ever get through the Metals, and reach Organic Products in future years, would take little notice of their industrial uses.

A change supervenes through the stimulus and assistance afforded by a Central Technical University. Thanks to a course of Popular Lectures on the Science of Daily Life, delivered on the Binary System,[1] our Apprentice sees the main landmarks of Physics and Chemistry taking up conspicuous positions in a series of connected facts;—he appreciates the interest of these facts, because they are expressed in simple and colloquial language, and supported by numerous specimens and experiments,—and he perceives at once their importance, because by well-chosen examples they are brought home to his daily wants. Science is no longer to his mind dry and indigestible, and he gladly avails himself the next year of an opportunity of attending a methodical series of about Twenty Lessons on Chemistry, not quite so easy as the previous Course, but still sufficiently clear and entertaining to claim his willing attention, even if there were not another motive directly connected with his industrial prospects. He is informed that by mastering a certain official Curriculum or select range of knowledge specially calculated to qualify him for his Trade, and the whole of which is as practical and intelligible as this Course of Chemistry, he may be able to pass a so-called Minor Examination at the nearest Technical College, and to obtain a Minor or Journeyman's Certificate as Dyer, which with the

---

[1] A full explanation of this System will be given in Chapter III., Section 3.

testimonial for good conduct promised him by his Master, will secure him employment wherever he goes. With the exception of some suitable test of practical efficiency, the Examination is to be by Papers written without book or assistance of any kind, and this troubles him, on account of the difficulty of expressing his thoughts in readable English; but he is assured that due allowance will be made for the unavoidable lack of literary training. He is further told that the Technical Examinations in question are conducted on a new principle specially devised for the benefit of Working Men.[1] He has in fact before him a detailed Syllabus containing the whole of the questions he is liable to be asked; and though they embrace seriatim the whole of the knowledge he is expected to possess, and he cannot tell which Questions the Examiner may select from each division of this long list, yet at all events he is secured against being asked anything calculated to puzzle rather than to probe his brains, or anything that is not included in the Elementary and Technical Text Books specially provided at a very moderate price.[2] The Junior Dyer's Hand-Book is well illustrated, and explains very clearly the rationale of all the *common* operations of the Trade, as well as those of the Cleaner and Scourer, generally connected with it; but in studying it he discovers to his dismay, that his Master's business is far from including all the appliances and processes of a complete modern Dyeing Establishment. The upshot is, that the local Institute and his Master's Establishment put together, cannot supply the whole of the instruction demanded by his Curriculum, and that at the conclusion of his Ap-

---

[1] See Chapter III., Section 5, and Chapter IV., Section 7.
[2] See Chapter IV., Sections 5 and 6.

prenticeship, he must spend a short time at the Technical College where he intends passing his Examination, in order previously to complete his Studies.[1] He determines for various reasons to do this at the model Technical College attached to the Central University, in the Museum of which, besides unrivalled special resources, he hopes to find a little remunerative employment in arranging illustrations of Dyeing Materials and processes, for being supplied at cost price to provincial and colonial Institutions.[2]

At the appointed time, our Candidate passes satisfactorily the Dyer's Minor Examination, including :—

A Grade 1 Course of the indispensable General Scientific Foundation.[3]

Grade 2 of Chemistry.

A cursory review of Textile Materials and Fabrics.

A certain insight into CHROMATICS, that is to say, the Science which founds on the solar spectrum a knowledge of primary and mixed colours, with their harmonies and contrasts.

A scientific review of the Dye House, its construction and apparatus, the materials usually dyed, and those commonly used in dyeing or for other cognate purposes, with indications of common adulterations, useful substitutes, &c.

A practical review of the various processes and operations generally included in the Dyeing Trade.

Rules of prudence to be observed for avoiding the

---

(1) The Dyer's Trade is peculiarly scientific. In the case of most of the ordinary Trades, an Industrial Institute ought to be able to supply the *whole* of the Preparatory Knowledge, though not unfrequently the Technical Knowledge will have to be completed at a Technical College.

(2) See Section 8, J., of the present Chapter.

(3) For a detailed Syllabus, see Chapter IV., Section 1.

unhealthiness or danger to which there is a tendency in some of the operations of the Dyeing Trade.
A few incidents of the History of Dyeing.
A suitable test of practical and manual ability.

Having obtained his Minor Certificate, our recognized Journeyman turns it at once to good account as a claim to satisfactory employment and first-class wages. But more than any actual profit, he enjoys the zest of doing everything with a perception of the "reason why," and his enjoyment is mingled with a confident anticipation of rising higher in his calling. He has before him the Syllabus of the Major or Master's Examination, and in conformity with its indications, he takes a Grade 2, instead of a Grade 1 Course for his General Scientific Foundation, and similarly he endeavours to push forward one Grade most of his other scientific attainments.[1] This is no more than every Technical College ought to offer him the means of accomplishing, or than he may possibly achieve by self-instruction. The third Grade thus assigned to his Chemistry only corresponds to the usual level of elementary works,[2] and is in fact what would be prescribed at a high class College to every beginner starting with a well-trained mind. Our Journeyman Dyer still lacks the advantage of understanding Greek derivations and the like, but the substantial knowledge of Chemical Bodies and their properties which he has acquired in mastering successively the two lower Grades, compensates for educational deficiencies, and makes smooth work of the Doctrine of Equivalents,

(1) For details see the Suggestive Summary of a Dyer's Curriculum in Chapter V., Section 2.
(2) See Chapter IV., Section 3.

whilst his acquaintance with the results of the various reactions, enables him to appreciate and enjoy the use of the Notations by which they are so graphically expressed.

Self-instruction in Science is mostly an arduous and ungrateful task for a beginner, especially for one deficient in general culture, but the case is different with the Artisan who has already gone through a regular Course of Studies, more particularly as the exercise of his trade supplies him with most of the technical illustrations he requires. It will probably be only for a short time previous to the Major Examination, that our Journeyman Dyer need repair to the University for completing the Technical portion of his attainments; that is to say for adding to his scientific knowledge of the *ordinary* Dyeing Materials and Processes, an equally thorough acquaintance with *new* ones not yet in general use, with *foreign* ones deserving of adoption, and generally with that extended circle of ideas, and more independent and originating style of thought, which ought to make the difference between a Master and his Men.—It need scarcely be said that all this intellectual expansion, rendered easy by the union of first-rate oral teaching with every resource that the best technological Collections and Laboratories can afford, will, independently of the honourable and gratifying distinction of a Major Certificate, very soon repay our Dyer for the time and outlay involved.

The Dyeing Trade is one of those in which higher positions can be aspired to than in many of the ordinary handicrafts. The Director or Foreman of large Dyeing Works should be a thorough Chemist, and possess a comprehensive knowledge of the various resources placed at his disposal by inventive and commercial industry, not only for ordinary times, but also for times when certain products fail, or rise to an exorbitant price, and must be replaced by substitutes. It is indeed to Men of this grade in the complex and scientific Trades, that we must mainly look for additions and improvements in the respective branches of manufacturing industry, and nothing will be spared at the Central Technical University, that can enhance their abilities, and secure for them an honourable recognition. Should our typical friend be of the right sort, a DIPLOMA OF EXCELLENCE presented with due ceremony, will some day attest the fact that he has satisfactorily passed the third and highest Technical Examination.

The foregoing brief sketch of a Dyer's progress, may, *mutatis mutandis*, serve to exemplify the stages of study, examination, and reward, similarly open to other branches of manual and commercial industry.

Commercial Trades, as stated in describing Category B, will to a considerable extent be associated with the producing and manipulating Industries. A Dealer owes it to himself and to his customers to understand well the nature, properties, and quality of the Articles he sells, and this generally involves a certain acquaintance with their origin and production or manufacture; though of course this knowledge, when spread over a great variety of Articles, need not by any means equal for each, that possessed respectively by those for whom they are specialties. Many circumstances will concur to render the participation of Merchants and Shopkeepers, and also of many of their respective Employees in the advantages of the Central University, alike expedient and economical. They will find just the information they want in many of the Courses delivered for the benefit of the more Technical Industries, and a large portion of the Museum will specially illustrate the ministration of Commerce to the wants and comforts of Daily Life.

Whilst domestic Book-keeping and the like will be included in the General Foundation of which every Student according to his Degree will be required to possess a certain knowledge, the more complex forms of Accounts will be specially taught for trade purposes, and other commercial subjects will be dealt with according to their importance, up to the highest spheres of Commercial Geography, History and Law.

SECTION 3.—ART INDUSTRIES.

If hitherto less mention of Art has been made than might have been expected, it is not by any means on account of any deficient appreciation of the part which it has to perform in many of our National Industries. It is perhaps as regards ornamental design, that the progress of other nations, and our own shortcomings, have been first and most forcibly brought under our notice through the number of foreign Workmen engaged by our employers of skilled labour. But, partly on account of that very circumstance, a great amount of intelligence and energy has already been brought to bear on this department of technical requirements, and the system of Art Schools centred at South Kensington, is so capable of meeting all the most urgent demands, that in most respects, and especially as regards the training of Designers, it will be the manifest interest, as it certainly will be the earnest desire of the promoters of Technical Progress, to strengthen and to profit by so useful an organization. It is hoped that a considerable stimulus may be given to the study of Drawing, by including it among the obligatory subjects in the Examinations of several Manual Trades, and no exertion will be spared to supplement existing resources, by measures calculated to secure with the least sacrifice of other useful studies, a more extensive and practical diffusion of artistic ability and refined taste, in the various departments of Technical Industry. It will not however be forgotten that the Working Man's

Education is a battle against time, and that as much harm as good might be done by blindly pushing into popular education, an amount of Freehand Drawing that must necessarily push out a proportionate amount of something else. There is a certain training of the eye and mind to understand what is meant by a painting, drawing, or print, to appreciate the difference between a good picture and a daub, and to be capable of selecting household ornaments not deserving of a place in the " chamber of horrors," which every sane and seeing individual should possess.[1] But manual ability at any branch of the Arts of Design, unless it be a natural gift in a very exceptional degree, is not to be acquired without much time and teaching; and accordingly it will be seen in the pages, specially reserved for the details of the proposed Instruction (see Chapter IV., Sect. 4), that particular care will be taken to reduce the *obligatory* artistic attainments to easy utilitarian standards; especially as regards Candidates for Minor Certificates. In many of the Chemical Industries, as for instance those of the Brewer, the Soap Boiler, and the like, the Arts of Design will be left out of the question. In many of the Mechanical Trades, the Minor Candidate will only be required to understand Drawing sufficiently to enable him to work from a drawn Pattern, whereas the Major Candidate will be expected

[1] I could name a Parochial School where a collection of Models, &c., supplied at reduced rates by the Science and Art Department, besides being used for a well-conducted Art Class, is displayed to good advantage along the walls. The incipient appreciation of Art conveyed to the minds of the children through thus having constantly some of its best types before their eyes, requires no comment. One cannot help reflecting on the further advantage that might accrue from similarly displaying chromo-lithographs, or other suitable reproductions of some of the choicest works of the Great Masters. It has been proved that if the supply of them were organized on a large scale, the cost would become very moderate.

to be able to draw one, and this in many cases will be made to include a certain amount of origination, as well as of artistic skill. Thus the Journeyman who aspires to a Major or Master's Certificate as Cabinetmaker, must show that in taking an order for a Sideboard or a Bureau, he is not tied down to a patternbook, but can sketch the proposed article either in perspective or isometrically, carrying out tastefully the instructions of his customer. But independently of obligatory attainments, sincere encouragement will be given to every Artisan favoured by talent and opportunity, to push forward an appropriate style of Art as a distinct "Subject," competing for Art Prizes if he has a chance, and aiming very high if competent judges pronounce him a genius.

In those Trades which come within the province of Art-workmanship, Design will occupy a foremost place both in the Minor and Major Studies, and Origination will receive that attention of which the experience of the Society of Arts shows that it is so much in need. It appears that as long as the supplying of Model Designs in the several branches of Art Industry formed part of the system of the Society's yearly special Examinations, there was every reason to be satisfied with the ability displayed by the respective Candidates in carrying them out; but that the attempt to raise our Art Workmen a step higher by inducing them to be their own designers, led to the results depicted as follows in the official Report of the Adjudicators.[1]

"In spite of the individual specimens of excellence to which we shall presently allude, we are bound to confess that the response made by Art Workmen to the Society's liberal invitation to compete for prizes

(1) See the Society's 'Journal' of the 5th March, 1869.

offered during the last session, cannot in our opinion be regarded as satisfactory. It will be remembered that the lists of subjects proposed differed materially from those of previous years—it having been considered well, as an experiment, to test the workmen's powers in the combination of original design with skilful workmanship, and in novel directions rather than to keep them in the groove of the reproduction of the best works of the past.* * * * Whether it is that the task recently set to the Art Workmen has been beyond their present powers, or as is more probable that they look with anxiety only to what affects their regular employment, possibly, in some cases, apprehending notoriety as a fault rather than merit in their masters' eyes, certain it is that the results of their labour, taken as a whole, are not such as we had hoped for, nor such by any means, as we think would have been made by French, or even Belgian workmen, had a similar invitation been addressed to them.—We do not necessarily attribute this to incapacity on the part of our Art Workmen as executants, but ascribe it rather to their want, in this case, of the directing and sustaining power which is supplied to them, in the course of ordinary business by the superior education and attainments of their masters, and of the artists and designers, from whose drawings, models, or suggestions, they may habitually work."

Now I perfectly concur in thinking that our comparative failure in the endeavour to elicit tokens of origination from our Art Workmen, is in great measure to be attributed to the simple fact that origination is not what their employers particularly wish them to possess; and I can perfectly follow out in my mind the hint given us by our Art Judges that "notoriety," or in other words distinguished excellence in design on the part of a workman, might not exactly please a master who required merely a perfect instrument, patient, opinionless, and not possessing any merits beyond those absolutely wanted. These are matters that must not be ignored, nor dealt with indiscreetly; but there are fortunately indirect means as safe as they are effective for overcoming the spirit of short-sighted jealousy and mutual distrust, and for establishing satisfactory relations between Masters and Men. One of the main points is that both parties should not merely be *told* what their mutual interests and duties are, but that their natural intelligence should by a suitable education, be trained into an habitual clearness of perception, and a logical way of reasoning, which may make them see for themselves what is right. Should adequate measures be adopted for securing to all ranks of the People a Schooling calculated to stimulate a healthy development of the intellectual faculties, we may expect on the one hand a more generous and far-seeing appreciation of the industrial value of Art, whilst on the other hand its adepts will soon show greater readiness and aptitude to turn to good account the artistic instruction placed within their reach. I am not a believer in panaceas, and do not argue that Science will make a lad an artist; there is on the contrary, a kind of science and a way of teaching it, more calculated to deaden than to improve the poetic sentiment which should pervade his appreciation of the chaste and beautiful in Art; but I firmly believe that Science so selected and so taught as to raise the youthful mind towards heaven in thankfulness for God's bounties, and at the same time to awaken it to an intelligent and lively use of them, is the most effective drilling that the mind can have for the development of its highest as well as of its most practical faculties. When Science of this kind shall have opened the understanding and elevated the aspirations of the Children of the People, we shall not see our soldiers hungry and shivering where a foreign campaigner would contrive to make a comfortable meal, neither shall we see our art workmen unable to compete with their neighbours in thought and origination.[1]

SECTION 4.—TRAINING FOR PROFESSORSHIPS.

Training for Technical Pursuits, may be considered as forming one great division of the educational operations proposed to be carried on at the Central Technical University and the affiliated Institutions. Another equally important division, will embrace the training of Professors or Teachers of various grades and denominations, for carrying out the scientific and technical propaganda of which this University is to be the centre. Considering that every thoroughly competent Professor sent out to a local Institution, will as it were, expand into a multiple of industrial power, it is obvious that this great division of the intended operations, will well deserve the fullest meed of attention and energy.— Every exertion will be made to secure by means of sufficient and well-administered Scholarships, and other inducements, a constant influx of young men with good abilities and a sound scientific foundation, who after completing their studies at the proposed University, may be sent out again in all directions to supply

---

[1] See also respecting Art Industries Sect. 8, G., of the present chapter, Sect. 4 of the 4th Chapter, and Sect. 14 of the 5th.

genuine Industrial Instruction, carrying with them the latest improvements in Knowledge and educational appliances.

It would be premature to attempt enumerating either all the educational levels to be supplied with Teachers, or all the Subjects which each stratum of Teachers should embrace. For the present a few general indications will suffice :—

—Attention has already been drawn to the important "Recommendation" of Her Majesty's Commissioners on Scientific Instruction, that suitable provision should be made at Training Colleges for the instruction of the Students in the Elements of Physical Science.[1] The urgency of giving effect to this recommendation, if anything like a general system of science-teaching is to be introduced in the Primary Education of the People, is as has already been indicated, but too obvious to those who have watched the educational proceedings of the last 12 years. The adoption of the Revised Code suddenly depreciated the value of Science in the eyes of the great body of Primary Teachers throughout the country, because every knowledge except that of the three R's, ceased to bring subsidies to their Schools. Even before that time, though Students at the Training Schools might gain certain advantages by learning Chemistry, Physics or other branches of Science, each branch was taken upon its own merits, without any reference to the others, or to the requirements of Daily Life. On the other hand, Domestic Economy was already included in certain Examinations for Females, but it did not rest on a scientific foundation, and the consequence was what might have been expected : a knowledge of facts

(1) See Chapter I., Section 2.

unsupported by a knowledge of principles, which the least shaking would upset, and rules without rationale, which would only guide through a straight path.

—Should the suggestions of Her Majesty's Commissioners be carried out on a liberal basis, and the provision at Training Colleges "for the instruction of Students in the elements of Physical Science" be construed into something like the range of select knowledge which has been described under the collective title of "Science of Daily Life," a vast impulse will be given to the studies of both sexes in that utilitarian direction, and valuable indeed will be the aid afforded by the Central Technical University, in supplying those Establishments with Instructors of the exact kind required. They will not only be fully versed in the select range of Elementary and Applied Science to be expounded to the Students, but they will have been specially trained for imparting to these the best methods of infusing genuine notions by degrees into the minds of Children, through an attractive use of cheap, simple and partly self-made apparatus; thus illustrating in the tuition that fertility of resources under difficulties which Science is particularly calculated to engender.[1]

[1] Besides the amount of Scientific knowledge which may be reckoned among the prime necessities of Industrial Life, there are certain general notions concerning the Earth and the Heavens of which the Child of the poorest Labourer should not remain ignorant. It is essential therefore that the special Training Colleges for School Masters, should afford them an opportunity of acquiring an easy insight into the leading facts of Astronomy, Geology, and Physical Geography, and it may be convenient that the Science Professors supplied to these Colleges by the Central University, should be competent to include these subjects in their teaching; but at the same time they must be able to impart the art of introducing matters like these briefly, strikingly, and as far as possible in the form of recreation.—It is consoling to find that Pedagogy, or the Art of imparting Knowledge, is beginning to receive in this country that attention which it has long enjoyed in Germany and elsewhere.

—Some years must of course elapse before these principles can have permeated the Teaching Power of the Schools of the People, and in the mean time, many Schoolmasters, Schoolmistresses, and other persons engaged in popular Instruction, may gladly avail themselves of special opportunities that will be offered them for imbibing the Science of Daily and Industrial Life at the various Institutions included in the proposed System, and more particularly at the Central University itself.

— Besides the Scientific Teaching Power required for the Schools of the People, we must take into account the Science Teachers demanded at a rapidly increasing rate by numberless Schools, Academies, Colleges, and Institutions for the Middle and Upper classes, having a commercial or otherwise a practical tendency.

— Next to Science at the School, we must consider Science at the Industrial Institute; that is to say, Scientific Instruction of an easy and attractive character, offered to Apprentices and Artisans in all parts of the Country, through the accommodation afforded by Industrial Institutes, or by any other of the Meeting Places for elevating purposes which it has been agreed to include, for brevity sake, under that name. The first step will be to train and send out special Agents, or Commissioners of Industrial Instruction, competent to deliver stirring Lectures in the Provinces on the practical value of Knowledge,—to organize on the best principles every variety of Working Men's Clubs, Institutes, Mutual Instruction Societies, Educational Museums and Reading Rooms, &c.—to initiate Evening Classes of various descriptions, and methodical Courses of Lectures on the plan adverted to in the next paragraph, and in short, to introduce every appropriate device for propagating attainments calculated to secure a substantial scientific and artistic foundation for local Industries.

—It will be seen by the account of a new form of Lecturing, fully described in Chapter III., Sect. 3, under the title of the "BINARY SYSTEM," that in order to follow up satisfactorily the foregoing measures of local organization, the Central Technical University must, besides ordinary Class Teachers, train and send out to the various Institutions, Readers and Demonstrators whose functions will be duly explained.

— The duties of the Central University towards Technical Colleges, will comprise in addition to the foregoing, the supply of a higher class of Professors including Technologists. It is essential that the Teaching Staff of Technical Colleges should unite to zeal and cleverness, that special training required for making each Institution harmonize with the others, and all of them with the Central and Provincial Universities. These qualifications should be pre-eminent in the person charged with the direction of a College, for on him will greatly depend the stimulus given, and assistance afforded, to the Industrial Institutes of the surrounding country.—The brains and hands required for the College Museum, must not be forgotten; nor must either the spur of zeal or the curb of prudence be wanting in the Chemical Laboratories; nor skilful supervision in the Technical Ateliers, which may assume so much importance if the introduction of the Cornell plan of alternating study with work, should be found successful; and so much more if the deficiencies of the present system of Apprenticeship should force us to seek a partial remedy in a special development of the Technical Colleges.

G

— We now come to the professorial staff of the Provincial and Colonial Universities, and of the Central Technical University itself,—to that body with many members, of which the influence for promoting Industrial progress and prosperity, will be immense, provided it be of one mind. It may be assumed that the affiliated Institutions of this grade, standing prominent in all the great Technical Divisions of the Empire, may in due time be able not only to train many Professors for the subordinate Colleges and Institutes around them, but to rear in some measure their own Staff; yet if the Central University makes good use of the superior resources resulting from its position, it will have no difficulty in maintaining the welcome ascendancy of its benefits. The best local men will feel that something is wanting to their scientific culture, till it has received the finishing touch at the Metropolis, and every possible inducement for resorting thither will be offered to them, for the sake of the centralizing influence which their improved and systematized abilities will harmoniously exercise in their several localities for the general good.

SECTION 5.—INDUSTRIAL AND COMMERCIAL ECONOMY.

Bordering on one side on Political Economy, and on another overlapping Social Economy, the department of enquiry designated by the above title, and which might also be styled the PHILOSOPHY OF INDUSTRY AND TRADE, must be managed with particular caution at an Institution like the Central Technical University, in order to avoid all semblance of class or party bias; but there can be no doubt that under judicious treatment, it may be made to afford additional claims to national respect and gratitude. Industrial Economy, as we will briefly denominate it, should be taught by Men of firstrate abilities to advanced Students, either aspiring to Diplomas of Excellence in important Industrial or Commercial Pursuits, or in training for Technological Professorships.—Then again its essential principles should be diffused by gentle permeation among all grades and denominations of Students, and through almost every branch of instruction, in the manner best calculated to exercise a harmonising influence among the varied elements of which the Industrial community is composed.—It may not be amiss to consider a few of the means by which that teaching, and that diffusion, may best be rendered safe and successful.—Repeated stress has been laid on the reputation for disinterestedness and impartiality which it is of vital importance that the Central Technical University should maintain, as an indispensable means for obtaining the confidence and co-operation of other Institutions. Accordingly no Professor will be appointed to a Chair of Industrial Economy, or International Commercial Law, who is not as well known for his discretion as for his abilities.

—The principles of mutual support and reciprocal benefits on which should be based the relations of Capital and Labour, will be advocated by the eloquence of facts, not of words. Well authenticated statistical details, somewhat similar to those given in M. Le Play's admirable work, "Les Ouvriers Européens," will afford materials for a parallel between the position of the English Workman and that of his continental brethren, and all other parts of the Labour Question will be discussed in the same earnest matter-of-fact style; not to justify foregone conclusions, but to result in suggestive ones. At the same time prominence will be given to every well-organized Institution, and well-conducted experiment, by which, in this Country or elsewhere, it has been sought to bring industrial difficulties to an equitable and satisfactory solution.

— Commercial Economy will, as far as the divergence of the two subjects allows, be treated in the same spirit as Industrial Economy. By enquiring with statistical accuracy into the commercial development, resources, and mutual relations of nearly all parts of the Globe, the causes of the rise and fall of certain commercial centres will become apparent with a power of instruction far above that of wordy warnings.

— The questions of Protection and Free Trade, with many others of analogous character, are more easily and consequently more frequently determined by dogmatic rules, but they will be more safely solved by a thorough investigation of the particular circumstances of the respective Countries, and the same process of enquiry will lead to the adoption of International Commercial Laws so constructed as really to be bonds of common interest and permanent good-will.

— As regards the gentle and gradual propagation of sound notions of Industrial Economy among the Industrial Community at large, this result will naturally follow when the wholesome leaven of Science is introduced into the minds of the rising generation. Though in small quantity, it will by a spontaneous process of fermentation, expand and vivify the intellectual faculties, and provided their exercise be discreetly directed by those who will have the opportunity of combining with the functions of a Teacher the influence of a Friend, we may look forward to better industrial times. The enlightened Employer will see more clearly how much the interests of his men are his own, and the clever Workman, alive and active in all the concerns of Daily Life, earnest and industrious in applying his attainments to his work, and intelligently thoughtful in turning them to the best account for the welfare of his home circle, will be ever ready to apply the same clear-headedness to his social relations, in a spirit at once conscientious and conciliatory.[1]

[1] The following is from a Letter which I addressed in March 1867 to the Secretary of the Labourer's Friend Society:—"It is through science that the working man may render his position more secure by increased ability to earn, whilst he adds to the value of his earnings by an improved and more thoughtful expenditure; and if master and man can both counteract high prices by knowledge and ingenuity, the chances are that they may thus be mutually enabled to live and to let live; and that by their united efforts, they may yet renew the best times of our industrial prosperity."

SECTION 6.—FEMALE TRAINING.

A praiseworthy desire is being manifested on all sides, for a due development of facilities for the training of Females in the various technical pursuits for which they are particularly well qualified, and it is evident that this department of operations deserves to receive a full amount of attention, in carrying out the main and affiliated portions of the proposed system of educational machinery. A few explanations may however be necessary in introducing this subject, to obviate misapprehension.

Nothing is easier than to show that a Workman is likely to give more satisfaction if his intelligence and skill are assisted by a good training in the practical elements of Science and Art, than if such assistance is withheld; but when an attempt is made to transfer the argument to the other sex, too many persons will try to stop the way with a barricade of hollow prejudice, strengthened here and there with real anecdotes of Girls whose heads were turned, or tempers spoiled with Learning. The fact is, that to the want of a discreet and judicious selection and adaptation of the subjects taught, is in most cases to be ascribed the failure of the teaching, and that in female as in male education, the prime condition of success is that it should be to the purpose. This remark extends even to those young Ladies who are in a position to be able to select pure intellectual refinement for the purpose of their study, and who can afford to attend fashionable Lectures in a fashionable neighbourhood. It would be a matter of great interest to review the numerous opportunities already offered them for acquiring a high standard of culture, showing to what extent the results are satisfactory or otherwise, what further Institutions are required, and generally speaking, in what directions a cultivated female mind may find that satisfactory consciousness of *time well spent*, in the hope of which those who attempt to soar above their natural sphere of action, are so often disappointed; but these questions might lead us too far from our present Programme, and moreover they are in a promising way of solution through the enlightened labours of the "National Union for Improving the Education of Women," organized under the highest patronage through the praiseworthy origination of Mrs. Wm. Grey.

Two chief classes of legitimate purposes meet our inquiry in the female as in the male department of the operations of the proposed University ; namely, direct training for Industrial or Commercial Occupations, and Training of Teachers.

A. *Training of Females for Industrial Pursuits.*

Several Female Occupations worthy of note will occur in the Analytical Review of Industries in general (Chapter V.), and a better opportunity than the present one will then present itself for canvassing various matters of detail respecting Female Employments. In the meantime it may be mentioned that the scientific instruction involved, is generally of a limited range and simple character.

B. *Training of Female Teachers.*

The introduction in the Primary Education of the People of select rudimentary notions of the Natural Sciences, applied to the familiar concerns of Daily Life, is not by any means an advantage to be confined to the Boys. On the contrary, as far as the comfort of the Home and the health of the Family are concerned, a tincture of common sense based on scientific principles, is peculiarly necessary for those on whom in the natural order of things will devolve the care of the household. A slight distinction may be made. For instance, the use of the Mechanical Powers is rather an affair for Men than for Women, but the many functions entrusted to the latter as regards food, clothing, cleanliness and the like, call loudly for a clear appreciation of what is right or wrong according to Nature's Laws, and this appreciation is most efficient when it has been acquired at an early age, and has thus become instinctive and habitual. Assuming these principles to be recognised and put in practice, a change will supervene in the training of Schoolmistresses and other female Teachers engaged in Primary Schools for Girls, even more marked than that in the case of Male Teachers, because hitherto Science has unfortunately been considered as altogether uncongenial to the female mind.

The means adopted will be the same for both sexes: —*Firstly* as regards the *present* Members of the Primary Teaching Community, Mistresses no less than Masters will have liberal opportunities offered them by the local Technical Colleges and other Institutions affiliated to the proposed system, for learning the "Science of Daily Life," and how to teach it. *Secondly,* the Central University will reckon amongst its essential duties, to provide the Training Colleges for Females no less than those for Males, with Professors thoroughly qualified to raise *future* Mistresses to the required scientific standard. There is no reason why some of these Professors, being intended for Female Colleges, should not themselves be Females.

The question of the assistance to be afforded by the Central University in reference to Schools of various descriptions for the Daughters of the Middle and Upper Classes, is much more intricate, and must in great measure be influenced by the operations of the National Union for Improving the Education of Women of all Classes.[1] In the meantime the following Courses of Instruction may be pointed out as likely to be included in that assistance, and to be followed up with special Examinations and Certificates.

— A Course on Female Pedagogy, considered both in its general bearings, and in reference to the various social levels. The Course should include the purposes to be aimed at, and the most appropriate means of attaining them, adding to the essential moral and hygienic rules and considerations, those appertaining to practical School Management and Discipline. The whole to constitute that professional knowledge which every person having charge of a Ladies' School ought to possess.[2]

(1) Address: Miss Louisa Brough, Secretary, Women's Education Union, 112 Brompton Road, S.W.
(2) Besides the valuable materials for such a Course that can be obtained from the above-mentioned Women's Education Union, I may point to the suggestive matter contained in a late Bulletin of the 'Belgian Ligue de l'Enseignement,' and especially to the highly interesting description of a Model School for Girls near Rorschach, by the Lake of Constance.

— An Exposition of the "Science of Daily Life," conformable in its range of Elementary and Applied Knowledge, and in its level of difficulty, to the Grade 2 Course proposed for Working Class Students, but altered in the selection of examples as well as in the style of language, to suit a high mental culture and habits of social refinement.[1]

— Detached Courses on interesting branches of Science, specially arranged, and being either further special developments of those included collectively in the foregoing range, or careful adaptations of others, such as Astronomy, Physical Geography and the like.

It is proposed that the Courses of Instruction indicated under these three headings, and such others as it may be found expedient to add, shall be held either in distinct Female Departments of the Technical Universities and Colleges, or at special affiliated Institutions, and that the most liberal facilities for regular attendance shall be offered, not only to Ladies engaged in or intended for the directing or teaching staff of Middle or Upper Class Female Educational Establishments of every description, but particularly also to Resident and Daily Governesses.—To establish in various parts of the Country, Female Scholastic Training Institutions on the most approved principles, where the above attainments would be judiciously associated with a selection of the usual ones, to provide for a constant supply of suitable Professors, and to organise reliable Examinations with Certificates of uniform Standards, would be a work which the Central Technical University with its ubiquitous affiliations would be particularly competent to undertake, and the satisfactory accomplish-

---

[1] It may be well to have also a more advanced Course corresponding to Grade 3.

ment of which, would be the means of vastly improving the chances of health and happiness among future generations. It is moreover through Science discreetly blended with moral influences, that young Ladies must learn the secret of crowning life's real enjoyment with the satisfaction of making ends meet.

There are a large number of Occupations, some approaching in character to those we have been considering, others shading off to Industrial Pursuits, which might be benefited by arrangements more or less similar to the foregoing. It will suffice to indicate by way of examples, those connected with Reformatory and Mission Work, with District Visiting and Poor Relief, with the care of the Sick, with certain forms of Domestic Service, &c.

SECTION 7.—PUBLICATIONS.

It has by this time become sufficiently apparent, that in order to be of real benefit, the proposed or any other National System of Technical Instruction, must acquire general acceptance and uniform interpretation for its scheme of Studies, its method of Examinations, and its standard of Awards. It has also become evident that a Central Technical University charged with accomplishing this result, whilst it will need, and must strive to deserve the hearty concurrence of local Institutions, must cheerfully support the chief onus of organizing the operations, and providing the educational Rolling Stock; the more so as it is only by means of several decided innovations, that Scientific Studies and Examinations can be popularized. Each of these innovations will successively be made the subject of special consideration in the following Chapters, but our present purpose of getting a general and connected notion of the *modus operandi* of the proposed University, requires that a few moments' attention should be bestowed on the literary or editorial portion of its labours.

Any INSTRUCTIONAL BOARD or Committee that may be charged with preparing the Educational Requisites, whilst the Architect is preparing the Building, will not have the easier task of the two. A great part of what will be wanted, in the literary line, will have to be created; especially the popular Text-books hereafter described (Chapter IV. Section 5), giving in two progressive Grades the Courses of GENERAL SCIENTIFIC FOUNDATION, Elementary and Applied, of which every Student will be required to possess a certain knowledge. They will be troublesome to compile on account of the difficulty experienced by high-fed minds in stooping to meet minds accustomed to a low diet. But when this task is done, the consciousness of the benefit conferred thereby on the teaching community, and through it on generations yet untaught, will be proportionate to the difficulties overcome.—In the higher Grades, existing Books, or selections therefrom, will be utilized under advice and guidance that will greatly add to their value. (Chapter IV. Section 3.)

Another literary task of greater magnitude will be to prepare two series of Industrial and Commercial Text Books; viz. a Junior Series of HANDBOOKS corresponding to the requirements of Minor Students, and an Advanced Series of MANUALS suited for Major Students; each series to be also printed for reference as a TECHNICAL CYCLOPÆDIA (Chap. IV. Sect. 6). The preparation of these voluminous works, and their publication with an abundance of superior Illustrations at a very reasonable price, to say nothing of keeping them up to the latest technical improvements by means of frequent Supplements, and of occasional new Editions, are labours of which only a powerful Body, sure of the general appreciation of its disinterested exertions, could afford to give the benefit to the country.

SECTION 8.—THE MUSEUM AND DEPARTMENTS CONNECTED THEREWITH.

A. *Visual Instruction for the Industrial Institute, the Technical College, and the Central University. Leading Principles.*

VISUAL INSTRUCTION, long valued in Germany under the name of *Anschauungs-Unterricht*, has deservedly obtained much favour in England, but still leaves great room for improvement. It is intended to be developed to the utmost at the Central University, both as regards its constant use, and also as regards the publishing of Diagrams and other Illustrations, more correct, better executed, and cheaper than it is possible for them to be supplied as a trade speculation.

It is in the MUSEUM DEPARTMENT of the proposed University that Visual Instruction will display its varied resources with peculiar advantage. Though its Collections will necessarily depend to a considerable extent on circumstances which cannot at present be foreseen, a number of points are ripe for consideration, involving principles and purposes susceptible of general application, which bring them within the compass of the present Chapter, though they may occasionally run somewhat into details.

The Central Technical University will have attached to it, a Technical College, and an Industrial Institute, models of organization and activity, and each having its Museum which will consist of Apparatus and Illustrations wanted from time to time for teaching purposes, and as far as practicable permanently displayed for the instruction of the Students. Adding to these repositories that of the University itself, we see that three distinct typical MUSEUMS will be offered for provincial and colonial imitation. They will be very different in their scales of development, but all three will have more or less for their nucleus and starting-point, the Illustrations required for a Course of Studies on the range of practical knowledge, *preparatory* and *applied*, which has been described as the "Science of Daily Life."

An INDUSTRIAL INSTITUTE of humble means may be obliged to content itself with a course of the simplest standard, namely Grade 1.—Of the apparatus, Diagrams, Chemicals, &c. appertaining to the preparatory division, only certain portions are suited for display; but a highly instructive little Museum may be formed at a small expense, and in a small space, with the Illustrations of the second division, representing as they do in methodical succession, the several Departments of DOMESTIC and SANITARY ECONOMY. (See Chapter IV. Section 1.) It is evident that an intelligent Student inspecting Class after Class of this ECONOMIC COLLECTION, with a printed Course of familiar Lectures in hand, will have as near an equivalent to the advantages of oral teaching as can well be desired, and particularly that it will afford him an excellent reminder if he has already heard the Course. Even in this humble Grade the origin and process of manufacture of many of the Articles will necessarily be illustrated; for it would be impossible otherwise to convey a sufficient idea of their nature and relative value for use. Should the Institute include a School of Design, many of the Art Illustrations required for its use, may as hinted elsewhere, be permanently displayed to good advantage.—Respecting other features of instructional or recreational development, I cannot do better than again refer to the papers published by the "Working Men's Club and Institute Union," 150 Strand.

In the Museum of a TECHNICAL COLLEGE, devised for a more comprehensive and advanced study of the requirements of Daily and Industrial Life, the Technological Illustrations will assume increased importance, giving as far as possible an insight into the rationale of all the most noted Industries.[1] Associated with this Groundwork of Practical Knowledge will be :—1stly. Illustrations of Art, pure and applied. They need not be very numerous, but should be carefully selected, so as to present a choice variety of marked types.[2]—2ndly. Any available materials susceptible of attractive and instructive display; such as: Illustrations of the Zoology, Botany, Geology, or Archæology of the neighbourhood;[3] Astronomical, Geographical and Historical Charts and Apparatus; Models of Machinery, &c.

---

(1) In each locality particular attention should be paid to the Illustrations of Industries prevailing in the neighbourhood.

(2) Some of the Indications given further on in reference to the Art Department of the University Museum, would equally apply to the case of a well-developed College Collection. The value of Chromo-lithography for reproducing at a cheap rate select works of the best Masters, has already been mentioned.

(3) "The museums of less important towns are generally very incomplete. They too often consist of specimens unconnected with each other, the gifts of travellers possessing little or no knowledge of natural history. When they are the results of the labours of some local naturalist or of

3rdly. Anything that may be found suitable for display in connection with the Mathematical or Classical Studies carried on at the College.

Generally speaking, it will be the policy of provincial Technical Colleges, to unite within their walls every Scientific, Artistic, and Literary Instruction of which their District may be in want; for where resources are limited, as in most country Towns, strength lies in the concentration of patronage. In a vast and wealthy Metropolis division of labour becomes possible and desirable. Whilst other branches of Education are attended to by other Institutions, each in the spirit of its calling, the special task proposed to be entrusted to the Central Technical University, that of organizing and maintaining an improved educational status of the Industrial Community, can only be accomplished by that practical and utilitarian spirit which characterizes Industrial Enterprise; and in its MUSEUM as well as in its Lecture Halls, nothing will as a rule be admitted, which has not a bearing on the technical success or physical well-being of the Industrial and Commercial Population.

As in the Institute and the College, the principle will be upheld of first making the proposed Studies point out whatever objects they require by way of illustration, and then displaying for the best advantage of the Public, as well as of the Students themselves, such of these objects as are suited for display. The more purely scientific Apparatus will probably be accommodated in or near the Lecture Rooms, but with the objects illustrating Science and Art applied to

some provincial society they are exceptional, and sometimes of great value." Fourth Report of H. M. Commissioners on Scientific Instruction, &c., April 1874.

Daily Life and Technical Industry, will be constructed an Educational Museum, intelligible and profitable alike to the Students specially concerned, and to the miscellaneous Visitors representing the thirst for knowledge of the neighbourhood. And here it must be distinctly remembered that as a *sine qua non* of the success of a University intended to be patronized by the Working Classes, it must be situated in a Working Class Locality, cheap, industrious, decent and not overdone with culture, where People are not fastidious about the cut of a coat or the material it is made of.

B. *Rules for Guidance in Collecting Materials.*

The majority of the old Public Museums have been the creations of opportunity rather than of deliberate design, and consist of gifts rather than of purchases. Consequently their growth is often distorted through wealth accumulated in one department, whilst another remains poverty-stricken, or is conspicuously absent, and there is an uncomfortable incongruity about them, because they have been enriched with collections formed with different views from those with which their first nucleus was produced. Then again a large proportion of the Museum Curiosities brought home by travellers from distant climes, are sensational rather than educational; or if they supply missing links in scientific series, they speak to the learned, and are mute to the Working Man. For these and various other reasons, if we submit to an impartial analysis the benefit supposed to be conferred on the masses by the frequent inspection of Public Museums of the old type, we find that it consists more in rousing the intellect than in feeding it, and less in mixing recreation with study, than in preventing its being mixed with vice.

The Museum of the Central Technical University will be got together on different principles. As soon as the general Schedule of the proposed Courses of Study has been made out, and the necessary Illustrations of every kind have been ascertained, these will be mentally digested together with the intellectual requirements of the surrounding Population, an eye being had at the same time to the available means and space. Thus a few distinct Collections will suggest themselves; each of these according to its nature and purpose will divide itself on paper into Classes, Sections, Groups and Sub-groups, and on this canvas, Science and Common Sense will assist each other in methodically inscribing the future contents of the Museum. To make the list for each Department sufficiently complete, the assistance of special Experts will be called in. To prevent any Department from expanding out of due proportion to the rest, special zeal will be restrained by general supervision.[1]

The WORKING PROGRAMME being completed, and a place devised for everything, the task will begin in earnest, of getting everything to its place. In the practical accomplishment of this, a certain discretionary power must be allowed, and considerable fertility of expedients called into action; for resources which one has had reason to reckon on may not always justify one's expectations, and on the other hand certain sets of articles may come in, rather beside the exact line, or out of due proportion to others, and yet too valuable

---

(1) Respecting suggestive Inventories and Working Programmes that may be inspected at the renovated Twickenham Economic Museum, see the Note at the end of the first Section of Chapter III.

to be refused; but on the whole, if the Programme is really the work of experience, forethought, and disinterestedness, and if the same principles preside over the expenditure of a liberal subsidy in realizing it, the proposed Museum will rapidly acquire sound educational efficiency, and in due season become an acknowledged model of its kind.

The willingness of Manufacturers and Tradesmen to come forward with their contributions on such occasions for the sake of the advertisement, will be turned to account as far as it is found to work in the right direction; but implicit confidence must not be placed on this resource, which is often too much relied on. Eagerness for increased publicity does not always proceed from consciousness of merit, and the best industrial gems are sometimes kept away by shyness, or the pride of worth. A certain amount of expenditure will be well employed in filling gaps left by speculative generosity, and in procuring for instructive censure at the Lecture Table, articles which if accepted as donations, would become entitled to silence. It is generally the etiquette of Public Museums to avoid the responsibility that might be incurred by explaining the merits or demerits of the various Articles; but the Central Technical University, strong in its unbiassed sense of duty, and in the practical abilities of its Professors, will make it a rule to afford, through the Labels and Handbooks of the Museum, no less than through the verdicts of the Laboratory and Lecture Hall, a plain and salutary guidance to all who seek it.

C. *Rules for Guidance in Arrangement and Display.*

1stly. The construction of the MUSEUM, like that of every other part of the University Buildings, will be in strict accordance with its purpose, and its general style of Decoration will be a lesson of neatness without pretence.

2ndly. The Stands, Glass Cases and other Fittings and Appliances of every kind, will be thoughtfully selected or contrived with a view to the following *desiderata* :—[1]

— Adaptation to space and light.

— Easy conveyance to the Lecture Hall of portions of the Collections not possessed in duplicate.

— Easy interpolation of unexpected additions, or exchange of unsatisfactory Articles or Diagrams for better ones.

— Convenience for keeping Pictorial Illustrations and other accessories near the Articles to which they appertain, and generally for compactly grouping together the often heterogeneous belongings of each Subject; references being made to those which on account of their bulk or other causes must be placed elsewhere.

— Convenience for close inspection of objects

---

[1] Various Illustrations of the devices and appliances adopted for these purposes at the Twickenham Economic Museum, are preserved in the depository erected on part of the site which it occupied before its destruction by fire on the 5th of April, 1871.

requiring it, and for showing a good amount of surface with small quantities, *e.g.* flat bottles, and glazed boxes with moveable bottoms.

— Exclusion of Dust and Smoke. Preservation from Insects, Damp, and other Causes of deterioration.

— Facilities for frequent Cleaning.

3rdly. All reasonable precautions will be taken against Fire, by appropriate modes of Construction and Materials, by a ready Water-Supply, and by good practical regulations.

4thly. The order of the Departments in each Division of the Museum, and of the Articles in each Department, will, as nearly as conveniently possible, be that adopted in the Courses of Studies, and in the corresponding Text Books; so that these may be used by the Students in their regular visits to the appropriate parts of the Museum, after the respective Lectures or Lessons.

5thly. In order to secure the full benefit of the foregoing suggestion, the arrangement of the Stands and other Fittings, and of the passages or gangways between them, will be such as to induce even the casual Visitor to take things in their proper order. He will be guided by suitable indications, and referred occasionally from one part of the Museum to another, as from one Chapter to another in a Book.

6thly. What is sufficient for an advanced Student, would be bewildering for a Beginner. Accordingly, certain indications, as for example Tickets of a particular colour, or bearing the words "commonly used," will point out to him for special notice the things that every one ought to be acquainted with, and a special Guidebook will tell him all it is most essential he should know about them, in so companion-like a way, as to leave the least possible room for regretting the absence of a Demonstrator.[1] It is evident that any intelligent person desirous of mastering a given range of scientific or technical information, who in successive visits to a Museum thus organized, attentively collates what he reads with what he sees, enjoys as near an approximation to, and as good a substitute for, a series of oral Lessons, as can well be devised. He has not the advantage of witnessing experiments, or of being able to ask questions, but he has that of adapting his visits to his leisure, and his progression to his mental strength.

7thly. Instructional Labels in bold type will court the Visitor's glance, and forcibly impress on his mind condensed abstracts of information particularly deserving his attention.[2]

(1) The plan of having the objects orally explained to successive groups of Visitors has been tried, but not found to answer where Visitors are numerous.

(2) When first I began organizing my Economic Museum, I intended to have everywhere in convenient proximity to the various articles displayed, Instructional Labels, giving the required information concerning them; and this plan was carried out in a portion of the Food Department; but I found that if adopted throughout, it would add too much to the difficulty, unavoidably great, of arranging a multitude of illustrations of all kinds within convenience of inspection, and moreover that the few persons who would take the trouble to read so much printed or written matter, would prefer taking it home with them in the form of a Hand-book. I consider that in the enlarged Food Collection instituted at the South Kensington Museum, and since removed to Bethnal Green, a wise course has been adopted in displaying only condensed abstracts of pithy information, which could be printed in large type and read at some distance. The Collection of these Labels, especially of those giving food analyses, is highly instructive, and has proved of great service to many local Museums, including my own, to which they have been courteously distributed.

D. *The Economic Collection.*[1]

The Collection of all others that will equally benefit the Public at large, and the University Students, will be that illustrating the Applications of Science to Domestic and Sanitary Economy. The Science of Daily Life, as already stated, has two Divisions.[2] The one consisting of a groundwork of ELEMENTARY SCIENCE is not easily taught by motionless display, and Visitors who wish to see its paraphernalia, will be recommended to see them in action at the Lecture Table. The other Division on the contrary, extending over the picturesque ground of APPLIED SCIENCE, is particularly adapted for supplying to the Public at large, in an attractive form, the principles of Health and Comfort, whilst at the same time, as the origin and progressive manufacture of many things must be gone into for understanding their nature and value, it supplies to the Technical Student, excellent Technological Practice. Accordingly this ECONOMIC COLLECTION, developed on nearly the same plan as that proposed for a College Museum, but on a vastly expanded scale, will receive every care and attention that may render it in all respects a standard for general imitation, and a national place of reference.

(1) The word *Economic* applied to a Collection or Museum, implies when taken in its fullest sense, that the contents illustrate Domestic, Sanitary, Industrial, Educational, and Charitable Economy.—A brief account of the Economic Movement, and of the formation of the Twickenham Economic Museum, will be found in the next Chapter.
(2) See the Syllabus in Chapter IV., Section 1.

E. *The Industrial Collection.*

Besides the amount of technological Illustrations which appertain to a complete Economic Collection, a great many others of less general interest, will be required in an Institution where Technical Industries will be constantly the order of the day, and where their study will be pushed by some to its utmost limits. It will therefore have to be considered to what extent the technological should be separated from the economic Illustrations, and transferred to a distinct INDUSTRIAL COLLECTION. Among various points that will have to be determined respecting the latter, will be the question whether new Inventions and Improvements should be exhibited for a time as such, in a special Department.

F. *The Commercial Collection.*

COMMERCE that supplies Daily Life with its necessaries, and manufacturing Industry with its Raw Materials, is well entitled to have a division of the Museum to itself, where many of these things may be exhibited over again, under the entirely different light thrown on them by a geographical classification; that is to say, by arranging them, not according to the purpose for which they are employed, but according to the Country or Region which supplies them.[1] Charts of

(1) The late Sir William Hooker, to whose enlightened administration Kew Gardens owe so many useful attractions, fully admitted the principle that differences of purpose, classification, and explanation, may make very distinct Museums with nearly the same materials. It may be remarked

Geological, Physical and Meteorological Geography, with others showing the consequent ranges of natural and artificial production, Tables of Commercial Statistics brought to the most striking form of visual expression, and in short everything that graphic illustration or printed matter can legitimately do, will be done to enhance the practical value of this Collection for all interested in commercial transactions.

G. *The Art Collections.*

The extent to which it may be desirable to develop the Art Gallery of the University Museum, will in some measure be in inverse ratio to the facilities which the proposed Establishment may possess for intercourse with South Kensington and other Art Centres, and in direct ratio to the desirableness of a local Collection.— For the present it may suffice to allude to the following points.

—The various Schools and high-art styles of Painting, Sculpture and Architecture, will be represented as thoroughly as circumstances may admit by typical examples, carefully selected for the instruction of the Public as well as of the Students. A critical Guidebook or *catalogue raisonné* will point out in detail the beauties and defects of each work, as well as the general character of each School and Epoch.—A thoughtful consideration will be devoted to the question of SUBJECT in Painting and Sculpture, and every endeavour will be made to refine not only the VISUAL TASTE, but also

that an Economic Collection will display a far greater number of articles, and go far more minutely into every interesting particular concerning each article, whilst a Commercial Collection will multiply the specimens of a staple article by a far greater number of places producing it.

the ÆSTHETIC SENTIMENT of the Working Classes. The principles of selection to which Art Galleries owe their origin and mode of growth, differ too often from those which would best conduce to popular improvement. The *commercial* value of a Picture, which exercises a greater influence on the appreciation of advanced Connoisseurs than they are always aware of, is to a great extent determined by other considerations than intrinsic merit, and the mere fact of rarity may make up for a multitude of sins. Sometimes admiration for a piece of foreshortening or some other technical difficulty, sought and conquered, carries all before it; at other times a style of outline which would make a false impression on the minds of our Designers, and if imitated by them would lead to stiffness, is not only pardoned, but viewed with interest as "*chasteness of contour*" in the productions of some of the early Italian Masters, for instance FRA ANGELICO and the BELLINI, because mental allowance is made for initiative genius. Similarly a ready excuse is found by an Amateur for the "*quaintness*" of certain old specimens of other schools, which if entrusted to Students not properly imbued with the spirit of the thing, would simply lead to caricatures.—There are certain associations and conventionalities in the classical refinement of the upper classes, which rightly or wrongly, have a potent influence on their appreciation of the moral element in works of Art; but those whose eyes and minds are yet comparatively in a state of nature, may perceive nothing humanizing in the most artistic representations of bodily torture, nothing edifying in the most masterly renderings of voluptuousness, and nothing elevating in a drunken scene, even though portrayed by a TENIERS or a BRAUWER. The subjects painted by the latter in particular are taken from low life, of the most unpleasing class. From the extraordinary skill displayed in the execution, the excellent colouring, the correct drawing, and the life and character of the design, they fetch a high price; but this appreciation of connoisseurs should not secure them a place in the Galleries of the People. Again, it would be of questionable expediency to exhibit without comment, the anachronisms of costume which disfigure certain productions of noted artists, as for example of BASSANO and even of REMBRANDT, or to display without explanation landscapes either deviating from nature, or copying from nature effects which are not what one would call *natural*, or violent architectural perspectives which are only right when the eye viewing them is at the proper *focus*.

—From the consideration of the various styles into which oil-painting divides itself according to the nature of the subject, whether History, Genre, Portrait, Animals, Landscape, Marine, &c., we pass to Paintings in Water-colour, and thence to a number of subordinate branches of the Arts of Design, the respective capabilities of each of which should be illustrated by types meriting study and imitation. The same applies to the various Polygraphic Arts, that is to say, to the Arts producing multiplied impressions or casts.

—Either incorporated with the foregoing or arranged separately, as may be found expedient, will be select Chronological Illustrations of the origin and progress of certain branches of Art.—If this Department is properly attended to, it may check the retrograde tendency observable in some branches after they have reached a certain degree of efficiency. An instance of this is afforded by Lithography. At certain times too the style of Polygraphic Art selected for a given class of subjects, say Prints of local Scenery, is less effective without being cheaper than one previously in fashion. This tendency might similarly be checked.

—Distinct in character, as exhibiting the practical means rather than the intellectual aspirations of Art, will be a Technological Series, showing for each notable kind of Artistic Productions the materials and appliances used, and exemplifying every successive stage of process.

H. *The Music Department.*

Allied to ART by poetical affinity and humanizing power, and at the same time supplying, as we shall soon see, materials of more than one kind for Visual Instruction, MUSIC next claims our attention.—It would be inexpedient that the Central Technical University should interfere with the work of the Academies of Music, that is to say, with the advanced teaching of professional Vocalists, Instrumentalists, and Composers; but it will be quite within its province,—*firstly*, to promote in general, as a matter of principle, the judicious and frequent enlivenment of industrial toil with intellectual and physical recreation; and *secondly*,—to promote in particular, by all reasonable means, the diffusion among the Working Classes of MUSIC, the most elevating of all recreations. Thus for instance we may imagine in a part of the Building not remote from the Museum, Class-rooms for vocal, instrumental and theoretical Music, to which the University Students of all denominations will be invited by instruction of the right sort on the most easy terms. A particular invitation will be addressed to persons preparing themselves for Science Professorships at Training Colleges, and also to Schoolmasters and Schoolmistresses who may, as mentioned elsewhere, come to the University for direct instruction in the Science of Daily Life, and the way of teaching it in Elementary Schools. Vocal Music associated with that instruction on the same principle of showing how to extend knowledge to the *maximum*, by reducing difficulties to the *minimum*, may through the instrumentality of these popular Teachers, become for the rising generation a source of manifold enjoyment, and a real *charm* against evil.—We return to the subject of Visual Instruction by entering a spacious Music Hall for choral, orchestral, and wind-band practice, along the walls of which are displayed Illustrations of the history of Music, and of the several systems of Notation and Tuition used in various countries, together with an assortment of good and cheap Instruments for the Million. Here may be seen the ingenious contrivances devised for beating time, and for the autographic notation of Music improvised on the Piano; also the appliances used in ordinary Music Printing. Here too, if not more conveniently located elsewhere, may be exhibited the interesting collection of illustrations of various kinds, wanted for the technical Instruction of Artisans engaged in the several branches of the manufacture of Musical Instruments.

I. *Miscellaneous Departments.*

The question of healthful bodily exercise of various kinds, of Gymnastics and Athletic Sports, Swimming, Drill and the like, has received sufficient attention of late years to produce many devices of which Models, Drawings, or other representations are well worthy of display. They will however find a place in the

Hygienic Department of the Economic Collection, unless they should be so numerous as to deserve being formed into a distinct Collection.

EMIGRATION has so much importance as a last resource for unsuccessful industry, that everything concerning it will be gone very thoroughly into by the Professors of Industrial Economy, whose business it will be to see that the peculiarly favourable opportunities offered through the relations of the Central University with all parts of the Empire, be turned to the best account for collecting reliable information of the kind that intending Emigrants require. Any useful guidance it may be possible to give to the swarm from the industrial hive, either by the organizing of special Lectures in the proper localities, by the publication of special instructions, or otherwise, will be quite within the circle of the Central University's initiation and supervision, though not perhaps of its direct and continuous action. Still more evidently within the range of its attributes, will be to promote the formation of Museums of Emigration at London, Liverpool, and perhaps at one or two other eligible places. The only question, and that merely one of convenience, which will depend in some measure on the site selected for the Central University, will be whether a Model Collection of this kind should be included in its own Museum, or be originated separately elsewhere.—A small collection of special requirements for Emigrants, &c., with indications of "Self-help for Emergencies," and contrivances of all kinds for lightening labour, as well as means of safety protection and convenience under the varied circumstances of travel, formed a part of my late Economic Museum; but I perceived that to do justice to so important a subject, a special Collection would be required, which I contemplated either as a Department of the East London Museum, or as the leading feature of a distinct establishment uniting every device for securing to Emigration the essential conditions of success. The following are the heads of the skeleton Programme which I drew up with a view to a " Museum of Emigration ":—

    1stly. The choice of the future Fatherland.
    2ndly. The Exodus.
    3rdly. The new Home.
    4thly. Instruction and recreation for self and the rising family.

The deviations from the ordinary routine of Domestic Economy which occur in the life of a Soldier, or a Sailor, and indeed in many other special pursuits, deserve being made the subject of particular attention in their respective special Institutions. An interesting example of one of these is afforded by the United Service Museum, where much care has been bestowed on collecting whatever might, under peculiar professional conditions, promote personal health and comfort.—It is in its behaviour towards Institutions of this description that the Central Technical University will be able to show the difference between obtrusive interference, and the welcome influence of disinterested assistance.

J. Means for propagating Visual Instruction.

In order to give activity and life to the dispensing of practical benefits and assistance by the Central Technical University to affiliated Institutions, far and near, it is proposed to organize among the Students and Junior Professors, an Association for carrying on by partly remunerated and partly gratuitous Labour, mental and manual, the following and other analogous operations, which are mentioned in connection with the Museum, because in or near it more than anywhere else, will the required resources of every kind be concentrated :—[1]

— To furnish on demand full Instructions as to the general principles of selection and classification which should guide the formation of local Museums illustrating Domestic, Sanitary, Technical, and Commercial Economy. For Institutions supplying sufficient data concerning the local industries and resources, suggestive Programmes and classified Lists of Desiderata will be specially prepared.

— To distribute Samples, Models or Drawings of the best and cheapest Museum Appliances, with addresses and prices. Also indications for turning to the best account every means of preservation and display, including makeshifts that may be available where better appliances cannot be had.

---

[1] An opportunity will thus be created for testing under favourable circumstances the principle of the Cornell University, previously adverted to, of mixing in the Studies of Artisans, a sufficient amount of remunerative work to help them through their term.

— To distribute Instructional Labels printed in bold Type, and to supply instructions and information for preparing Guide-books.

— To prepare for being supplied at a very low scale of charges, complete Sets of Illustrations of the Science of Daily Life. Some of these Sets will be got up in the simplest and cheapest style of material and workmanship, in order to meet the limited resources of Primary Schools and small Industrial Institutes; others in a rather more expensive style for Institutions of a higher class.

— Particular attention will be paid to cheapening the production of models and apparatus illustrating Mechanics and Machinery, on account of their utility for technical instruction, and their usual expensiveness.

—Particular encouragement will also be afforded to the simplified production of Models showing the best construction of the Premises, Fittings and Appliances used for Technical Processes.

— Production of Models for instruction in Crystallography, Stereometry, and Drawing from the Solid.

— Production at reduced rates of Phototypes, Chromolithographs, photographic Magic Lantern Slides, and every other successful means of Illustration for Instructional purposes.[1]

— Distribution at cost price of educational Prints in various styles, and of certain other requisites for Art Schools.

— Distribution and Exchange of Specimens and

---

[1] For an account of the highly successful use made of the Magic Lantern and its congeners in scientific, historic, and other instruction, see 'Les Mondes,' edited by the indefatigable Abbé Moigno, rue Bernard-Palissy, No. 11, Paris. A Catalogue of the slides, &c., may be obtained of the manufacturers, J. and A. Molteni, rue du Château-d'Eau, No. 44.

Samples of all kinds for Museum or Class-room Instruction.

— Gratuitous agency for the purchase and transmission of Articles appertaining to scientific and Industrial Instruction.[1]

— Active Correspondence with home, colonial and foreign Institutions, on all matters relating to educational and industrial progress.

Various other measures might be suggested as calculated to facilitate the spread of genuine practical knowledge, and as likely to be very acceptable to provincial Institutions with limited means; but the foregoing enumeration is sufficient to show the thoughtful as well as disinterested spirit, which will prevail throughout the operations of the intended University.

(1) To this might be added any means by which the distribution of surplus specimens by the British Museum and other wealthy collections might be facilitated. "The authorities of the British Museum should be empowered to dispose, by gift, in favour of local museums of any specimens which may be ascertained to be duplicates."—Fourth Report of H. M. Commissioners on Scientific Instruction, &c.

CHAPTER III.

FACTS IN SUPPORT OF PROPOSALS.

SECTION 1.—ECONOMIC EXHIBITIONS AND MUSEUMS.

HAVING in the two preceding chapters sketched the general physiognomy of the proposed System of Technical Education, and shown the adaptation to each other of its several features, it is time that I should enter into the origin and rationale of some of these, adducing facts in support of those proposals of which the expediency may not be self-evident. This kind of justification is particularly required in respect of the principles which have guided me in determining what Scientific Knowledge Working Men want, and how they may acquire it. It is necessary to show that these principles have not been biassed by a connection with or predilection for any particular department of Science, but have resulted from a concourse of circumstances peculiarly favourable for studying in various forms, popular ignorance, its consequences, and its remedies.

For more than thirty years of the earlier part of my life, I dwelt alternately in France, Switzerland, and Italy, and though hampered by infirmities resulting from several severe accidents, I could so far identify myself with the languages and usages of widely differing localities, as to be able to compare with advantage the intellectual and physical condition of their respective Working Populations. Thus I became more and more convinced :—*firstly*, that much practical benefit would accrue to all nations, if they would adopt on principles of good fellowship, a cordial interchange of useful notions, habits, and contrivances; and *secondly*, that still greater advantages might accrue from bringing the united exertions of science and inventive industry, to a direct and constant bearing on the requirements of the Million.

Considerable difficulties however stood in the way of any scheme for realizing these views. Everywhere ignorance and habit kept back improvement, or only allowed it to progress slowly in given grooves, and engendered an unthinking prejudice against new devices, and especially against those of the foreigner. Then as regards the willingness of Men of Science and Manufacturers to stoop as it were to minister to the wants of the Million, the kindly spirit or the pious sense of duty which might have prompted this, were too often found wanting.[1]—Even among the Clergy

(1) The following remarks on this subject are from a Memorandum which I brought out in the spring of 1855 :—

"If we look at the usual Exhibitions of Industry in this and other Countries, we see there articles of ornamental and external beauty, or of elaborate and costly workmanship, rather than of real and durable utility to the Million; well-paid additions to the luxurious enjoyments of the rich, rather than ill-paid contributions to the comforts of the poor. Or if we take patents as a criterion, we see that by far the majority of patented inventions and improvements are brought out with a view to emolument or renown, and that they are accordingly addressed to those privileged classes which deal out money and reputation.—Manufacturers, tradesmen, and mechanics, work their inventive abilities in a business-like manner. Men of Science, more specially so called, have not yet, generally speaking,

whose true interest it was to be as in former ages the foremost pioneers of useful knowledge, too many were ignorant enough not only not to care for Science, but even to regard it with apprehension.

These facts first conveyed to my mind by my continental experience, were afterwards more strongly impressed on it by a special knowledge of the wants and resources of British Workmen, acquired during an apprenticeship of many years as Member of the Committee of the Labourer's Friend Society, under the chairmanship of the illustrious philanthropist, Lord Shaftesbury. Thus arose my idea of advocating in this country and on the continent, the formation of public collections of the objects best calculated to meet the wants of the Million ; collections which while they taught the rationale of kind, quality, and adaptation to purpose, and brought into prominence the best existing resources, might powerfully stimulate and safely guide the further exertions of Science and Industry. It was obvious that these collections might be in the form of permanent Museums, and that in this way could best be developed their instructive character ; or that they might simply be in the form of temporary Exhibitions, and that this was the plan best

attained to such a standing in this Country, as to be able to lay aside the thoughts of themselves and their families. Even amateurs, though their position may raise them above pecuniary considerations, are seldom above the temptation of bringing out that which may be conspicuous, in preference to that which might be quietly useful. In short, the homage of knowledge and ingenuity is very naturally paid to those who can best give an acceptable return ; and whilst every appliance of Science and Art is called into requisition to meet the fastidious fancies of the rich, and to supply novelties for the cravings of fashion, the requirements of the humbler classes are comparatively overlooked.—The sum of this is, that we must do all we can to make it suit the pecuniary interest or the personal feeling of inventors and producers, to attend to the wants of the Industrial Population."

suited for starting the movement, offering as it did to Exhibitors a more attractive chance of publicity and distinction.

These views after having been sanctioned by the Council of the Society of Arts in 1852, obtained in July 1855 the active support of the Philanthropic Conference held at Paris. The following resolution was adopted at one of its sittings :—

" La Réunion émet le vœu, conformément aux vues exposées dans le mémoire de M. Twining, qu'il soit constitué dans les divers pays un musée économique permanent, où seront réunis et classés tous les articles destinés à l'usage domestique et qui se distinguent par des qualités d'utilité, de solidité, et de bon marché, ainsi que les procédés et les appareils qui se rapportent à l'hygiène et à l'assainissement des habitations, des ateliers, etc."

A special Committee was at once appointed for selecting throughout the Universal Exhibition, then open, the articles most conspicuous for cheapness, usefulness, and good quality, and thanks to the friendly assistance of the Commissioner-general, M. Le Play, a Gallery of Domestic Economy was inaugurated on the 15th of September. Its success was such that M. Augustin Cochin expressed himself as follows in a Report prepared on behalf of the Committee :—

" Désormais, aucune Exposition universelle ne doit avoir lieu sans qu'un large espace soit réservé à l'exhibition spéciale des objets utiles au bien-être physique ou au développement intellectuel des classes les plus nombreuses de la société."

The principle thus initiated gained everywhere such

ready acceptance, that nearly every International Exhibition held on the Continent since that time, has devoted a department to means for the improvement of the condition of the Working Classes, and that several important Exhibitions have been held in various countries for that distinct purpose. Among these are particularly deserving of mention the Domestic Economy Exhibition organized on an extensive scale at Brussels in 1856, by the illustrious Belgian philanthropist, M. Ed. Ducpetiaux, and the great Amsterdam Exhibition of a similar character, held in 1869 through the exertions of Baron Mackay.

There is one point to which I particularly wish to draw attention as connecting these details with our present purpose; namely, that though the domestic and sanitary requirements of the Working Classes were originally thought by some too vague in their circumscription to constitute a manageable subject, yet whenever in the various countries, men devoted to the welfare of their respective Industrial Populations have taken it up earnestly and methodically, there has been in their several ranges of ideas a very satisfactory conformity with each other, and with the original Programme, in all essential particulars.

Whilst however Temporary Exhibitions of the foregoing kind are, when well directed, of immense advantage for giving to manufacturing industry an impulse in the required direction, and for inducing the spontaneous juxtaposition of articles which it would be otherwise very difficult to get an opportunity of comparing with each other, yet I felt all along that the most practically useful lessons would be those taught by Permanent Collections, organized on more strictly educational principles, so that one might not only see the things to be adopted or avoided, but learn at the same time the "reason why." It was in this spirit that I began in 1856 to form at the House of the Society of Arts, the permanent and decidedly educational collection of the things appertaining primarily to DOMESTIC and SANITARY ECONOMY, and subsidiarily to Industrial, Educational, and Charitable Economy, which from its being thus devoted to ECONOMIC KNOWLEDGE, took the name of ECONOMIC MUSEUM.

The Food Department was exhibited in embryo at South Kensington in 1857, and gave rise to Dr. Playfair's admirable Food Museum.[1] The next year it was transferred to the Polytechnic Institution, where other classes were added. It was evident however that my infirmities precluded any satisfactory development of the collection at a distance from the place of my abode, and accordingly it was removed in 1860 to a capacious building erected for the purpose close to my residence at Twickenham. My intention was, after sufficiently working out its various departments, to consign it to some suitable London Institution where, besides imparting instruction to numerous visitors, it might serve as an inducement and a convenient pattern for the establishing of analogous collections in all parts of the country.—My hopes were foiled in a melancholy manner through the destruction of the Museum by fire on the 5th of April, 1871; but together with many valuable Manuscripts, one thing still more valuable has remained to me, namely, the EXPERIENCE acquired through diligently devoting so many years to what may be called the PRACTICAL MANIPULATION OF

---

(1) See the fifth Report of the Science and Art Department of the Committee of Council on Education, 1858.

THE REQUIREMENTS OF DAILY LIFE. That experience bears *inter alia* on the following points :—

1stly, the determining, selecting, and classifying of the departments of Economic Knowledge needed by the bulk of the population;

2ndly, the selecting or devising of the Articles, Models, Diagrams, and other representations best suited for the visual illustration of that Knowledge, and also of the best, cheapest, and most compact appliances and contrivances for displaying the same.

3rdly, the devising of the most effective and acceptable means for diffusing that Knowledge among the Working Population at a distance, supplying at the same time a substantial Scientific Foundation.

The discussion of these points in their direct relation to the duties of the proposed Central Technical University, will be the subject of the following Sections.[1]

---

(1) A Building has been erected on part of the site occupied by the former Museum, where persons interested in the propagation of Practical Knowledge among the Industrial Population, have at their disposal the following resources:—
— A Library of Reference.
— Inventories, Working Programmes, and other manuscript documents, calculated to facilitate the formation of Economic and Commercial Collections of various calibres.
— A Show Room containing a selection of the most appropriate Stands, Cases, Glassware, Labels, and other Museum appliances.
— A Laboratory, Workshop, Artists' Room, and Repository for preparing and storing Sets of Illustrations required for circulation with gratuitous Courses of Popular Lectures. (See Section 3 of this Chapter.)
— Programmes, Questionaries, Candidates' Cards, Prize Certificates, and other items connected with the holding of Artisans' Examinations in Elementary Science. (See Section 5 of this Chapter.)
The Building still retains for purposes of correspondence the name of " Twickenham Economic Museum." Enquiries and other communications to be addressed to the Superintendent Wm. Hudson, Esq.

SECTION 2.—RANGE OF ECONOMIC KNOWLEDGE REQUIRED BY THE INDUSTRIAL POPULATION.

It would be difficult to imagine a more interesting occupation, or one better calculated to give a thorough insight into the requirements and resources of the Working Population, than that of collecting, examining in all their details, selecting, and arranging for visual study, the infinitely varied articles which come within the scope and powers of display of a full sized Economic Museum, either by actual presence, or through the proxy of a Model or a Drawing. Neither is any pursuit likely to tempt the studious Economist into more diversified experiments of classification. He tries successively the theoretical and the practical principle, the arrangement according to origin, and according to purpose, and most probably ends, as I have done, in adopting a compromise between them; occasionally sacrificing in some measure the strict logic of scientific affinities, to that intuitive appreciation of common sense impressions, which tends to make classification pleasant to the mind, and easy to the memory. Another expedient which I found very effective in keeping selection and arrangement in a practical direction, was that of assuming that I was lecturing to mixed, but chiefly Industrial audiences, on the means by which they might hope to prolong life, and to promote its rational enjoyment :—The discourse thus originated in my mind would naturally be about what was wanted for supplying a Guide-book to the proposed

Museum, whilst the required Illustrations of every kind would be exactly suited for forming the Museum itself.

The subject which naturally presents itself first in reviewing my experience of an Artisan's requirements, is one of such paramount importance that a few words of special consideration will demand no apology. It is the HOME to which his thoughts recur even as he toils, and where they fain would nestle in an imagery of comfort and repose, but are too often met with a cold and cheerless reality that drives them to the Public House. It would be irrelevant to enumerate the many aspects in which Dwellings Reform has presented itself to me during the years in which I made it the subject of my special study.[1] There are however one or two points which must be particularly kept in view in arranging this portion of the studies of Artisans, and the corresponding Economic Illustrations, especially the model Economic Collection at the Central University.

DWELLINGS REFORM for being carried out on anything like the scale required by vast and urgent necessities, demands the mutual co-operation of the Capitalist and the Working Man. The latter can do much by investing his savings in bricks and mortar under well devised and wisely managed organizations; but he can do more by an intelligent appreciation and acceptance of accommodation provided on an extensive

---

(1) A number of manuscript documents are available for perusal by persons specially interested, at the Library in the new Museum Building, together with a selection of Books bearing on this subject, and a more comprehensive list of them. These publications may be divided into two Categories: those which chiefly aim at bringing into relief prevailing defects and errors with their consequences, and those which deal with the various proposed forms of improvement.

scale by that wholesome mixture of good-will and fair returns which may be called *remunerative philanthropy*. Often, the good-will has been there, and the capital has flowed abundantly; but the result has been a failure for want of judicious application. Often, on the other hand, we have seen endeavours as judicious as they were generous, baffled by actual want of *appreciation* on the part of those classes for whose direct benefit they were made; not to say by stolid apathy or wanton prejudice. Nothing of all this would have occurred, if the minds of rich and poor had been clarified with that scientific acumen which it is now sought to introduce. It will be shown in the proper place, that the Second Division of even the Grade 1 Course of General Scientific Foundation, prescribed as the minimum of Science for the Industrial Classes, will afford a good opportunity for impressing on them in these times of Typhus and Cholera, the life and death importance of Domestic Hygiene; an opportunity which will largely increase in the second Grade Course. Then, besides serving to support oral teaching in the Lecture Hall, the numerous illustrations of Dwellings Reform publicly displayed in the University Museum, will address their Visual Instruction to those who build, as well as to those who inhabit. First, attention will be roused by startling representations of the ill-situated, ill-contrived, crowded, and squalid habitations which both in Town and Country still bid defiance to the laws of health, notwithstanding all the arguments of Science and the eloquence of Philanthropy; then will follow, partly displayed, and partly in portfolios for reference, a carefully classified Series of plans, elevations, working drawings, models, &c., of all the most successful designs for the erection of improved Dwellings in Town and Country, as well as for the adaptation of existing ones. From this Central Repository, organized with such enlightened impartiality as to deserve the co-operation of the various Bodies and Individuals most eminent for services rendered to the cause, will be supplied abundantly to the provinces, such Illustrations as are susceptible of multiplication, and best calculated to give impetus and safe guidance to the movement.

After these preliminary remarks due to the important branch of economic inquiry through which I was led to the study of the rest, I will proceed to give methodically concerning my late Economic Museum, such details as are necessary for understanding the Educational System to which its development has gradually led me. This Synoptical account may at the same time afford convenient guidance to persons charged with directing the arrangement of the corresponding portion of the University Museum, or with preparing smaller Economic Collections for other Institutions.

The Twickenham Economic Museum was divided into 9 Classes.[1] The Library of Reference which formed the ninth, had a room to itself; the other 8 classes, with the exception of cumbersome articles which were placed in an annexed space, were contained in a Hall about 80 feet long by 30 feet wide, divided lengthwise by rows of Stands, into 4 parallel gangways or passages.[2] Along both sides of these

---

(1) A Programme furnished with a Plan may be had on application.

(2) The Stands which were made of deal in 5 and 10 feet lengths, were so constructed as to supply both wall and counter space. Specimens may be seen with the other appliances in the present Display Room. If used in a Museum adjoining a Lecture Hall, they might be mounted on Castors.

passages, the Exhibits of every description were arranged in due order of sequence, so that in walking up along the left side of each passage, and down along the right side, the Visitor would naturally inspect the several Classes, with their respective Sections and groups, in due order of succession, as one reads the Chapters and Paragraphs of a Book.

Synopsis

Of the chief Series of Illustrations in the late Twickenham Economic Museum.

Class I.—Economic Architecture.

PRELIMINARY DEPARTMENT.—Homes of the People in former times and in foreign Countries. Existing lurking places of pestilence and demoralization.

CLASSIFIED SERIES OF MODELS AND DESIGNS.(1)

GROUP 1.—Urban Dwellings on the External Gallery System.
GROUP 2.—Urban Dwellings on the Internal Corridor System.
GROUP 3.—Urban Dwellings on the system of Flats, with two or more Tenements on each floor.
GROUP 4.—Urban Buildings of a mixed character.(2)
GROUP 5.—Renovated and adapted Town Houses.
GROUP 6.—Model Lodging Houses.

(1) This Department was one of the most complete. The models made of Cardboard or Wood, were of very simple workmanship, being merely intended for the instruction of persons unaccustomed to obtain an idea of a Building from a Plan. They have been entirely destroyed, but some of the Designs, though partly burnt, are still susceptible of being used by Architects, and an Inventory has been preserved supplying various details; also forms for collecting Statistics of Model Dwellings.

(2) Among the plans that have been preserved, is one for a large Building, partly on the external gallery and partly on the internal corridor system, intended to unite at moderate rents, apartments for Families and Club Chambers for single Artisans, with every desideratum in the way of health, comfort, and educational advantages. A description reprinted from the 'Builder' is also available for inspection.

SECT. 2.] *SYNOPSIS OF THE ECONOMIC MUSEUM.* 127

GROUP 7.—Suburban Dwellings; comprising Small Houses or Cottages arranged in rows, and suited for the suburbs of a Town or the streets of a Village.

GROUP 8.—Rural Dwellings of every description, forming several sub-groups.

GROUP 9.—Select Foreign Dwellings. *Cités ouvrières*, &c.

GROUP 10.—Habitations contrived for peculiar purposes.

 Sub-group a.—Dwellings adapted for special circumstances of climate or locality.
  b.—Emigrants' Dwellings.
  c.—Moveable Dwellings, Caravans, &c.
  d.—Tents.
  e.—Camp Huts.
  f.—Barracks.
  g.—Guard Rooms and Shelters.
  h.—Living accommodation on board of Ships, Barges, &c.

GROUP 11.—Homes, Asylums, Hospitals, &c.

 Sub-group a.—Homes for Sailors, Servants, &c.
  b.—Alms Houses and Asylums for the aged and disabled.
  c.—Hospitals and Infirmaries.
  d.—Unions.
  e.—Dormitories for the Houseless Poor.

GROUP 12.—Buildings for Educational purposes.

 Sub-group a.—Refuges and Asylums for destitute Children.
  b.—Reformatories.
  c.—Schools.
  d.—Mechanics' Institutes, Working Men's Clubs, &c.
  e.—Educational Museums.

GROUP 13.—Public Drinking Fountains.

GROUP 14.—Public Baths and Wash-houses.

GROUP 15.—Farm Factories, Granaries, Dairies, Slaughter-houses, Markets, &c.

GROUP 16.—Soup Kitchens, Cooking Depôts, Eating and Coffee-houses, &c.

GROUP 17.—Applications of Sanitary construction to Factories, Workshops, and other places of crowded resort.

GROUP 18.—Mortuaries.

DETAILS OF CONSTRUCTION.

Models, Diagrams, &c., showing the best modes of Construction of House-Drains, Foundations, Wallwork, Partitions, Floors, Roofs, Stairs, Chimneys, Dust-shafts, Sanitary Conveniences, Doors, Windows, &c.

Class II.—Materials for Building and for Furniture.

Illustrations of the Geology and Chemistry of the Rocks.—Educational Series of the most useful varieties of Stone. Commercial Series of the chief Building Stones.—Lime, Mortar, and various Cements and Stuccoes. Concrete and Artificial Stone. Asphalt and its substitutes.—Slag and other available igneous compounds.—Bricks of various kinds, including hollow bricks, glazed bricks, fire bricks, &c.—Tiles for paving and roofing.—Slate.—Miscellaneous Roofing Materials. Metal Roofing.

Growth of Exogens and Endogens. Botanical Series of Woods of all kinds for constructive and decorative purposes. Commercial Series of the chief kinds of Timber, with indications for selecting them according to purpose, and using them to the best advantage. Preservation of Wood and other materials from dry rot, decay, and combustion.—Iron as a Building Material.—Varieties of Window Glass and processes of manufacture.—Whitewashing and Distempering. Painting in oils; materials and processes eligible or the reverse. Staining, Graining, Stencilling, &c.—Paperhangings; block and cylinder printing.

Class III.—Fixtures, Furniture, and Household Utensils.

Educational Series of the Metals most commonly used, and their Alloys, with Illustrations of mining, smelting, casting, laminating, perforating, wire drawing, nail- and screw-making; welding, brazing, soldering, tinning; bronzing, lacquering, japanning, &c. Building Ironmongery; including fastenings of every description, and a priced series of articles required for a Model Dwelling.—Appliances for water supply. Filters.—Appliances for gas supply.—Appliances for warming, cooking, &c.(1)—Select serial examples

(1) Grates, Stoves, and other cumbersome articles were collected for more convenient display, in a special part of the building, considered as an Annexe to this class.

of household ware and furniture supplied by the Furnishing Ironmonger and Cutler, the Earthenware, China and Glass Dealer, the Cooper, Turner, Brushmaker and Basketmaker, the Upholsterer, &c. These series were intended to illustrate among other points the following:—processes and stages of manufacture; various ways in which things are made in this country and abroad, and the relative advantages; various instances in which the stuffed furniture, carpets, and the like, prevailing in the Dwellings of the Working Classes, might be replaced by other appliances with a great advantage to health, and without increased expenditure or sacrifice of real comfort.(1)—There were also in this Department, estimates illustrated with actual specimens representing gradational assortments of Hardware, Earthenware, and Glass Articles, adapted to 3 or 4 Grades of requirements and means, from those of a Labourer, to those of a Skilled Workman, Foreman, or Clerk with first-rate emoluments.(2)

Class IV.—Textile Materials, Fabrics, and Clothing.

Educational Illustrations of Spinning, Weaving, Bleaching, and Dyeing with its allied processes; fast and fugitive colours.—Textile Materials as produced, and in various degrees of preparation for the Loom; including:—Hemp, Flax, Jute and other analogous Fibres. Cotton. Silk of various kinds. Wool. Hair.— Furs.— Leather, its various modes of preparation and chief varieties.

Matting. Carpeting; Druggets; Felts.—Horsehair Fabrics.— Fabrics for curtains, bedding, house and table use, and for all

(1) The subject of Furniture has by some been considered beyond the capabilities of illustration of an educational Museum; but my experience enables me to affirm without hesitation, that if due pains be bestowed on the selection of points to be elucidated, and on the choice of means of illustration, a great deal of attractive visual instruction, possessing considerable hygienic importance, and borrowing valuable suggestions from foreign countries, can be satisfactorily got together in a moderate amount of museum space.

(2) Typical series of this description, with estimates appended, are when prepared conscientiously and with much selective care, well calculated to check wasteful expenditure, by showing that cheapness need not necessarily preclude efficiency or tastefulness. It is however necessary to revise such collections from time to time, on account of the changes which supervene in the price of articles, and the rate of wages.

descriptions of under and outer, summer and winter Clothing.—Trimmings, Fastenings, and other miscellaneous accessories and ornaments of Dress.

India-rubber and Gutta-percha; their preparation and applications. Waterproof Fabrics of various descriptions, including Macintoshes and the like; Leather Cloths; Oilskins, Painted Tablecovers; Floorcloths; Kamptulicon, Linoleum, &c.

ARTICLES OF APPAREL selected as being suited for certain branches of industry, or as conspicuously uniting convenience and appropriateness with cheapness and durability; including:—Haberdashery, Hosiery, made-up Apparel; Head coverings, Foot coverings (*chaussures*) and the like, with illustrations of the respective processes of manufacture.— Priced Lists of articles forming complete outfits suited for various circumstances of age, sex, occupation and means.(1)

The best safeguard against an excessive subserviency to the dictates of Fashion in matters of Dress, is the possession of clear scientific notions as to what is best suited to a given purpose, and clear artistic notions as to what under given circumstances may be most becoming. There is perhaps no more agreeable or more effective way of acquiring this enlightened appreciation, than an impartial study and comparison of the costumes used under various circumstances in various parts of the world. Under this impression I displayed along the whole length of the Clothing Department, an appropriate series of cosmopolitan prints, and gladly availed myself of the willingness of the Science and Art Department, for exhibiting the actual working-class costumes brought from China by Dr. Bowring, which I associated in geographical sequence with dresses of Peasants and Artisans from other Countries.(2)—Equally instructive in a Hygienic point of view, was a collection of Clothing for Infants used in different Countries of Europe, which had been displayed at the Brussels Economic Exhibition of 1856.

(1) A number of these lists, extending upwards from the requirements of a Lad beginning his Apprenticeship, or of a Girl first going to service, were prepared for me at the establishment of Messrs. Silver, and included complete outfits for Emigrant Families bound for distant regions of widely differing climates.

(2) The best part of these Collections having been deposited for the sake of greater publicity at the Crystal Palace, were destroyed in the fire which consumed the " Tropical " end of that Building. The remainder shared the destruction of my own Museum.

Class V.—Food.

This Division of an Economic Museum surpasses perhaps all others in interest, owing to the vast and varied scope it affords for practical improvement through scientific research, and to the peculiar suitableness of most of the information arrived at, for being conveyed or supported by Visual Instruction; without which indeed many important details would be almost unintelligible. This led me to the adoption of various new Museum appliances and devices, including the use of Instructional Labels offering in close proximity to the respective articles, concise summaries of useful information. It was thus that I endeavoured to make easy and profitable the study of Food Economy, even to those who had not had the advantage of a previous acquaintance with the scientific principles on which it depends.

I have throughout been convinced that the plan adopted by some of our most eminent men, of reducing food articles to their *ultimate* Constituents in computing the value of Dietaries, is unsatisfactory, as so much depends on the properties of the *proximate* Constituents of which those articles are composed. There would in fact be no difficulty in proving that the amounts of Carbon, Hydrogen, Oxygen and Nitrogen, which grouped in the form of certain proximate constituents produce a wholesome meal, would, if otherwise grouped, produce indigestible or unpalatable articles, and even poisons. I have accordingly adopted the plan firstly, of endeavouring to give the Student as clear notions as possible of the Proximate Constituents of animal and vegetable Food, their chemical nature, and their nutritive value, and then of reviewing seriatim all the ordinary Food substances, applying as far as possible to each, the test of a Proximate Analysis.[1]

The following heads of classification are mainly selected from those of the original Working Programme of my Museum.[2]

---

[1] The present Display Room contains a small collection designed to illustrate these principles. It comprizes, firstly, introductory Placard Labels giving outlines of the philosophy of Nutrition, secondly, a series of Proximate Constituents with full explanations of their properties, and thirdly, visual Illustrations of Proximate Analyses of the Cereals. For these Analyses the metrical system of weights has been adopted.

[2] A copy of this rather voluminous Manuscript is kept for inspection. It may prove convenient as constituting a tolerably full list of food articles,

PRELIMINARY DEPARTMENT.

Elementary notions of the Chemistry of Food and Nutrition.—Composition of the Human Body.—Proximate Constituents of Food, with Instructional Labels giving their Ultimate Analyses, and leading properties.—Transmutations of organic matter by natural means: saccharine, vinous, and acetic Fermentations; putrefaction. Transmutations by artificial means.—Classified List of processes for the Preservation of Food, with specimens and drawings.—Synopsis of Foods naturally unwholesome, or rendered so by various causes.—Synopsis of common admixtures and adulterations, with indications for detection.(1)—Culinary Science; Rationale of its operations. How may be made palatable that which is wholesome and cheap.—Diet and Dietaries.(2)

FOOD AUXILIARIES.

WATER; Its extensive prevalence and its value in Food;(3, qualities which fit it for drinking, culinary purposes, keeping, (as in sea voyages) washing, &c.; Tests, and means of purification.—SALT: its value in Food;(4) leading varieties.

FOOD SUPPLIED BY THE ANIMAL KINGDOM.(5)

Zoological Synopsis of esculent species.—Composition and nutritive value of the several parts of Animals, such as Bones, Flesh, Blood, Fat, &c. Gelatin; its culinary uses.—Diseased or tainted Meat.

elaborately classified, and also as serving to illustrate the plan recommended in Section 8 of the previous Chapter of forming a Museum according to the indications of a thoughtfully prepared yet sufficiently elastic " Programme," and not of leaving the relative development of the various departments to the uneven bias of chance donations or of contributions made in an advertising spirit.

(1) Independently of this synoptical table, the chief respective adulterations were shown near each article throughout the class.

(2) I found it practically convenient to display the dietary tables in this part of the Class; otherwise I might have preferred reserving this subject for the end of the Food Department, as a Corollary of its teaching.

(3) See Dr. Lankester's Food Lectures, p. 127, and Dr. Carpenter's Manual of Physiology, 4th ed., p. 101.

(4) See Dr. Lankester's Food Lectures, p. 37, and Dr. Carpenter's Manual of Physiology, 4th ed., p. 102.

(5) In the greater part of this Section of my Museum, pictorial illustrations were considerably preponderant, on account of the difficulty of preserving actual specimens in a state suited for close inspection.

SECT. 2.] SYNOPSIS OF THE ECONOMIC MUSEUM. 133

MAMMALIA: Domestic and Wild, including foreign available resources.—BIRDS, (land and water) classified like the foregoing.— REPTILES.—FISH: fresh water and sea; indigenous and otherwise.— SHELL-FISH: Crustacea and Mollusca.—EGGS: various kinds; modes of preservation.—MILK; various kinds of Preserved Milk; Butter; Cheese; processes and appliances.

FOOD SUPPLIED BY THE VEGETABLE KINGDOM.(1)

Botanical Synopsis of Food-supplying Plants.

THE CEREALS, including all notable products, preparations and processes, and giving particular prominence to Wheat, flour, bread, biscuits, maccaroni, &c.; Rye; Barley; Oats; Rice; Maize, &c. Causes of harm to plants and products. Adulterations.—Esculent Leguminous Seeds, or PULSE: Peas; Lentils, Beans; Haricots; Gram, and other foreign kinds deserving of attention.—ROOTS and BULBS.—VEGETABLES of which the stalks, leaves, or tops are chiefly eaten.(2)—FRUITS: eaten as vegetables; wild; cultivated; imported in a juicy state; imported in a dry state, &c.—Cryptogamous plants, chiefly MUSHROOMS, available for Food.—SECRETED and EXTRACTED PRODUCTS: Glutens, Starches; Gums; Saccharine substances, including Honey.—CAKES and CONFECTIONERY, and substances used for adulterating them.—CONDIMENTS, including: Flavouring Herbs and Aromatics; Stimulants and Spices; Vinegar and Pickles Sauces; Fats and Oils.—Tobacco and other Narcotics.

BEVERAGES AND DRINKS.

BREAKFAST BEVERAGES; including Cocoa and its preparations, Coffee, and Tea, with their respective adulterants and substitutes.— FERMENTED LIQUORS including Beers, Wines, and their various adulterants and substitutes. — DISTILLED LIQUORS. — REFRESHING DRINKS.

FUEL AND OTHER HOUSEHOLD STORES.

Considering that Stores for warming, lighting, and clothing are, no less than alimentary stores, articles of actual *consumption*, and

---

(1) Including as far as possible dried specimens or pictorial representations of the plants, and proximate analyses of the products.

(2) In my Collection this Group was divided into Sub-groups which may be respectively typified by the asparagus, the celery, the lettuce, the cabbage and the cauliflower.

not merely of use or wear, and at the same time considering the similarity of the arrangements and appliances required for their respective display, I was induced to unite the whole in one class with two divisions, which was accordingly headed with the inscription "FOOD, FUEL AND OTHER HOUSEHOLD STORES. The Appliances for warming, lighting, and laundry purposes, were allotted to Class III., (Fixtures and Furniture,") the cumbersome ones being relegated to an outlying annexe of that Class. All this proved practically convenient; nevertheless, if I had to form another Economic Museum with ample space and resources, I should prefer uniting the materials and the appliances for warming, lighting, and cleaning, in a distinct class, which would become the Sixth.—In the following brief indications, the things which would be removed into the proposed new Class from Class III. are enclosed in square brackets.

WARMING. Coal; Chemistry of its formation, and Geology of its position in the Earth's strata. Coal mines; their wonders and their dangers. Safety Lamps. Varieties of Coal in common use, and their relative qualities. Fire lumps.—Coke.—Peat.—Wood as an article of fuel.—Charcoal.—Materials for lighting fires.—Appliances for ignition.—[Grates, ranges, stoves, kitcheners, &c., adapted to the wants of the Working Classes; including Gas Stoves and their atmospheric burners.]

LIGHTING.—Flame; its nature, its varieties and their illuminating power.—Gas; illustrations of its manufacture and concomitant products. [Burners and Accessories.]—Liquid and solid materials of all kinds for lighting purposes. [Lamps, Reflectors, Shades, and the like.] (1)

CLEANING. — Soap of all kinds ; materials and processes, auxiliaries and substitutes. Means for the removing of stains, for brightening and polishing, &c. [Laundry fixtures and Furniture.]

(1) It is optional to take lighting after or before heating, and there are indeed good reasons on either side. For instance it is convenient to describe the manufacture of Gas in connection with lighting, before adverting to its use for heating purposes; whilst on the other hand it is desirable to take the history of Coal, before the manufacture of Gas. Doubts of this nature are constantly occurring in an Educational Museum, and must be solved according to circumstances.

Class VI.—Sanitary Science.[1]

Public Works for protection against inundations, &c., for the improvement of unhealthy districts, for water-supply, sewerage, &c. Appliances for the ventilation of dwellings, for preventing inconvenience from damp, smoky chimneys, noxious effluvia, &c.—Disinfecting agents and appliances.

Nursery appliances. Means for promoting a normal development of the frame in infancy and youth, and the healthy maintenance of its functions throughout life; including appliances for judicious exercise and rational sport, for swimming, &c.—Baths, and other means for promoting a healthy condition of the skin.—Orthopædic Apparatus. Crutches. Artificial limbs. Dentistry. Means of relief for deafness, defective vision, (especially myopia and presbyopia,) and weakness of the eyes. Means of occupation for the blind, and specimens of their work.—Appliances for the sick chamber. Means of comfort for invalids. Ambulances.

Household remedies.—Educational series of drugs commonly used, with Botanical Illustrations. Medicinal herbs.—Drug adulterations. Injurious articles likely to be mistaken for medicinal substances.

Antidotes and treatment for poisons and venoms. Prints of poisonous plants, venomous reptiles and insects, noxious fish and mollusca, animals in a rabid state, &c.—Means of safety from, and destruction of beasts of prey, house and field vermin, &c.—Means of safety or relief from the effects of excessive heat or cold, from asphyxia, drowning, shipwreck, lightning, fire, dangers, and discomforts in travelling by railway or otherwise, and accidents to which children are particularly liable.

Prevention of the injuries and diseases which specially attach to industrial occupations.[2]— Articles of which the manufacture is injurious. Eligible substitutes.

---

(1) In 1861 I arranged the Synopsis of this Class of my Museum, as a suggestive Programme for the use of the Sanitary Committee appointed with a view to the then approaching International Exhibition. A copy remains for inspection.

(2) In 1854 I induced the Council of the Society of Arts to appoint a Committee for the investigation of this subject under the name of "Industrial Pathology," and among the papers now open to inspection, is a Synopsis kindly prepared for me by Mr. John Simon.

Class VII.—Home Education.—Self Instruction.— Recreation.

Home education of children. Instructive toys. Educational prints and cabinets.—Adult self-instruction and scientific recreation. Synopsis of the Sciences with Illustrations of their respective natures and purposes. Cheap and serviceable apparatus for in or out-door studies. Formation of Herbaria, &c.

Illustrations of the Arts of Design and Polygraphic Arts, showing the processes employed, and the respective results.—Instruction in the principles of taste, in outline, colour, and subject; select prints, figures, and other articles for Cottage Decoration.

Music, vocal and instrumental, in its popular applications. Modes of instruction and notation adopted in various countries.

Class VIII.—Miscellaneous Articles, not referable to the foregoing Classes.

Scientific appliances for household use, including clocks, barometers, thermometers, scales and weights, measures, &c.—Stationery in all its departments.—Miscellaneous Household Requisites. Toilet articles.—The Housewife's Work Box. Female handwork of all kinds.—The Cottager's and Emigrant's assortment of Tools for carpentering, shoemaking, &c. Garden and Field implements.— Seeds of varieties suited for horticulture and small husbandry. Resources for barren localities.—Appliances for locomotion and the conveyance of burdens. Contrivances of all kinds for lightening labour.—Special requirements of Travellers, Emigrants, &c. Self-help for emergencies.

Samples of Museum Fittings and Appliances, with estimates for the use of persons desiring to form Economic Collections on any scale of development.

Class IX.—The Economic Library.

This Class, besides supplying every desirable information concerning the subjects illustrated in the body of the Museum,

SECT. 2.] *SYNOPSIS OF THE ECONOMIC MUSEUM.* 137

furnished knowledge which could not be taught by visual display; but which was a necessary complement to a full course of domestic and sanitary studies. It contained books, pamphlets and documents, (British and Foreign) on matters of domestic, sanitary, educational, and social economy, arranged for convenience of reference, and particularly intended for the use of persons engaged in the organization of Provident and Charitable Institutions, and of Clergymen, Medical Men, Schoolmasters, and others entrusted with the bodily welfare, or intellectual guidance of the People.(1)

The comprehensive nature of the foregoing Synopsis may raise the question,—how could so vast a range of subjects be illustrated in an area which including the Library and the annexe for cumbersome objects did not exceed 100 feet by 30? The chief secret lay in a careful selection of representative Types; but I must confess that in many departments the articles, and especially the diagrams, were packed more closely than would be proper in the University Museum, and making allowance for additional desiderata, I calculate that its Economic Collection should occupy twice the space above mentioned, even reckoning that several series may be transferred to other Collections.

Besides wishing to facilitate the formation of the Central and Provincial Economic Collections, I have in giving the sketch of my Museum, been actuated by a desire to show the groundwork of experience on which is based the SYLLABUS of ECONOMIC KNOWLEDGE prescribed for Artisan Students in the next Chapter (p. 179). I trust that from the fact of my having laboured for so many years to represent, fairly and evenly, all the essential features of this range of subjects

(1) Part of a descriptive Catalogue of the Library was printed for private circulation, and a copy is available for inspection.

138 *RANGE OF ECONOMIC KNOWLEDGE, ETC.* [Chap. III.

in a moderate space, carefully selecting and methodically packing together every available kind of visual illustration and explanatory labelling, it almost necessarily follows that the requirements of our population in general, and of the Working Classes in particular, underwent time after time, such a sifting, classifying and scrutinizing, as to supply peculiarly suitable and reliable materials for the preparation of such a Syllabus.

SECTION 3.—FORM OF INSTRUCTION FOUND MOST SUCCESSFUL.

I had not long been engaged in the practical development of my Economic Museum, before I became aware that the visual instruction it was intended to convey, would require more explanation, either oral or printed, than I had originally supposed, and that subsidiary educational measures would be required to complete its practical utility. The object to be had in view was not merely to point out to the Working Man what to choose and what to shun, or to store his memory with authoritative advice, but to give him an opportunity of acquiring that knowledge which might enable him to reason for himself, and to select with judgment what might be most· likely to suit his requirements and his means. For this purpose it was necessary to look beyond the outward surface of things, to enquire into the peculiar properties of some, to trace the origin of others, and the processes which had transformed the raw material into a manufactured article.

Unfortunately, in whatever manner I attempted to convey this instruction, a serious obstacle stopped the way. That obstacle was the almost total absence of scientific knowledge among the bulk of the community, and their consequent inability to understand the true scientific reason why one thing might be objectionable, and another preferable. It was for instance of little avail to show designs on the most improved principles for Dwellings in town or country, to persons to whom those principles were a mystery; who were ignorant of the Physical Laws on which Ventilation is chiefly founded, and who did not indeed know enough of Chemistry and Physiology to be able to understand why Ventilation should be necessary at all. Or how could I hope in displaying rational articles of apparel, to convince those who were not acquainted with the normal structure of the Human Frame, that its organs are often displaced, and their functions impeded through a blind subserviency to the dictates of fashion in matters of Dress. Or again, in what other than chemical language, unintelligible to the many, could the Curator of my Museum, in showing a Visitor through the Food Department, explain to him the reason why Wheat is more nutritious than Potatoes, and Peas more nutritious than either, or why he would find Beef Tea more strengthening than Calf's Foot Jelly, and fresh Meat more wholesome than Salt Junk; or how he might detect the commonest adulterations of Food or Drink, and the like. An Artisan and his Wife, visiting the admirable Food Department then at the South Kensington Museum,[1] might be struck and interested amazingly by some of the sensational illustrations and labels, but they would be so much at sea in all that relates to the chemistry of nutrition, that they would

---

[1] Now at the East London Museum, Bethnal Green.

scarcely venture to alter one item of their daily fare in accordance with a scientific dietary.

I am sorry to say that the deficiency of scientific training, and the consequent inability to appreciate the value of scientific guidance, has often met me among Schoolmasters, and even among Divines, on whom I had particularly reckoned for encouragement and assistance in propagating among the poor, intellectual means for physical improvement, and among the rich, notions of judicious and discriminating benevolence. On the other hand I am happy to be able to state that from many enlightened members of the Clergy, I have received the most gratifying tokens of sympathy and support. I hope indeed that the Representatives of all religious denominations will ultimately concur in the conviction, that the frame of mind suited for imbibing spiritual instruction, is far more likely to be found among the intelligent inmates of a healthy home, than among those whom ignorance has degraded to a torpid state of misery. I further hope that agreeing in this important principle, they will also acknowledge themselves to be the most fitting instruments for giving it practical effect; that as ministers of good will towards all men, they will eagerly promote the practical enlightenment in matters of Daily Life, which would be an unquestionable boon to all, and that they will readily support every *bonâ fide* endeavour to promote wholesome and humanizing knowledge, seeing that the same ignorance which engenders vice and callousness, opens the way both to superstition and infidelity.

More vitally important however than the question of patronage and assistance to my instruction, was that of its acceptance by those I wished to teach. Whichever way I turned, I found myself face to face with the serious doubt—" Can the Working Classes be induced to imbibe in any way, the fundamental truths of which they are, unawares, so much in need"? Can they for instance be persuaded to listen earnestly to a series of methodical Lectures on subjects of a kind commonly considered rather the reverse of attractive? The fact is that much had been done to give the audiences at Mechanics' Institutes a false and unfavourable idea of Science. When genuine and educational, it had often been in substance too high and dry, and in form too didactic and technical; so as to require on the part of the artisan an amount of preparatory knowledge which he could scarcely be expected to possess, and which indeed it seemed to be nobody's business to give him. When amusing, it was rendered so by experiments more calculated to be admired than understood, and it consisted mainly of sensational bits, picked here and there from the range of scientific knowledge, and which without the general context could scarcely produce the clear impressions required for practical use.

To test the possibility of remedying this state of things, I composed in 1864 a detailed explanatory Syllabus of six connected Lectures respectively entitled as follows:—

1. " The Alphabet of the Science of Common Life; or, a first peep into the mysteries of Health and Comfort."
2. " A good Home, and what belongs to it."
3. " Furniture and Clothing; and Health as affected by them."
4. " Food: its purposes, principles, and resources. How to make meals palatable, wholesome, and cheap. Beverages."

5. "Fire: what it is and how to make the best of it. Contrivances for Ventilation."
6. "Good Health, and how to keep it."

Each lecture having been sketched out in detail was provided with an ample assortment of specimens, diagrams, and apparatus, prepared at the Twickenham Museum, and suitably packed for circulation. Arrangements were made for the delivery of the Course at ten places of Meeting of the Working Classes in the Metropolis during the ensuing winter. The six lectures were entrusted to three competent persons, who severally undertook to adhere as closely as might be to the substance of the Syllabus, clothing it in their own language, and who acquitted themselves of the task in a very creditable manner.

The result proved not only that the plan of prefacing the study of Domestic Economy with an introductory account of the sciences on which it depends, was a step in the right direction, but that there was no occasion to condemn myself to the awkward task of squeezing that account into limits which it was ready to burst in all directions. I found to my great delight, not only that illiterate audiences could perfectly imbibe and appreciate the elements of Science, when presented to them in a simple and palatable form, but that there was no reason why these elements, instead of being confined to one Lecture, should not be spread over several. This was highly satisfactory, as regards the main *principle* of teaching Science to the People; but as regards the particular *manner* in which I had made the attempt, it was obvious that however detailed and explicit might be my Syllabus, it could not, when distributed in parts to several lecturers, sufficiently bind and control them to secure throughout a uniform treatment of the different subjects. Again, if I entrusted the whole range to one lecturer, he must indeed possess consummate abilities to extemporize a series of fluent and taking Lectures on materials prescribed by me, in accordance with my particular views; and in fact a professor competent to do this, could scarcely be expected to submit his talents to my control for any sum that I could afford to offer him. The higher his attainments, the more tedious it would be for him to be engaged in teaching my industrial Friends the mere "Alphabet of Science." The abundance of his higher knowledge would at times be thoughtlessly poured out to the detriment of the particular simple facts which I had thoughtfully selected; the ardour of his mind would ever and anon break the line of connected purposes, and carry him off into the realms of speculative fancy, his facile eloquence would dilate on favourite themes, and three quarters of the allotted time would be consumed before one half of the allotted matter was got through; so that the remaining half would either be cut and maimed, or galloped over at a headlong pace.

These conclusions, resulting from various practical experiments, are here related in the spirit of a mariner who points out the shoals he has encountered, not only that others may avoid them, but also that he may the more be trusted when he points out the safe passage through which he has sailed at last.

Having found that there was no alternative left but to write out the proposed Lectures in full, I prepared by the autumn of 1866 under the title "SCIENCE MADE EASY," a first instalment of the proposed Scientific Foundation, comprising five familiar Lectures on the Elements of Physics and Chemistry. In doing this

I endeavoured to carry out to the best of my power the instructions which I had prepared for others, as to the manner in which the subject matter should be selected and arranged, and as to the style in which it should be expressed. Here however I was met by another ominous question: would Working Men listen to a *read* lecture? As a rule the reading of a discourse of any kind only answers when there is nothing to do but to read, and is a failure when the delivery of the text is broken by experiments, writing on the black-board, or any other kind of visual demonstration. The expedient for overcoming this difficulty which happily occurred to me, was very simple, but proved effective beyond my most sanguine expectations. It consists in the joint action of a Reader and a Demonstrator, and may therefore be named the BINARY or DOUBLE-ACTION SYSTEM.

Wherever a specimen is to be shown, a diagram to be pointed to, or an experiment to be performed, a cross (✗) in the text warns the Reader to make any pause that may be required. The Demonstrator who has before him a full list of instructions, with every device for enabling him to be ready at the right moment, does the needful, and the reading is resumed without the least embarrassment or loss of time. I am the more induced to lay some stress on the remarkable success which this plan has obtained, because I believe it to be a new one, and to be susceptible of becoming a valuable addition to the present very deficient resources of popular Instruction. A Reader cannot be expected to deliver a written or printed lecture satisfactorily, if he has to go to and fro between his text and his diagrams, or his apparatus. On the other hand a Professor who can deliver a whole educational course in a concise yet easy style, without further guidance than a few notes, and who at the same time is a skilful experimentalist, is an expensive luxury, even in London, and almost unobtainable in most provincial localities. But on the contrary there is scarcely a country town where the vicarage or the school cannot supply a good Reader for a philanthropic object, whilst the Doctor or the Chemist of the place will be sufficiently up to the performing of any amount of chemical or other experiments, involved in expounding the Science of Daily Life.

It will not everywhere be possible to secure the services of so first rate a Reader as Mr. Henry Ellis, whose professional co-operation I have been fortunate enough to engage ; nor will it be easy to find a Demonstrator so dexterous at manipulation and clever in the performance of experiments, and at the same time so competent to answer scientific questions, and so well qualified for undertaking the general management of Circulating Lectures, as my Secretary Mr. Wm. Hudson. The peculiar smoothness and homogeneity with which their joint action progresses as rapidly as that of the most fluent and experienced extempore professor, will not be attained to by many provincial amateurs, but neither will this degree of efficiency be expected by provincial audiences, or required for making this mode of teaching Science a welcome expedient in local Institutes, and in many ways a valuable resource for Schools.—Be this as it may, there can be no question as to the service which this mode of delivery, specially contrived for sending out *identical* Instruction in various directions from a given centre, may render to the proposed system of Technical Training. With a vivifying action, resembling that of the Heart in

L

the Human Frame, the Central Technical University may thus dispense well elaborated Knowledge throughout the United Kingdom and the Colonies, and secure everywhere that uniformity of Studies so necessary for establishing uniform Standards of Certificated Merit.

Another educational contrivance on which I considerably reckon, is the plan of open-handed Examinations, to which I have been led by the success of my Popular Course; but before describing it, I must dwell for a moment on the circumstances of this success itself, in order to show that it has afforded repeated and incontestible *proof* of that aptitude of the Working Population for receiving scientific knowledge, which has been adopted as an essential element of calculation in framing the proposed System of Technical Training.

SECTION 4.—APTITUDE OF THE WORKING POPULATION TO RECEIVE APPROPRIATE SCIENTIFIC INSTRUCTION.

It may be well to explain that the above-mentioned Course entitled "Science made Easy," which has now accomplished its eighth season among the London Institutions, consists of nine Lectures embracing those Elements of Natural Science which are most essential for understanding the rationale of the Things of Daily Life. They are strictly methodical and progressive, leading an uneducated person through a facile acquaintance with the outward forms and properties of Bodies, to an enquiry into the Physical Forces by which they are governed in rest or motion, and into the theory and practice of the Mechanical Powers. Two Lectures are thus occupied. A third concludes Mechanical Physics with an exposition of the principles and familiar applications of Aërostatics and Hydrostatics, and a few notions of Acoustics. Lecture IV., under the title of Chemical Physics, deals with the most practical phenomena of Light and Heat. The fifth and sixth Lectures are devoted to a glance at the most important features of Inorganic and Organic Chemistry. Lecture VII. gives such outlines of Natural History as everyone ought to know something of, and finally two Lectures devoted to Human Anatomy and Physiology, turn to account the previous instruction in a compendium of all that is most essential to be known for promoting a healthy condition of the system.

Even this very brief summary may suffice to show that my Course, though I have sought to make it palatable to the Working Classes by a familiar and cheerful style, and by abundant illustrations, is nevertheless a *bonâ fide* endeavour to unite all the scientific data most essentially necessary as a ground-work to the study of the requirements and resources of Daily Life. This earnest character becomes particularly apparent on reading the Syllabus of the 9 Lectures, contained in the Programme distributed among the Working Population of the Districts where the Course is to be given. It is nearly identical in range with the Syllabus of Elementary Knowledge which will be found in Section 1 of the next Chapter.—Such being the character of the Course, earnest, methodical, and having practical usefulness for its constant aim, let us now see in what manner it has been received year after year, not by select students, but by miscellaneous audiences admitted *gratis* in various parts of the Metropolis, including some of the most densely populated and least refined.

Referring for fuller information to "Science for the People;" and for further proofs to the testimonials and letters of thanks in my possession, I will proceed to quote the Reporter of the *Daily News*, whose article entitled "Science among the Costermongers," appeared in that Paper on the 5th of November 1868. It describes a visit to Mr. Orsman's Mission Hall in Golden Lane, where my Course was then being given for the second time.

"On Tuesday evening last, a person passing down Golden Lane, a long, narrow, and poverty-smitten thoroughfare, leading from Barbican into Old Street, might have observed numerous individuals belonging, as their costume unmistakeably indicated, to the Costermonger class, silently making their way, amid piles of empty barrows and heaps of decaying vegetable refuse, towards a building, not very pretentious in its external appearance, situated at the rear of the City Baths. The locality is not a very inviting one. * * * * Entering the building above mentioned, a curious and suggestive spectacle met the eye. In a large room having a spacious platform at one end, and encircled by a strong and commodious gallery, were crowded together some 400 or 500 men, women, and children, belonging, for the most part, to the poorest classes. Many, perhaps the majority, were members of the street-trading community, the rest of the audience consisting of labourers, artisans, workmen's wives, factory girls, shop boys, street arabs, and the like. And for what purpose were they thus voluntarily collected together? * * * * They had met for the purpose of listening to a Scientific Lecture on Chemistry, * * * * eagerly listening to a Lecture, in which were explained the various properties of Oxygen, Hydrogen, Carbon, Sulphur, and other non-metallic Elements; the phenomena of combustion; the decomposition and recomposition of Water, and the like. * * * * In previous Lectures, those present had had explained to them the conditions of matter, laws of gravitation, mechanics, aërostatics and hydrostatics, light, heat, and other elementary portions of Scientific Knowledge; but instead of becoming wearied with the formidable mass of technical teaching with which they were threatened, the number of hearers was found to increase considerably with each successive Lecture. The order maintained was admirable. * * * *"

More conclusive however than could be any isolated instance, is the general tone of orderly earnestness with which the Course has been attended throughout its London career, from Paddington to Holloway and Spitalfields, from Chelsea to Shadwell and Stratford, and from Nine Elms, Lambeth, and Bermondsey, to Woolwich Arsenal. In order nevertheless to prevent disappointment on the part of those who may be prompted to engage in similar undertakings, as auxiliary to the proposed systematic training of the people, it is right I should point out a few conditions of success which must not be disregarded. Thus whilst it is essential to select Institutions where Working Men have acquired the habit of congregating in considerable numbers, it is scarcely less so that there should be some intelligent and active person on whom reliance can be placed for superintending the machinery of publicity. As for my own Course, it succeeds best where an energetic Minister or Missionary has intelligently induced, and still actively keeps up habitual gatherings of the people in large numbers, by mixing innocent recreation with religious and secular instruction. There my Lectures are quite at home; for it has been my constant endeavour to leaven them with a spirit of Christian morality, and to show how religion and science may, and always should, go hand in hand.

As a rule I have found, as might be expected, that audiences consisting of men and women of the lower social strata, such as are called in by the invitation of *open doors* in a poor locality, require to be presided over by a person who, besides being a good man of business, possesses the art of making them feel at their ease; whereas when the chair is filled by any one to whom they look up as to a condescending patron, a kind of reverential awe seems sometimes to chill their interest, and check the liveliness of their appreciation.

The ability to preside at such a Meeting is not a privilege reserved for gentle birth and classical education. At one of the places where my Course was given in the first season, the chair was taken by one who neither stumbled over his own words, nor treated the Meeting with a flow of eloquence when it was not wanted, but could always when required, speak cleverly and to the purpose, and conduct an evening's proceedings with tact and courtesy. Both he and the audience showed more than once that they were not novices in parliamentary routine. Now the place of Meeting in question, was the Hall of the Old Pye Street Working Men's Club, instituted through the generous and enlightened exertions of MISS ADELINE COOPER, and occupying the Ground Floor of a Model Lodging House adapted to the requirements of one of the poorest localities in Westminster. The benches were compactly filled with Hawkers, Costermongers, and Labourers, and their clever and efficient Chairman was one who earned an honest livelihood by uniting the trades of Glass and China Mender, and Knife and Scissors Grinder.—To give an idea of the discriminating eagerness with which such audiences hail the advent of knowledge, provided it be of the practical kind they require, I will mention that for some time Science had got into evil odour in Old Pye Street, through causes of the usual kind. I accordingly resolved to give Miss Cooper's friends only one Lecture by way of experiment, and Lecture IV. was selected, as being of fair average interest. They were so well satisfied, that they not only asked to have the whole Course, but desired that the Lecture already given, should be repeated in its proper place.

Generally speaking about three hundred Working Men are a sufficient audience for purposes of steady and continuous instruction, and are indeed as many as can conveniently enjoy the portable Illustrations and Experiments of a circulating Course like mine; though for the purpose of exciting a taste for Science, much larger numbers may be convened. In the vast hall of the Lambeth Baths which, thanks to the untiring energy of the Rev. G. M. Murphy, is appropriated in winter time to a continuous weekly rotation of devices for uniting instruction and amusement, the numbers have repeatedly reached eight or nine hundred, and on one occasion considerably beyond a thousand. I am bound to say that even these multitudes admitted with open doors, in a neighbourhood which used at one time to be looked upon as one of the lowest in the scale of refinement, have maintained the most decorous and attentive silence, only broken by hearty applause, or well-timed manifestations of hilarity.

I trust that the direct bearing of the foregoing details on the subject of establishing a National System of Industrial Training, is self-evident; for certainly, as I said in a letter to the Secretary of the Labourers' Friend Society, such facts prove that there are in our working population, sterling qualities of great promise, germs of thoughtful improvement, which only want judicious fostering and disinterested guidance, to produce results of infinite value for their physical and social welfare.

SECTION 5.—EXAMINATIONS FOR INDUSTRIAL STUDENTS.

In the Autumn of 1868, when preparations were being made for the third session of delivery of my Course at the Lambeth Baths, the Rev. Mr. MURPHY [1] suggested the expediency of testing by means of Examinations, the actual amount of knowledge imbibed and retained by those men of Lambeth who seemed so earnest in their desire for Science. I the more readily adopted this suggestion, as it afforded me an opportunity of putting to a practical test, certain innovations which I had long been anxious to see applied to the furtherance of scientific instruction among the People. The vast importance of Examinations in general, and the deficiencies of most of the existing plans, had become strongly impressed on my mind in serving on the Educational Committees of the Society of Arts, and in watching from its beginning the development of the system of *Open* Competitive Examinations due to the

(1) Now a Member of the London School Board.

initiation and indefatigable perseverance of my late friend and colleague, Mr. Harry Chester. By *Open* Examinations, I mean those not confined as in a College to Students taught by the Examining Body, but open to all Candidates duly qualified. It is more particularly to such Examinations that the following considerations will apply, inasmuch as it is desirable that the system of allowing Candidates free option as to the way of acquiring prescribed attainments, should be introduced as far as possible in the operations of the proposed Central Technical University.

Examinations whether *competitive*, that is to say, offering for *competition* a limited number of Prizes or Rewards, or *competential*, that is to say offering a certificate of *competency* to any number of Candidates whose attainments come up to a specified mark, may subserve various purposes. They may stimulate and test individual industry and abilities, with a view either to reward merit for its own sake, or to determine fitness for certain duties; but they may, and should if possible, further have a more special educational aim. When established in strict accordance with the carefully ascertained requirements of certain categories of Students, they may afford in addition to a powerful *incentive*, a highly valuable *guidance* to their respective studies. It may be all very well for those who are well educated and well to do, and can afford to regard study as the mere solace of leisure hours, or as a wholesome drilling of the mind for an indefinite purpose, to have a set of competent Examiners ready to question them on any branch of knowledge to which they may take a fancy, without considering whether or not it may be likely ever to render them any direct and practical service; yet even among Students of this description, laxity of purpose should not be encouraged, nor should they be left to the misleading influence of fancied aptitudes or futile predilections. With Artisan Students and others who have to carry on the struggle of mind against matter, we have a much stronger case. To them knowledge is a necessity, without which their incomings are likely to be precarious, and their outgoings are sure to be unsatisfactory. And woe to them if some friendly guidance is not at hand, to point out the knowledge of the right sort, and the way to get at it. Ignorant of the distinctive characters and bearings of the different branches of Natural Science, unacquainted with the differences of treatment which each branch has received from different Authors, unable to pick their way through works which contain vastly more than they want, and liable through lack of previous culture to find confusion and dismay in etymologies which give a lift to the classical Scholar, they must not be entrusted with their own guidance in the pursuit of knowledge, whilst at the same time they cannot afford to pay for tutorship or *coaching*. Sometimes indeed the soundest and most friendly advice, tendered to Artisan Students from the level to which their employers belong, is apt to be viewed at first with diffidence; but let it be based on facts that speak for themselves, and on considerations that find an echo in their own shrewd common sense, and they will soon cheerfully accept its guidance, for they know very well that ignorance cannot be its own guide to knowledge, and that they who most require to learn, can least afford to waste time and opportunities in learning what is not to the purpose.

It is naturally to the Programme of Examinations yearly published by an Examining Body, that one looks for the means of guidance required by the respective Students. I used to turn to the "Syllabuses," hoping to find in each a succinct exposition of what the respective Candidates were expected to learn, followed by satisfactory indications of Books, or parts of Books, recommended for their use. I scarcely like to say how often I was disappointed. Instead of a congruous series of sketches, each clearly defining a distinct course of study, and yet all linked together by unity of design, I saw evident signs that each Examiner had been left to his own devices. His own individual conception of his Subject was more or less fully or faintly indicated, without reference either to the plans of his colleagues, or to that of his predecessor, and a student could only surmise the probable level of his questions, from a knowledge of his temper and antecedents. As for indications of Text Books they were often inadequately given, or the works named contained so much beside or beyond the mark, as to be bewildering without the aid of an adviser competent to pick and choose.[1]

At that time so little regular attention had been paid to the different intellectual requirements and capabilities of Students differing in social level, educational advantages, and intended pursuits, that it was perhaps excusable if Examination Programmes displayed rather the indefinite unstable and tentative character of an experiment, than the homogeneous solidity of a well developed system, or if they were deficient in those legitimate means of guidance which earnest but inex-

[1] Every department or range of knowledge in which an examination of a given level is to be held, with a view to the requirements of a given Category of Students, should be embodied in a special Text Book giving just what they want. To supply a reflex thereof in the form of a detailed Syllabus is then a comparatively easy task.

perienced Candidates are entitled to expect.—Had there been a permanent Committee of Supervision or Board of Moderators, uniting the lay with the professorial element, and possessing due authority for framing and maintaining Systematic Examinations distinctly intended for a certain educational level, say, that on which the lower portion of the Middle Class, and the upper portion of the Waged Class, may be said to meet, they would perhaps have adopted Measures of Improvement founded on Conclusions to the following effect :—

— That the subjects to be included in an Examination Scheme, whether scientific or otherwise, should always be selected in accordance with the requirements of the class of Candidates in view.
— That a distinct CURRICULUM or Course of Studies, should be agreed upon and marked out for each Category of Students, and that the required Text Book or Books should be selected, or if necessary specially produced.
— That each Curriculum should be represented in the Examination Programme, by a SYLLABUS giving all necessary details, and referring to the respective Text Books or portions thereof.
— That sufficient unity should prevail in the degree of explicitness and general character of the various Syllabuses, to stamp them as parts of one and the same scheme.
— That unity of purpose should similarly govern the Examinations throughout, so as to maintain even in different departments, conformity of guiding principles, and uniformity of standards.
— That in the Questions to be proposed at a given

Examination, there should be no undue preponderance given to topics that might happen to be favourite ones with the Examiner.—This kind of favouritism besides being liable to be taken advantage of by Candidates who happen to be aware of it, tends when continued year after year, to disturb the proper relation between the importance of each matter, and the amount of attention bestowed on it.

— That due care should be taken to defeat the cleverness of Candidates, or their *Coaches*, at guessing what questions are most likely to be asked, and every exertion used to put down the practice of *cramming*.[1]

— That far more satisfactory means exist for accomplishing the foregoing objects than rendering the Examination Syllabus vague and reti-

---

[1] The following passage is from a discourse pronounced by his Grace the Archbishop of Canterbury, on the 18th Nov. 1873, on the occasion of distributing the Prizes in connection with the Oxford Local Examinations at Margate:—"I believe that those who are preparing for competitive examinations are very well acquainted with a system now too common in this country, which goes by the name of 'cram.' A great deal of what appears as most successful results in many examinations now held throughout the country is of very little value. I have been told, for example, of an eminent tutor who has a vast number of pupils because his pupils are almost sure to be successful in some of these examinations, and who has this mode of enabling his pupils to exhibit before the examiner the writing of essays on any subject of history. The tutor, from long experience and careful study of the questions set in past times, ascertains almost with mathematical precision that one of ten subjects will be set for an essay at the coming examination. He in his study carefully prepares ten essays on ten subjects, such as the history of Queen Elizabeth, the influence of the French Revolution on the civilization of the world, and other subjects. Having prepared these, day by day he requires his pupils to commit these ten essays to memory. They enter the competition. To their delight they find a familiar subject in the essay proposed, and in a few moments a beautifully prepared essay, such as fills the examiner with surprise, is prepared by the youth of 14, and he is awarded the required number of marks."

cent, and then when the examination time is come, pouncing on the Candidates with out of the way questions, having scarcely anything to recommend them beyond their being perfectly unexpected, or depending for an answer rather on the fortuitous direction than on the general extent of knowledge.[1]

— That the questions should as much as possible be distributed over all the Departments or Sections of a Science, in such a manner as to encourage an even distribution of attention over the whole of its ground-work and essential features.[2]

— That of the Questions put for each Science at each Examination, some should be more comprehensive in grasp, or more difficult in character, and others the reverse; so as to enable Candi-

---

[1] A few examples may serve to illustrate my meaning:—At a geographical Examination a Candidate was asked about a River in Asia, of which the only known account was one recently published in a Book of Travels which *happened* to have been read by the Examiner.—A question on the numerical estimation of the Conductivity of Heat was repeatedly put, for an explanation of which the ordinary Text Books were for some time vainly consulted. It was afterwards learnt that the Examiner had discussed the matter in an article written for a certain expensive Cyclopædia.—Candidates in Geometry were asked to prove with a piece of paper and a pair of scissors that the three angles of any triangle amount to two right angles. The cleverest mathematician might be less able to do it than a child who had seen it done.

[2] The following is from a Memorandum which I addressed about 10 years ago to the Secretary of the Society of Arts.

" What most of our young men assuredly want, and may hope to acquire in the limited time and with the limited means which they can devote to preparing themselves for the Society of Arts Examinations, is a clear and logical perception of the scope and nature of a Science, and a competent knowledge of its ground-work, classification, leading facts and ordinary processes; so that whatever further information they may have opportunities for picking up through books, lectures, or practice, may be at once rightly understood, and brought home for permanent keeping to its proper Chapter of the memory."

dates possessing a considerable amount of knowledge, to make it proportionately manifest, whilst the less fortunate would at least be afforded opportunities for turning to the best advantage, such amount as they might have acquired.

— That a clear understanding and agreement should be come to among the Examiners, and between them and the Board of Moderators, as to the total number of marks obtainable in a given Science, or other subject,—as to the number of marks respectively required for obtaining given Prizes or Certificates, and,—as far as practicable with the aid of select typical examples, as to what should be considered *good answers* to Questions of stated grades of difficulty.

— That every endeavour should be made, year by year, to work the appreciation and measurement of knowledge and abilities, and their representation by numerical Marks, into a system not only logically uniform as regards one Examining Body, but also logically harmonious as regards the *ensemble* of Examining Bodies.

— That though a gradual rise of Standards, conformable to the progress of Science and to its more general diffusion, may be allowable from year to year, yet there should if possible, be nothing beyond this; especially no sudden rise unannounced in the Programme, and due to a change of Examiner or a lack of Prize Fund; in short no change due to any other cause than the benefit of the Examinee.

— That as regards Elementary Examinations, precise Regulations should show as clearly as possible what is expected of the Candidates, not only as regards isolation from assistance and the like, but also as regards the fair English in which their knowledge is to be expressed, and the fair legible hand in which it is to be conveyed to paper.

— That the yearly Programme issued by the Examining Body, should unite in a well-considered order of sequence, the Syllabuses agreed upon by the several Examiners in concert with the Board of Moderators, and at the same time supply such prefatory information, and such a full statement of regulations, as may clearly establish the proposed purpose, educational level, and *modus operandi*; and that the said Programme thus carefully prepared, should remain *unaltered*, subject to *yearly revision* for correcting any errors of omission or commission that may become apparent, for introducing any practical improvements pointed out by experience, and for keeping pace with scientific progress.

Several of the foregoing conclusions appear to have been more or less adopted of late years, but they were untried at the period to which I am alluding, namely the autumn of 1868, when the approaching delivery for the third time of my Course on the Elements of the Science of Daily Life, induced MR. MURPHY to ask for Examinations. The opportunity for a practical experiment was an excellent one; but at the same time I perceived that to ensure success among a class of Students considerably below the Society of Arts Candidates in social position and educational culture, it was necessary, besides enforcing several of the above principles, to clear away by a still more searching reform, every obstacle, real or imaginary, which might deter my Artisan Friends from entering the lists. To them the very idea of a scientific Examination had in it something awful, and it would have been almost ridiculous to invite them to such an ordeal, if I had not seen my way to a plan of operations somewhat subversive of the usual mystery of an Examination, but which was nevertheless in reality as safe to the Examiner as it was reassuring to the Examinee.

The manner in which the experiment was conducted is described as follows, in a Letter which I addressed to Mr. MURPHY at its conclusion, and which was read by him at a crowded meeting held at the Baths on the 20th February, 1869, when LORD SHAFTESBURY distributed the Prizes to the successful candidates:—

* * * * "Looking carefully through the eight Lectures into which I had endeavoured to compress the scientific facts most wanted in Daily Life,[1] I divided the subject matter of each Lecture into a convenient number of parts,[2] and prepared for each part a question so framed, that anyone answering it might fairly and freely show to what extent his intelligence and his memory served him; and of course being competent to answer all the questions would be tantamount to having in one's mind the gist of all the Course. You were kind enough to read publicly the whole of these questions at the opening of the Lecture Season, and as a copy of them was deposited at the Baths, together with a copy of each Lecture, and moreover, as each question had attached to it the maximum number of marks which an answer to it might gain, the whole scheme of studies and rewards was spread open from

---

[1] A second Lecture on Chemistry has since raised the number to nine.
[2] There were mostly ten to twelve parts in each Lecture.

the beginning. Each student knew what he was recommended to learn, and was offered the convenience of learning it without the annoyance of conflicting books. Each candidate knew what questions he was liable to be asked, and what each satisfactory answer might be relatively worth in marks."

"When the Examination time came, I selected with the assistance of Mr. HUDSON, who is thoroughly conversant with matters of this kind, two of the questions on each Lecture, one easy, the other more difficult; making sixteen questions for the whole Course. Each candidate who having satisfactorily passed a preliminary Test on the 22nd of January, was admitted to the definitive one on the 5th of the present month, found on taking his assigned place at the Examination Table, a copy of those sixteen questions before him. The Candidates were informed that they were not *expected* to take up more than *one* question on each Lecture; but such was the zeal with which they set to work, that in the three hours allowed, several of them went through nearly the whole range; and this in such a manner as perfectly to justify the sanguine expectations raised in my mind, by the remarkable earnestness with which the Lambeth Baths' audiences, young and old, had throughout listened to my Course. The number of Candidates has been small, but quality has made amends. Altogether, the appreciation of scientific knowledge, the power of imbibing and retaining it, and the power of expressing it in writing, evinced by the Working Men of South London, are facts most encouraging to the advocates of Industrial Instruction; whilst they reflect the highest credit on the enlightened Pastor under whose guidance this intellectual elevation has been attained."

The success of my plan of OPEN-HANDED EXAMINATIONS in Lambeth, induced me not only to repeat the experiment at the Baths when my Lectures were delivered there for the fourth time, but also to try it in other parts of London. The result having been everywhere most encouraging, I applied the plan to a somewhat more advanced Course of 8 Lectures on Inorganic Chemistry for Industrial Students, delivered likewise on the Binary System, at the St. Thomas's Schools Hackney.[1] Every trial has fully confirmed the conviction entertained from the beginning, that the plan of converting a *reticent* SYLLABUS into an *open* QUESTIONARY, *i.e.* a Catechism *minus* the Answers, is not only an expedient suited for overcoming the diffidence of uneducated Students, but might become with proper management almost equally advantageous in Middle-Class, and other analogous Examinations, whether conducted by the Examiners personally or from a distance. At all events it has become obvious that in a general scheme of Popular and Industrial Training this plan would afford a peculiarly convenient means of establishing and maintaining a uniform Standard of Examinations for each Grade of each Industry in all parts of the country.

The following is a *précis* of the plan of proceedings, taken from the Examination Programme which has stood the test of actual service for several years with only a few trifling alterations. It consists of 8 quarto pages printed in double columns, except the Title-page, which gives concerning the Popular Course to which

[1] The prizes were delivered to the successful candidates, all *bonâ fide* Working Men, on the 18th day of April, 1873, by Charles Reed, Esq., M.P. (now Sir Charles Reed, Chairman of the London School Board), supported by the Rev. J. A. Picton, whose valuable patronage much contributed to the success of the Course.

it belongs, a few particulars necessary for understanding certain references made thereto. Copies of this Programme and of a similar one prepared for the Industrial Course of Inorganic Chemistry, will be freely supplied on application, by the Secretary of the Economic Museum, Twickenham.

PROGRAMME OF EXAMINATIONS.

PLAN OF PROCEEDINGS.(1)

1.—Those essential Elements of Scientific Knowledge which it is most desirable that every Adult of the Working Classes should possess, are methodically arranged and explained in an easy and familiar manner in the present Course of Lectures.

2.—For the purpose of the Examinations each Lecture has been divided into a convenient number of parts, large or small according to the nature of the subjects; and a Question has been prepared for each part, embracing as far as possible the main substance or gist thereof, and so framed that a person answering it may be fairly prompted to show to what extent he has understood and retained the matter referred to.

3.—The whole of the Questions, which amount to about one hundred for the nine Lectures, are included in the present Programme. It is from them only that will be selected those to be put to the Candidates.

It is in this plan of publishing from the beginning in the Programme the Questions which the Candidates may have to face at their Examination, that lies my greatest deviation from the ordinary practice. It is unquestionably a startling innovation, and yet a perfectly safe one. It will be seen further on, (Article 11) that the Examiner selects two questions out of the 10

---

(1) A difference of Type will distinguish a few remarks added to draw attention to the points most likely to be found serviceable.

or 12 allotted on an average to each Lecture in the Programme. Which these two may be the Candidate cannot possibly guess. He has the satisfaction of feeling that he cannot be asked any question which he has not a fair opportunity of preparing himself to answer, but such preparation must, as far as his abilities allow, embrace the gist of the whole Course (see Article 5) and this is all that can be desired.

4.—The Questions, besides being consecutively numbered for each Lecture, have appended to them other numbers ranging from 3 to 25, which indicate their more or less comprehensive and difficult nature, and show the *maximum* number of Marks which a Candidate could obtain at the Examination by answering them in a thoroughly efficient manner.

5.—It is obvious that a Candidate, in order to be competent to answer *thoroughly* any of the Questions which the Examiner may happen to select, must master the whole of the Elements of Practical Science as embodied in the course; but far less than this degree of efficiency will suffice for securing to the Candidate a satisfactory Certificate.

In Examinations held like mine, simply to test and recompense the earnestness of a desire for scientific instruction, and the ability to benefit by it, the average level of capabilities of the class of Students dealt with in each locality, has of course to be leniently considered in determining the *minimum* of Marks sufficient for securing a Certificate; but in adapting the present plan to a National System of Technical Examinations, much care must be bestowed on determining, and everywhere uniformly maintaining in the several subjects and grades, the proper proportion of that Minimum to the total of attainable Marks.

6.—The Series of Questions, with maximum numbers appended, remains the same from year to year, subject to any improvements which it may be found expedient to introduce, and to any alterations or further development of the Course. It will be freely circulated previously to and during the beginning of the delivery of the Course in each locality, together with the Syllabus of the Lectures, so that the Working Men of the district may know at once what they are wanted to learn, what questions they will be liable to be asked if they become Candidates, and what may be the value of each satisfactory answer.

7.—Till the Course is printed, such facilities will be afforded as convenience may allow, for the inspection of a manuscript copy.[1]

8.—Persons desiring to become Candidates in the several localities where the Lectures and Examinations may be organized, will be expected to conform to the following regulations, which will be entrusted to a Committee of Supervision.

Regulations.

    a.—Candidates must be at least 18 years old, and strictly of the Working Class.—In localities where it may be found desirable to admit Female Candidates, the minimum age required for these will be 16.

    b.—They must furnish particulars of Name, Address, Age, and Employment.

    c.—Each Candidate approved by the Committee will receive a Card, on which he (or she) will at once inscribe his (or her) own name.

    d.—As a rule no Cards will be issued after the 4th Lecture, and no Candidate will be examined who has not attended at least 6 of the Lectures, and each time signed the Attendance Book.

    e.—No Examination will be held unless there are at least six eligible Candidates.

9.—The Examination will be held by Mr. Hudson, the Examiner, in the room and on the evening appointed in concert with the Committee of Supervision, such evening to be if possible within one month after the conclusion of the Course.

10.—At a Friendly Meeting of the Candidates, held some time before the Examination, they will be asked by the Examiner to answer in writing a few very simple questions concerning the

---

(1) In conformity with this announcement, a manuscript copy of the Lectures is deposited at each locality for about six or eight weeks previous to an Examination.

Sect. 5.] *PROGRAMME OF EXAMINATIONS.* 167

Lectures, and those only who do this satisfactorily will be admitted to the Examination.

This "*Friendly Meeting*," intended to serve the purpose of a Preliminary Examination, has admirably succeeded, not only in separating from the Candidates a few who would have stood no chance in competing with the rest, either through want of knowledge or through want of ability to express it in writing, but also in giving those who remain a clear idea of the actual nature and requirements of an Examination. In fact it has been noticed that in the time between the Preliminary and the Final Examination, remarkable progress has been generally made, evidently the result of earnest application.

11.—Previously to the Examination, two questions on each Lecture will be selected and marked on the Programme by the Examiner. The one will be more easy, and the other more difficult or comprehensive, in order to give each Candidate an opportunity of adapting his attempts to his abilities.

12.—On the evening appointed for the Examination, the Candidates will find before them on taking the places assigned to them at the table, besides the necessary writing materials, copies of this Examination Programme on which the 18 questions are duly marked. Three hours will be allowed them for preparing in presence of the Examiner, and without any assistance, plainly written answers to as many of these questions as they may choose; but they will not be *expected* to take up more than one question for each Lecture, and even six good answers may earn a Prize. Minor formalities will be duly explained by the Examiner.

I wish to draw attention to the fact that the Candidates are encouraged to spread evenly their Studies throughout the Course. In dealing with a class of Students having enjoyed greater educational facilities, the answering satisfactorily at least one question per lecture, might be not merely *expected* but *required*.

168 INDUSTRIAL EXAMINATIONS. [CHAP. III.

13.—The Examination Papers, signed by the respective Candidates, and left by them on the Table, will be brought to the Economic Museum for being carefully examined, and each more or less satisfactory answer will be rated at a proportionate number of Marks.

14.—Each Candidate whose total number of Marks reaches the amount fixed beforehand as necessary by the Examiner, will receive a Certificate.

Here follows a set of rules concerning the Prizes, which are omitted as having in view circumstances of a local nature, and essentially different from those likely to occur in carrying out the scheme of Industrial Education contemplated in the present Work. Then comes the last Article which stands as follows.

16.—The results will be made known to the Candidates if possible within a fortnight after the Examination, and a convenient evening will be appointed in concert with the Committee of Supervision, for the distribution of the Prizes and Certificates.

I may here mention that delay in making known the awards of Examiners has been known to create considerable inconvenience and dissatisfaction, and should be carefully avoided in the system proposed to be centered in a National Technical University, in which it will be further desirable to make known to the Candidates, as I have done in my small way, the number of marks which they have respectively obtained; also if possible the number of Marks respectively allotted to them in each subject in a Composite or Trade Examination.

EXAMINATION QUESTIONS.

The remainder of my Examination Programme, forming five double column pages, is filled by the QUESTIONARY, which contains 103 Questions for the nine Lectures; being not under nine nor over fourteen for any one of them. The Questionary in the Examination Programme of the Industrial Chemistry Course, contains 81 Questions for its 8 Lectures, ranging for each from 8 to 12.[1] The insertion of these Questionaries would add too much to the already considerable space devoted to this subject, but both Examination Programmes, together with the Syllabuses of the respective Courses, will be freely forwarded to persons interested. I trust that the more the working details of this "OPEN-HANDED SYSTEM" are looked into, the more evident will become its peculiar suitableness for giving shy and untutored merit a fair chance of success. Feeling perfectly secure against abstruse or out of the way Questions, since none can be asked but the carefully and clearly worded ones of a published Questionary, Candidates will set to work with a satisfaction, and present themselves at the Examinations with a self-possession, that could scarcely be secured by any other means. They cannot possibly guess which questions the Examiner will choose from the Questionary, but being told that those Questions will be evenly selected from its various parts, they will naturally be induced to spread throughout the range of their studies, that uniform attention which is so desirable. At the same time it is obvious that the comparatively permanent character given to the Questionaries, will repay a great amount of care bestowed on their first preparation, whilst periodical revisions will secure their being

---

[1] It will be found that there is in this system a tendency to involve long answers, and it will be well in examining advanced Students to accompany some of the Questions, especially in Chemistry, with appropriate skeleton or blank Schedules that can be rapidly filled in with the required details.

kept up to the level of the time. Lastly, I may mention the great saving of trouble it will be to the conscientious Examiner, to have a well-digested series of Questions spread out before him for his selection.

It is evident that all these advantages cannot fail to tell with peculiar force in an Educational Scheme which, like the one proposed, encompasses in its comprehensive embrace large portions of the Industrial Population comparatively untouched by previous devices for Technical Improvement, and which essentially depends for its success on the popularity of its Tests and the uniform fairness of its Awards. The open-handed System will indeed be equally convenient in those cases where the necessity for oral questioning, or any other cause, may demand the personal presence of the Examiner, and in those where it may be preferable to follow the plan adopted by South Kensington, the Society of Arts, and other Bodies, of examining from a distance; that is to say, the sending out of Examination Papers to the various localities for being answered in writing under responsible supervision, and the sending in of these answers to the Central Body for Examination and Award.[1]

---

[1] See the Programmes of Examinations published yearly by the Science and Art Department, and the Society of Arts.—At the London University the higher Examinations are conducted under the direct supervision of the Examiners, and to a certain extent Oral Examinations are included.

CHAPTER IV.

RANGES AND GRADES OF INSTRUCTION. — PUBLICATIONS. — EXAMINATIONS.

SECTION 1.—THE FIRST GRADE OF GENERAL SCIENTIFIC FOUNDATION.

THE chief object in view in this and the two following Sections, will be to define the various departments or ranges of Instruction that belong more or less to Technical Training, and to indicate certain GRADES or Levels in each, introducing conventional Titles, which may be conveniently used in the next Chapter, in sketching out suggestive Curricula of Studies for various Industrial Employments.

It has been shown in the previous Chapters:—1stly, that a certain amount of the insight into Nature's Laws which Science affords, is indispensable to every Working Man who is anxious to foster his health and strength, and to make a judicious use of his resources;[1] 2ndly, that a certain amount of scientific insight is equally in-

---

[1] Several years ago when I was advocating among my friends at University College the policy which has since led to the institution of a special Professorship of Hygiene, I adopted the terms "BIONOMY" or "PRACTICAL BIONOMY," and "BIONOMIC KNOWLEDGE" for briefly designating the Rules of Guidance through Life afforded by an insight into the Natural Sciences. I have since repeatedly found these terms so con-

dispensable for the satisfactorily carrying on of almost every technical Industry.—Fortunately the range of scientific Knowledge necessary for the first purpose and which may be called the *Science of Daily Life*, is at the same time exactly suited to serve as a GENERAL SCIENTIFIC FOUNDATION for Technical Studies.

It is unnecessary to repeat what has been said in Chapter I. respecting this fact, and the educational facilities which will result from it for establishing on a sound scientific basis the Technical Training of our Artisans; but it would be difficult to overrate the importance of rendering this fundamental support thoroughly effective. Hence the care that will be bestowed in this and the following Chapter, on showing in detail how it may be initiated and developed, and how from it, as from a main stem, may be made to spring the various branches of Technical Knowledge.

It has been seen in Chapter III. how in the formation of a comprehensive Economic Museum, I was led to study and classify the Applications of Science to every department of Domestic and Sanitàry Economy, for the purpose of selecting and displaying in methodical arrangement the best representative articles in each; and how in order to render this course of Visual Instruction in APPLIED SCIENCE intelligible to those for

venient, that I have been strongly tempted to use them in the present Volume. I feel however that I am already proposing innovations intrinsically calculated to startle those whose studies in Educational Economy have not been kept up to the requirements of the times, and accordingly I think it prudent to avoid rousing opposition by strange names; the more so as I have seen names coined merely for convenience' sake, made a pretext for ridicule, and used as a handle for throwing overboard proposals with which it would have been otherwise difficult to find fault. It will be time enough to give the title of BIONOMY to the scientific guidance which the Industrial Classes so much require, when it shall have quietly made its way into their favour under the name of "SCIENCE OF DAILY LIFE."

whose benefit it was designed, I found myself obliged to supply a key to it by preparing an Introductory Course of 9 Popular Lectures on ELEMENTARY SCIENCE, which has for several years been freely delivered at the London working-class Institutions. It is with materials thus thoroughly sifted and tried, that has been framed the following Syllabus of a GRADE 1 STANDARD COURSE. The *Elementary* Division nearly corresponds to the above Introductory Course, and is separated into corresponding parts. In the *Applied* Division, I endeavour to epitomize the Economic Knowledge which my Museum was intended to illustrate, reducing it to that minimum which every Working Man ought to possess, and which in fact it would be more difficult for him to do without than to learn.

SUGGESTIVE SYLLABUS
FOR A GRADE 1 STANDARD COURSE OF
GENERAL SCIENTIFIC FOUNDATION.

FIRST DIVISION.—ELEMENTARY KNOWLEDGE.[1]

MECHANICAL PHYSICS.

NATURE AND PURPOSE OF THE COURSE :—Definition of the SCIENCE of DAILY LIFE. Importance of its study for securing health and comfort. Necessity for a preparatory knowledge of the Elementary Sciences on the application of which it is founded.

[1] The Elements of the Natural Sciences readily take colour from the examples by means of which they are illustrated. In the Course of Instruction here indicated, every example should as far as possible be borrowed from Domestic and Industrial Concerns; so that even in this Elementary Division, may already be recognized the utilitarian character of the "Science of Daily Life."

FORMS OF BODIES :—Introductory explanation of the terms commonly used in describing recti- and curvilinear plane and solid Forms.

FIRST PRINCIPLES OF PHYSICS :—Varieties of Attraction. Contending Forces resulting in the three conditions of Matter, solid, liquid and gaseous.

DISTINCTIVE PROPERTIES OF BODIES :—Compactness, porousness; hardness; brittleness, toughness, malleability, ductility, tenacity; flexibility; elasticity; sonorousness; opacity, translucidity, transparency. Crystallization.

WEIGHT :— Specific Gravity of Solids.[1] Comparative Density of Liquids; Hydrometers. Comparative Density of Gases; Balloons.

REST AND MOTION :— Inertia; momentum; centrifugal force.

GRAVITATION :—Increasing velocity of falling Bodies. Prevention of Fall. Centre of Gravity. The Pendulum.

THE MECHANICAL POWERS, shown in action or *dynamically*, rather than *statically*, in order to afford a lively and impressive demonstration of the inverse ratio of POWER and SPEED :[2] The Lever. The Wheel and Axle. The Pulley. The Inclined Plane. The Wedge. The Screw.—The Roller. Wheel Carriages.

Notions of AEROSTATICS and HYDROSTATICS: Phe-

(1) It has been found expedient to introduce Specific Gravity among the distinctive properties of Bodies, leaving the mode of determining it as regards insoluble Solids, to be alluded to in Grade 1 of Hydrostatics, and more fully explained in Grade 2.

(2) Some of the continental Writers on Mechanics adopt this mode of explaining the Mechanical Powers; for instance, A. Privat Deschanel in his well-known "Traité Elémentaire de Physique." See Part I. of the English translation by Prof. Everett.

nomena connected with the pressure of Air and of Water, and rationale of their most useful and interesting applications. The Diving Bell, the Syphon, the Common and the Forcing Pump, the Barometer, &c.

Notions of the production and transmission of SOUND. Echo.

CHEMICAL PHYSICS.

LIGHT: its production and transmission. Reflection, Refraction, and their practical applications. Decomposition of the Solar Ray; Phenomena of Colour.

HEAT: Expansion of Solids. Expansion of Liquids; Thermometers. Expansion of Gases; artificial Ventilation.—Practical Phenomena connected with the transmission of Heat by Conduction, Convection, and Radiation.—Changes of the condition of matter produced by Heat and Cold. Interesting and important phenomena connected with the boiling of Water, and the generation of Steam. Production of Cold by evaporation.

Notions concerning ELECTRICITY and MAGNETISM.[1]

INORGANIC CHEMISTRY.[2]

Difference between Mixture and Combination. Simple and compound Bodies. Synthesis and Analysis.—The most important of the non-metallic Elements, and

---

[1] There is no positive reason why these two subjects should not, to a certain extent, be here illustrated; but Dynamic Electricity involves a previous knowledge of Chemistry, and all things considered I think it will be found best to adopt, as I have done in my Popular Course, the plan of reserving the whole group for the Second Division, bringing in Static Electricity in connection with the subject of "Safety from Lightning," and Magnetism and Dynamic Electricity, in mentioning the services rendered by the Mariner's compass, and the Electric Telegraph.

[2] The following extract from the first of the two Lectures (Nos. V. and

of the Metals.[1]—Combustion and its results.—Groups of Compounds having common properties, such as the Acids, the Alkalies, the Salts, &c. Notable Operations, such as fusion, distillation, solution, neutralization, crystallization, precipitation, &c. The whole illustrated by useful examples.

VI.) devoted to Inorganic and Organic Chemistry in my Popular Course, may serve to indicate the spirit in which it is proposed that this Science should be taught in this first Grade of Instruction.

"To those who are aware of the comprehensive nature and vast importance of Chemical Knowledge, it may at first seem almost ridiculous to attempt to condense such a Science into two Lectures. But you must bear in mind that in the present instance, the purpose in view is merely to impart a few fundamental notions of the *Bodies* which Chemistry deals with, of the *manner* in which Chemistry deals with them, and of the *terms* which Chemists commonly use for designating those Bodies and explaining those operations. You will be surprised to find how easy this knowledge can be rendered by methodical arrangement, by the avoidance of all unnecessary technicalities, and by a few simple Experiments. At the same time I am sure that you will be pleased at discovering how many new features of interest are displayed on all sides by the Things of Daily Life, when you look into the chemical rationale of their origin and of their use; and I doubt not that you will become more willing, as well as infinitely more able, to undertake in earnest the study of Domestic and Sanitary Economy."

I may add to the foregoing that a considerable development of the acquaintance with Chemistry thus initiated in the first Division of the Course, will be attained in an easy and agreeable way in reviewing its applications to Domestic and Sanitary Economy in the proposed Second Division, which on its side gains by being interspersed with Chemical Experiments.

[1] The Non-metallic Elements included in my Popular Course are 4 Gaseous ones: Oxygen, Hydrogen, Nitrogen and Chlorine, and 4 Solid ones: Carbon, Sulphur, Phosphorus and Iodine. As regards the Metals, I have found it best to reserve the chief account of them for the 2nd Division of the Course, where it forms a suitable introduction to the Hardware Department. This, like many other parts of Chemistry, gains in interest for Grade 1 Pupils, by being brought into immediate connection with its practical applications.

ORGANIC CHEMISTRY.

Difference between Inorganic and Organic Bodies, and between Ultimate and Proximate Constituents.—Proximate Analysis illustrated by that of Flour, Milk, &c.—Review of various GROUPS of Organic Bodies, represented by their most common and useful Types:—Fibres, Starches, Gums, and Sugars. Oils and Fats. Resins, Balsams, and Essential Oils. Fermentation; [1] Alcohol, Ether, and Chloroform. Vegetable Acids. The Alkaloids. Colouring Matters.—Albuminoid and Gelatinoid Constituents of Food.

OUTLINES OF NATURAL HISTORY.

The three kingdoms of Nature, Purposes of Geology and Mineralogy.

Purposes of BOTANY.[2] Wonders of the Structure and Functions of the various parts or Organs of Plants, and of the mysterious cycle of Germination, Development, and Reproduction.—Differences between Endogens and Exogens, Flowering and Flowerless Plants and other distinctions of leading importance in Classification and Diagnosis.—Separation of the Vegetable Kingdom into Orders, Tribes or Families, Genera and

---

[1] Fermentation should be here very lightly touched upon, and further explained in connection with Fermented Liquors in the 2nd Division.

[2] In my Popular Course, Botany only occupies at present a small portion of a Lecture chiefly taken up with Zoology, which thanks partly to Patterson's Diagrams, proves an attractive as well as instructive subject. If I can succeed in giving Botany an acceptable form, I intend carrying out at the first opportunity the arrangement here indicated, by which the number of Lectures will be raised from 9 to 10; but this will not in the least prejudge the question of the length of the future Official Course, for which I can only presume to offer suggestive indications.

Species. Convenient practice of distinguishing Plants by their Generic and Specific names.

Review of some of the most important Orders, illustrated by useful types.

Purposes of ZOOLOGY. Chief points on which its classification is commonly made to depend. Review of its leading divisions with the aid of Patterson's Diagrams, or some other series of Illustrations, bringing into relief the most instructive features and incidents of the Animal Kingdom.[1]

HUMAN PHYSIOLOGY.

The bony framework of our system.

Review of the chief Organs and of their Functions: the Brain and the Nervous System. The Organs of Motion. The Blood; its composition, its circulation, its heat producing functions. The Lungs and Respiration. Effects of respiration in crowded dwellings. Importance of Ventilation.

Necessity for Nutrition. The organs and functions of Nutrition. The Senses: Sight, Hearing, Smell, Taste and Touch. The Skin: its functions, and the care required for preserving their healthy action. The Nails. The Hair.

Advantages to be derived from the study of Physiology, and manner in which it should be conducted.

---

(1) It is to be regretted that Patterson's Diagrams are in descending progression, whereas it is more convenient and more useful to begin with the lower orders and proceed upwards, a plan which the same author himself has adopted in his 'Zoology for Schools.'

SECOND DIVISION.—APPLIED OR ECONOMIC KNOWLEDGE.(1)

The Matter of this Division is not, as that of the preceding one, separated into parts sufficiently equal to form as many Lectures, but arranged so as to present a number of Subjects or Groups of Subjects which besides receiving collectively from the Student the general attention due to all, may be conveniently singled out for further special study.

THE WORKING MAN'S HOME: Causes which influence the sanitary character of a SITE, and render it healthy or unhealthy, or better suited for one temperament or occupation than for another. Drainage, Sewerage, Water-supply and other questions relating to Public Hygiene, which everybody should understand. Hints respecting the selection of a Dwelling in Town or Country.

BUILDING DESIGN AND CONSTRUCTION: Necessity that occurs in populous towns for building houses many stories high. The external gallery system, the corridor system, and other contrivances. Adaptation of existing urban houses. Dwellings suited for suburban localities. Rural Cottages; how they may be made comfortable and healthy with the least difficulty and expense.

(1) This part of the Syllabus is mainly an adaptation to the requirements of Junior Students, of the range of Subjects reviewed more extensively, and from a somewhat different standpoint, in the Synopsis of an Economic Museum given in Section 2 of the preceding Chapter, to which reference may be made for additional details if desired. I can further place at the disposal of the persons charged hereafter with preparing the "Official Courses," several Drafts of Lectures containing suggestive matter relating to Dwellings' Improvement, and other hygienic questions.

Notions such as every one should possess concerning the difference between *Running up* a House, and building it substantially; also concerning the shape and arrangement of Rooms, and the various contrivances for the avoidance of damp, noxious effluvia, vitiated air, smoke, danger from fire, &c. Best construction of Staircases, Floors, Roofs, Windows, &c.

BUILDING MATERIALS : Stone; notions respecting the geological formation, chemical composition, and practical value of the leading kinds.—Mortar, good and bad. Cements. Concrete. Artificial Stone. Asphalt. —Bricks and Tiles; their manufacture and chief varieties.—Slate and Roofing Materials in general.

WOOD : Sources and properties of the chief kinds used for building and for furniture. How to select and use Timber to the best advantage. The Dry Rot; how to preserve Wood and render it fire proof.

Manufacture of Window Glass.—Useful notions concerning the materials and processes of House-painting and other modes of preservation and decoration, including paper hangings. Principles of Taste and appropriateness in Cottage Decoration.

Applications of Chemical and Mechanical Science to the production and elaboration of the most useful and familiar of the METALS, including the salient features of their progress from the Ore to the manufactured article. Review of the most interesting points in Household Ironmongery, with hints for selection and repair. Culinary Utensils, Domestic Hardware, and Cutlery; insight into their manufacture, and hints respecting them, borrowed from Science or from the experience of other countries.—Manufacture of Earthenware and Porcelain, Bottle-glass and Flint-glass, Mirrors, &c.—A glance at the trades of the Cooper, Turner,

Brushmaker and Basket-maker. Furniture as it often is in the dwellings of the poor, and as it ought to be.

TEXTILE MATERIALS, FABRICS AND CLOTHING : Practical value of an insight into the resources which Science and Industry supply for this department of our daily wants. Preliminary notions about spinning, weaving, bleaching, and dyeing, fast and fugitive colours, &c.—Chief Textile Materials; their origin and nature, raw condition, and stages of elaboration for the loom.—Useful information concerning Textile Fabrics, from coarse Matting up to those used for household purposes and apparel, bringing into relief the good qualities or defects which most concern working-class Families.—Leather; its manufacture and chief varieties. Furs.—Gutta-percha; its use for soles and other purposes. India-rubber; waterproof fabrics of various kinds, and hints concerning their use for apparel.

HABERDASHERY and GARMENTS of all kinds, including foot and head coverings; interesting features of manufacture; adaptation to weather, occupations, and means. Glance at the rational points in the Costumes of other nations.

FOOD : Brief recapitulation of what has been said in the First or Elementary Division concerning Nutrition in treating of Human Physiology, and concerning proximate and ultimate Analysis in treating of Organic Chemistry. Importance of basing as far as possible the appreciation of the various articles of animal and vegetable food, on a knowledge of their Proximate Constituents, and of the value of these for Nutrition; careful review of the chief Proximate Constituents, for the purpose of carrying out this principle.

Causes of deterioration of Food, and means of preservation. Prevailing forms of Adulteration and Fraud; means of detection.[1]

WATER; impurities, how ascertained and remedied. SALT.

Educational Review of FOOD RESOURCES supplied by the ANIMAL and VEGETABLE KINGDOMS, Proximate Analyses being given wherever conveniently practicable.[2]—CONDIMENTS; their use and abuse.

BEVERAGES: Cocoa, Coffee, Tea, and their substitutes. Fermented and Distilled Liquors.[3] Refreshing Drinks.

GENERAL CONSIDERATIONS respecting PROVISIONS: Ignorance and bad management which are too often displayed in the purchase of Provisions, as well as in their use. Rationale of Culinary Processes. Improvements that may be borrowed from foreign nations. How wholesome and palatable meals may be prepared with cheap materials.

Remarks on DIET, that is to say on the selection of Food and the arrangement of meals, on rational principles, according to individual requirements, and available supplies.

(1) Special adulterations will be mentioned in connection with the respective Articles.

(2) There is generally speaking no difficulty in training even illiterate minds to the appreciation of Food Analyses, provided that at first either actual quantities are shown, or representative Diagrams on the plan introduced by Dr. Lyon Playfair; and provided also that numerical data be reduced to the simplest approximate figures. A pound avoirdupois is a very popular standard, but the various fractions of ounces stand to it in no clear proportion, and I have found best to adopt percentage Analyses, taking 100 grams as the standard. The opportunity thus afforded for popularizing the metrical system is in itself an advantage.

(3) For Grade 1 Students, it may be best to place here the rationale of the Saccharine and Vinous Fermentations, in connection with the processes of Malting, Brewing, and Distilling. For Grade 2 Students, it may be best to place the theory of these as well as of the Acetic and Putrid Fermentations either in Organic Chemistry, or in the general considerations at the beginning of the Food Department.

HEATING: The Coal Mine; its tales of olden time; its treasures, how hardly won, and how ignorantly wasted. Review of the best kinds of cheap Fuel, and of the Grates, Ranges, Stoves, and other appliances best calculated to turn them to useful account for warming and cooking.

LIGHTING: Review of the most important Materials and appliances; how to obtain at little cost a quality of Light congenial to the eyes, convenient to manage, and free from danger.

CLEANLINESS: Manufacture of Soap; review of its varieties and substitutes, and generally of the best articles and appliances for washing, brightening and polishing.—Disinfectants.

PERSONAL HYGIENE: Brief History of the Human Frame from infancy to old age, showing how much may be done to foster its development, and maintain its healthy action even under difficulties, and in short how much it lies in our power to enhance the enjoyment of life, and to prolong its duration, by a constant attention to the united dictates of hygiene and morality. —Hints for warding off incipient infirmities, and for relieving the sufferings, and increasing the resources of the infirm.

SAFETY OR RELIEF:[1] Advice concerning injurious articles likely to be mistaken for nutritive or medicinal

[1] Much good may be effected in this direction by simply drawing attention to existing means of guidance, such as the following:—
- 'Diagrams of Edible and Poisonous Fungi,' with Text by W. G. Smith (Hardwicke).
- 'Poisonous Plants,' by the Christian Knowledge Society.
- 'Noxious Plants,' by Darton and Hodge.
- 'Relief for cases of Poisoning.' Chart by Wm. Stowe (Churchill).
- 'First Help in Accidents,' by Ch. Schaible (Hardwicke).
- 'Plain Rules to be observed in case of Illness or Accident,' by Dr. R. Druitt (S. P. C. K.).
- Cox's 'Companion to the Sea Medicine Chest, &c.' (Simpkin and Co.).

ones, and articles of food which become unwholesome at certain seasons or under certain conditions; poisonous Mushrooms and other productions; recognition of Hydrophobia. Antidotes and other means of immediate relief for cases of poisoning, and for venomous stings or bites. First help in sudden accidents, including Asphyxia and Drowning. Rescue from Fire or Shipwreck.[1] Precautions against Lightning, prefaced with a brief account of the phenomena of Static Electricity. —Hints for avoiding the detriment to health, or the danger to life or limb, incurred in various industrial occupations.

Useful notions concerning various SCIENTIFIC APPLIANCES in common use :[2]—Gunpowder and Fire-arms. —The Steam Engine.—The Mariner's Compass and the Electric Telegraph, prefaced with brief explanations concerning Magnetism, Dynamic Electricity, and their mutual influences.—Clocks and Watches. Alarums.— Scales of various kinds, and how to ascertain their correctness.— Weights and Measures of length and capacity; advantages of the metrical system.

MISCELLANEOUS HOUSEHOLD REQUISITES: Manufacture of paper, ink, pens, pencils, &c. Pins, needles, and the like. Rope and string, glue, &c.

RECREATION: Attention to be drawn to useful and recreative studies in Science, Art, and Music; rational

---

(1) Many lives might be saved through propagating a knowledge of the various means of rescue most frequently available.

(2) The possibility of admitting into a Grade 1 Course of Popular Instruction, subjects like those here enumerated, will necessarily depend on the pains bestowed to reduce each principle to its simplest and clearest expression, and to illustrate each appliance by models or drawings of its most representative types, pointing to appropriate sources for the further gratification of the interest which it is sought to awaken.—Barometers and Thermometers have been explained in the First Division of the Course.

amusements and healthful games and exercises, including drill and swimming.[1]

FRIENDLY ADVICE: Household management and the keeping of accounts; Evils of the Tally and Pawning Systems; Co-operative Stores and the like; Savings Banks, Benefit Clubs and Mutual Provident Societies; Co-operative Building Societies; present and deferred Annuities, Life Assurance, and various other forms of PROVISION FOR SICKNESS, OLD AGE, and DEATH.

REMARKS ON THE FOREGOING SYLLABUS.

It must be borne in mind that the object of this Grade 1 Course is to gain a hold of minds in a very incipient state of culture, to raise them to a sound practical acquaintance with many useful things, and at the same time to endow them with a capability for learning more. Accordingly all classifications of the subject matter of Physics which involve previous mathematical studies of any kind are set aside, and the Pupil, whether youthful or adult, has his attention first directed to nearly the same sort of "Objects," as are displayed to Infants in a Kindergarten. His training in mental gymnastics begins in the notice he is induced to take of the leading plane and solid forms, in connecting with them names which will be frequently used, and in learning something about angles and their measurement, the relation of the circumference of a circle to its diameter, and other little items of information, as easy and entertaining as they are useful. After these preliminary explanations intended to obviate future trouble and

---

[1] The advantages of swimming as a branch of education cannot too strongly be urged.

stoppages, Physics proper is introduced with a few manifestations of attractive force which diversify the instruction, and then the illustrated Glossary of useful terms is resumed by reviewing the chief Physical Properties of Bodies.[1] Weight taken as one of them brings the mind to the subject of Inertia, from which a natural reaction leads it to motion with its consequences. From this time the attention is almost unceasingly attracted by phenomena *in action*, which thus become as clear to the uninitiated Student, as mathematical proofs are to one duly trained.[2]

Some apology may appear necessary for assigning so comprehensive a range to the General Scientific Foundation of the studies of Artisans. To require that they should be able to answer any reasonable set of questions on the whole of the facts and principles thus embraced, may seem exaggerated and unnecessary to those who have not investigated the requirements of Industrial Life, and altogether illusory to those who are not aware how welcome intellectual food is to Working Men, provided it be rendered conveniently available, and properly suited to their mental palate. Now as regards necessity, it must be remembered that the Artisan should know the art of spending as well as that of earning, compassing with his mind the rationale of the things used by himself and his family, as well as of the things he manipulates for the use of others. Then as regards his ability to take in useful information, I need only refer to the manner in which the " Popular Course " above referred to has been constantly received

(1) In my Course delivered at London Institutions, I have found it expedient to place the Glossary of Forms after that of the Properties of Bodies, and in some cases it might be omitted altogether.

(2) A few similar remarks on the mode of teaching Chemistry are reserved for the 2nd Grade.

SECT. 1.] *GENERAL SCIENTIFIC FOUNDATION.* 187

for the last seven years, and still continues to be received by the Working Men of London.

An account has been given in Chapter I. of the manner in which the Scientific Training of the Working Man may be initiated in his Primary Education, and in fact fresh "signs of the times" are constantly pointing that way; but it will at the best be a long time before that agency is in full operation. Even then the scientific instruction of a large proportion of working-class youths and adults will have to be completed, as all has now to be supplied, by Class Lessons, Free Lectures and the like, supplemented by private reading. For these purposes, the best *printed* form to be given to the Instruction in question, is doubtless that of a familiar Discourse divided into Lectures for delivery on the Binary system, but susceptible of easy subdivision for Class Lessons. The work should be fully illustrated with woodcuts for home reading.[1] As regards the amount of matter, I reckon that any competent and painstaking Author who may be selected by the University for the purpose, will be able to embrace in 9 or 10 Lectures all that it is essential to include in the First or Elementary Division of the above Syllabus. As for the Second or Applied Division, I am less competent to judge, having as yet only written fragments for it; but I do not think that it need extend beyond ten or twelve Lectures. Thus we may reckon that Grade 1 of the whole range may be taught in about 20 Lectures, or about twice as many Class Lessons such as Artisans might be expected to attend during their Apprenticeship.—If we suppose this amount of Instruction to be spread through the

---

[1] For further particulars see Section 5 of this Chapter.

evenings of two winter seasons, we have before us something not unreasonable, and which leaves plenty of margin for collateral or subsequent studies.—I am now dealing with the question as it will have to be dealt with as long as the Education of the People remains what it is. When Elementary Science approaches a proper development in Primary instruction, the Studies of the young Artisan will receive a proportionate lift. On leaving a well-conducted Primary School, say at 14, he will already be tolerably well grounded in Grade 1, and the opportunities for attending Evening Classes during his Apprenticeship, may be devoted at once to the further studies prescribed by the Curriculum of his Trade.

In dealing with Artisan Students, especially as regards the lower grades of instruction, Class Teaching will be preferred to the more formal plan of teaching by Lectures. These will as a rule be of one hour's duration each, taking three quarters of an hour as a minimum, and an hour and a quarter as a maximum. This supposes them to be delivered on the binary system, which is more rapid than an ordinary single handed delivery. Class Lessons, which will generally be on the latter plan, will for about half the matter, take rather more than half the time. Moreover they will be followed by an Examination on the subject just treated of, or preceded by an Examination on the subject of the previous discourse. A Lesson is assumed to be more colloquial than a Lecture, and if the Teacher fancies that his Pupils do not quite understand him, it is not out of order to invite questions.[1]

(1) Mr. J. C. Ward of the Geological Survey, in delivering his 9 Lectures on Natural Philosophy, employed 3 evenings to each Lecture; viz.: one for actual delivery, one for colloquial explanations with repetition of some experiments, and one for the examination of the Pupils.

SECTION 2.—THE SECOND GRADE OF GENERAL SCIENTIFIC FOUNDATION.

The Range of studies will be essentially the same as in Grade 1, but there will be a general expansion of the Subjects; inconsiderable in some, as for instance in Physiology, but very considerable in others, as for instance in Chemistry. The mode of treatment will throughout the Course be much more earnest and circumstantial, though care will still be taken to reduce whatever has to be said, to its simplest expression, and to avoid all unnecessary technicalities. To afford practical guidance in matters of every-day life, will still be a constant aim, but the introduction of additional *technological* Illustrations wherever opportunities present themselves, will give a decidedly *industrial* character: so much so that whereas the previous may be denominated the POPULAR COURSE, the present may be entitled the INDUSTRIAL COURSE.

It is unnecessary to give a fresh Syllabus; but the leading headings of the previous one will be repeated for the convenience of introducing a few remarks.

FIRST DIVISION.—ELEMENTARY KNOWLEDGE.

MECHANICAL AND CHEMICAL PHYSICS.

THE difference in the tone of instruction between Grade 1 and Grade 2, may in some measure be assimilated to that between the intellectual capabilities of a School-boy and of an Apprentice. Accordingly, we may now review at once the standard Properties of matter in the order commonly assigned to them, whereas previously it was necessary first to train the mind to the appreciation of some of the most visible and tangible forms of Bodies, before it could be considered competent to realize the abstract conception of *Matter*. But even now we must scarcely reckon on any acquaintance with Geometry or Algebra. The consequence is that instead of laying down the abstract laws of the physical forces, and then working them out with geometrical figures or algebraical equations, introducing experimental demonstration as a secondary affair, we must in most instances blend with the enunciation of each law its actual and visible working, and be perfectly content if we can thus make principles sufficiently clear by their practical results, for being readily and rightly applied whenever there may be occasion.

Working Men have as a rule a quick perception of what can practically benefit them. Hence it has been a frequent subject of remark at the London Institutions that the illustrations of the MECHANICAL POWERS in the "Popular Lectures" though less showy than some other parts of the Course, have particularly riveted the attention of industrial audiences. But then it must be remembered that they were always exhibited in actual operation, and explained *dynamically*, not *statically*. Even in this 2nd Grade the applications of the Mechanical Powers will not be pursued beyond the simplest and most common forms of Machinery. More advanced Instruction in this line will take the name of MECHANISM. (See p. 217.)

The subject of Light will be considerably expanded,

though Polarization, Fluorescence, Diffraction and the like, will be reserved for Grade 3 of Chemical Physics, and Spectrum Analysis probably transferred to that Grade of Chemistry.—The title CHROMATICS will be given to the Technical Knowledge of Primary, Secondary and Tertiary Colours, and of the rules of their Harmony and Contrast.[1]

The subject of Heat will be considerably enlarged.

As regards Electricity and Magnetism, a plan may still be adopted similar to that recommended for Grade 1.

CHEMISTRY—INORGANIC AND ORGANIC.

As already hinted, a very considerable expansion is necessary in this subject, as compared with its treatment in the previous Grade. In fact Grade 1 which professes to embrace only that which if possible every Working Man should know, irrespectively of the nature of his employment, must necessarily be confined to what almost every man may have a chance of taking in. Accordingly only the most essential outlines of Chemical Knowledge should be given. The case is different as regards the Grade 2 of Chemistry now before us, which is designed to be the regular standard of Chemical Studies for Apprentices and Artisans whose Pursuits are of a chemical character. It will indeed be frequently prescribed for Students for whom Grade 1 of other parts of the General Scientific Foundation may suffice. Accordingly it will present whatever it is essential that they should know, in a practical and

[1] See Section 4 of this Chapter (p. 217). In introducing the term "Chromatics," I have the authority of Chas. Brooke; see Chapter XIX. of his 'Natural Philosophy.' (Churchills.)

sufficiently circumstantial form, and yet in such easy language as perfectly to keep within the compass of their intellectual culture. Thus to make up for educational deficiencies is in truth no easy task, but a very necessary one. In Physics even of the 2nd Grade we found our Pupils still incapable of following out abstract Laws by means of geometrical figures or algebraical equations. So likewise will these same Pupils be quite unable to compass with their mental grasp a system of EQUIVALENTS, or to draw benefit from the working of SYMBOLIC FORMULÆ. They would turn with dismay from one of those pages of hieroglyphics in which the experienced Chemist sees depicted a series of the most interesting phenomena, and which to his mind's eye, bring a whole laboratory within a nut-shell. Yet they will perfectly comprehend the nature and properties of Chemical Bodies if *properly illustrated*. Progressing step by step they will take in first elementary, then compound ones: and provided everything be as much as possible brought home to their outward perception, they will soon sufficiently understand chemical action and reaction, to comprehend in an approximate way any processes not too intricate, and not involving articles yet unknown to them.—This last consideration to which attention has been previously drawn, points to the expediency of reserving for the end of the Chemical Course, many processes of production, mutation, and analysis. Laboratory Practice conducted methodically *under strict surveillance*, becomes here especially an important element of Instruction. It should be made provision for in all Technical Colleges, and as far as possible in the local Industrial Institutes.[1]

[1] Laboratory Practice should be the *illustration*, not the *forerunner* of Chemical Knowledge. In the former case it impresses safely and mar-

By means of these various expedients, any Working Student whose intelligence is naturally good, though not sufficiently cultivated to allow of his appreciating or being benefited by a scholastic chemical training, may be turned out fit for undertaking intelligently any ordinary technical manipulations of a chemical character. This is satisfactory as compared with the present prevailing ignorance; but the prospect is still more cheering if we consider the facilities created by this amount of knowledge for learning more. The main difficulty which at first stood in the way of the comparatively untutored beginner, that of connecting in his mind the hazy conception of a set of Bodies strange to him, with a set of Symbols still more strange, and of attaching to these in his memory an ill-appreciated set of arithmetical equivalents, has in great part ceased to exist when he has made actual acquaintance with the Bodies in question, has in a measure followed out the mutations for the expression of which Symbolic Notation is used, and has learned to conceive the advantage of graphically expressing these mutations by a few compact signs and numerals. Accordingly, the Course should not be concluded without such an initiation in the latest standard edition of the System, and its practical working, as may enable the intelligent and willing Student to pursue the next stage of his chemical progress with the aid of any of the most approved Hand-books of Elementary

vellously on the memory, the logic of connected facts; in the latter case it gives a taste for desultory information, inspiring the tyro with a self-complacent idea that he knows much more than he really does, and without supervision at the first, it is unquestionably dangerous. I cannot help wishing that Professor Bloxam, instead of rendering "Laboratory Teaching" easy and attractive for the ignorant, had reserved the benefit of his talent for Students capable of understanding the rationale of the operations entrusted to them, and of appreciating the intelligent caution required in handling chemicals.

Chemistry.[1] It is essential that the initiation here referred to, should be as much as possible divested of all personal or controversial bias; that a sincere endeavour should be made on the part of those entrusted with the preparation of the Text Books, to select from the crowd of innovations in theory and nomenclature, those which have the best chance of permanency, and that to a lucid exposition of these, such conscientious and discreet advice should be added, that the Student, having been previously well fortified with a substantial knowledge of chemical facts, may be able to keep his mind clear amid the confusion of conflicting authors, and cull information from books that teach differently from the way in which he has been taught.

It must be borne in mind that in the system of Industrial Instruction which it will be the business of the Central Technical University to establish and propagate by precept and example, all measures will be

---

[1] The following is taken from W. Stanley Jevons' 'Lessons in Logic,' p. 59 (Macmillan), 1872:—"There is no worse habit for a student or reader to acquire than that of accepting words instead of a knowledge of things. It is perhaps worse than useless to read a work on natural history about Infusoria, Foraminifera, Rotifera and the like, if these names do not convey clear images to the mind. Nor can a student who has not witnessed experiments, and examined the substances with his own eyes, derive any considerable advantage from works on chemistry and natural philosophy, where he will meet with hundreds of new terms which would be to him mere empty and confusing signs. On this account we should lose no opportunity of acquainting ourselves, by means of our senses, with the forms, properties and changes of things, in order that the language we employ may, as far as possible, be employed INTUITIVELY, and we may be saved from the absurdities and fallacies into which we might otherwise fall. We should observe, in short, the advice of Bacon— *ipsis consuescere rebus—to accustom ourselves to things themselves.*"

Among other authorities for making the study of Combining Weights or Equivalents, Symbols, and Notations, follow instead of precede a visual acquaintance with the bodies to which these principles apply, I may name Prof. Roscoe. See his 'Chemistry,' one of Macmillan's Science Primers. 1872.

concerted, and every part made to dovetail with its neighbour. Adopting this principle, the Technical College, or even the more humble Industrial Institute where the aspiring young Artisan may be making his way through Grade 2 of his Studies, will offer him at the same time, besides the indispensable complement of Drawing, Music, etc., an easy course of Mathematical Lessons, calculated to improve his mental acumen, and greatly to facilitate regular physical, chemical, or other ulterior scientific attainments.[1]

NATURAL HISTORY.

The tendency of the Sciences to split as they expand, will be particularly noticed in dealing with the divergent studies of the Three Kingdoms. In each. however the difference from Grade 1, will consist rather in the more manly tone of treatment, than in the addition of new matter, which will be chiefly reserved for the review of the respective Products as they occur in the several Departments of the 2nd Division, or in the Technology of the various Trades.

HUMAN ANATOMY AND PHYSIOLOGY.

Here particularly the change from Grade 1 will less consist in enlarged proportions given to the subject in hand, than in the more effective mode of treatment rendered possible by increased preparatory knowledge, and riper age. Even in Grade 1, it was expedient to go rather fully into Physiology as the foundation

[1] Respecting Metallurgy as an offset of Inorganic Chemistry, see p. 217.

of the Laws of Health; but it is not much mixed up in the special development of Technical Knowledge, except where called to the front by dangerous or unhealthy Trades.

SECOND DIVISION.—APPLIED OR ECONOMIC KNOWLEDGE.

In consequence of the greatly increased amount of chemical knowledge that can now be brought to bear on this range of subjects, as compared with the analogous part of the Grade 1 Course, not only can the *rationale* of the various departments of Domestic Economy be duly explained, but the manufacturing processes, an insight into which is so essential an element of domestic knowledge, can be treated of in such a manner as to afford at the same time an excellent general notion of INDUSTRIAL TECHNOLOGY; so much so, that certain portions of this Division will frequently be recommended for special study to Artisans respectively concerned, who may content themselves with Grade 1 of the remainder of the course.

The concluding subjects of the Grade 1 Syllabus, in which Domestic Economy borrows a leaf from Social Science, are matters that interest Daily Life in general, rather than Technical Life in particular, and demand but little progressive expansion.

REMARKS ON GRADE 2.

This Course, like the preceding one, must be specially prepared under the direct auspices of the Central Technical University. It is only thus that we may hope to see the necessary amount of pains bestowed by those who possess the necessary amount of abilities, in reducing the whole required range of knowledge to a uniform state of condensation at a uniformly low level.[1] There are indeed, especially as regards Grade 2, certain works giving in an appropriate style, one this, and another that portion of the information required; but the multiplicity of books would be alike inconvenient and expensive for Working Class Students, and it is unquestionably expedient to substitute for fragmentary and disjointed information, a connected and methodical arrangement of matter, and a uniformly progressive style of expression. Moreover it is necessary that the University should have entirely under its control, these STANDARD COURSES on which its influence for raising the condition of the Industrial Population will so much depend.—As nevertheless it may be desirable to try the proposed system of Trades Examinations with the aid of temporary means of instruction before the official Text Books can be got ready, I will presently indicate a few publications of which certain parts may thus provisionally be used.[2] None have been indicated for Grade 1, because the more humble the instruction, the more difficult it is to find reliable scientific Books that stoop low enough. Even as regards the present Grade, it will be necessary to make a careful selection of the Books most conveniently avail-

---

(1) The manner in which the publication of both Courses may best be made to subserve lecturing purposes and home study, will be explained in Section 5 of this Chapter.

(2) The difficulty of finding scientific Books of which the whole may be used by Artisan Students, arises partly from the fact that few have been written for their special use; but partly also from the uneven character of many of the Books intended for beginners which for reasons that have been explained elsewhere, contain here and there hard bits that stop the way of the tyro, sticking up like granite boulders in the midst of rather soft alluvium.

able, and then to prepare as a temporary substitute for the Standard Course, a detailed Syllabus thereof referring to the Books in question for each subject or fraction of a Subject.[1]

The only data I possess for computing the time which Students may require for a Second Grade Course of the General Scientific Foundation, are those which I have obtained by writing a course of "Chemistry for Industrial Students" which approximately corresponds to this Grade, and which, when revised, will probably consist of 18 to 20 Lectures.[2] It may be objected that this number, equivalent to about twice as many Lessons on the one subject of Chemistry, seems to indicate for the whole of Grade 2, an amount of study out of comparison with the amount of available time; but for reasons already adverted to, the expansion of Chemistry in the second Grade as compared with the first, is quite exceptional. As far as I am able to judge, the remainder of the first Division of Grade 2 need not exceed 16 to 20 Lectures.[3]

Let us then suppose about 35 to 40 Lectures for the first Division of Grade 2, and 20 to 25 for the second or Applied Division; making 55 to 65 for the

---

[1] I can show to persons interested, a manuscript example of the kind of Syllabus of Reference here mentioned. It was framed some years since in connection with my proposal for special examinations of School Teachers of both sexes at South Kensington, of which an account is given in 'Science for the People,' pages 55 to 58.

[2] The Manuscript, which was nearly complete at the time of the destruction of my Museum by fire in April 1871, was fortunately saved, but the Illustrations, of which there was a chest full for each Lecture, were destroyed. Those of the inorganic portion, comprising eight Lectures, have however since been replaced, and thus far, as mentioned in the preceding Chapter, the Series has been satisfactorily tried.

[3] The computation on which this amount is based, is not, like the amount for Chemistry, derived from actual experience. It is accordingly only given for what it may be worth in the following Table. Two

whole of the second Grade, as against 20 to 22 for the whole of the first Grade; and if these amounts, which would be doubled for Class Lessons, strike us as rather excessive, let us take into account the following extenuating circumstances:—

a.—As stated in describing Grade 2 of Chemistry, that subject by itself will in many cases be prescribed for Minor Candidates, *i.e.* Candidates for a Journeyman's Certificate, with only Grade 1 of the rest of the General Scientific Foundation. In some instances, only a portion of Chemistry will be pushed beyond Grade 1; such as Inorganic Chemistry for Workers in Metal, Porcelain, Glass, etc., for most of whom a slight notion of Organic Chemistry will suffice, the theory of the Fermentations for Candidates from the Brewery or the Distillery, the rationale of the Formation and Destructive Distillation of Coal for those engaged in Gas or Paraffin Works, and so on. The same principle of special selection will apply to certain portions of other departments, as for instance the portions of Chemical Physics treating respectively of Light and Heat. It will also as before stated frequently apply to certain departments of the 2nd Division.

b.—Major Candidates, *i.e.* Candidates for a Master's

Lectures are marked for Grade 1 of Natural History in conformity with the Syllabus given in the preceding Section:—

	1st Grade. Lectures.	2nd Grade. Lectures.
Mechanical Physics	3	6 to 8
Chemical Physics	1	2 to 4
Natural History	2	4 or 5
Anatomy and Physiology	2	3 — 3
	8	15 to 20
Chemistry	2	18 to 20
Total	10	33 to 40

Certificate, for whom chiefly a Grade 2 of the whole range of General Scientific Foundation will be prescribed, will as a rule have abundant time for acquiring it. Supposing the young Artisan to pass his Minor Degree at or soon after the conclusion of his Apprenticeship, several years may be assumed to elapse before he can aspire to pass his Major Degree, through which time he can conveniently spread the additional studies required. In fact an interval of two years would suffice for these.[1]

c.—By means of the Text Books which it will be the duty of the Central Technical University to render as near an approximation as possible to an oral discourse, the Artisan Student will be able to make considerable headway by himself in certain parts of the Range assigned to him. This particularly applies to various portions of the 2nd Division, of which it is to be hoped that actual Illustrations corresponding to the Standard Text, will whenever practicable be exhibited in the form of an Economic Museum.[2]

d.—As already stated, when Primary Education is properly organized, the young Artisan will arrive at his Apprenticeship well grounded in Grade 1 of his General Scientific Foundation. Till then a sort of compromise must be made, and he may possibly have to begin his Grade 2 without the advantage of a previous preparation. Generally speaking however the Tyro who takes up a Science entirely new to him, should, especially if unassisted by previous general mental culture, acquire first of all some idea of its nature and

---

(1) It is true that at the first starting of the system, many Masters who have never passed any science examination at all, will have to get through the two degrees with as little interval as possible.

(2) See Chapter II., Section 8, A.

leading features, as given in a Grade 1 Course. This will notably facilitate the understanding and retaining of the facts which a more regular study, say that of Grade 2, will next impress on his memory, still leaving deeper mysteries for a subsequent 3rd Period of research, when he will be armed with still more practised powers of investigation.

The following are the Publications referred to at p. 197, as susceptible of supplying in the hands of competent advisers, a temporary substitute for the proposed Grade 2 Course.

[LIST 2.]
SUGGESTIVE EXAMPLES
OF WORKS PROVISIONALLY AVAILABLE FOR
GRADE 2 STUDIES.

A.—ELEMENTARY.

Mackenzie's Series of Science Primers. From –/2 each. (Hardwicke.) (1)
Gleig's School Series. From –/9 upwards. (Longmans.) This Series includes some of Tate's well-known little Scientific Works.
Macmillan's Science Primers. 1/–.
S. P. C. K. Manuals of Elementary Science. 1/–.
Some of Colling's Elementary Science Series. 1/–.
,, ,, Heywood's Class Series of Text Books on Science. 1/– (Manchester, and Simpkin, Marshall & Co., London.)
,, ,, Murby's Science and Art Department Series of Text Books. 1/–.
,, ,, Cassell's School Series. From 1/6 upwards.
,, ,, .Weale's Series. From 1/– upwards.

(1) It must be remembered respecting epitomes of this description, that condensation into a small compass for the sake of a low price, generally tends to render Science less easy than it can be made in more explanatory Primers.

Some of Baker's Circle of the Sciences.
„ „ Chambers' Useful Handbooks. -/6.
„ „ Blair's Works. (Whittaker.)
„ „ Timbs' light scientific Works.
„ „ the Illustrated Works published by Jarrold.
„ „ the Christian Knowledge Society's series of Educational Scientific Works.
„ „ the Educational Trading Company's Books.
Parts of Elementary Natural Philosophy, by J. C. Ward. 3/6 (Trübner & Co.)
„ „ Cyclopædic Science Simplified, by J. H. Pepper. Cr. 8vo. 9/- (Warne.)
„ „ the Popular Educator. 6 vols. 6/- each. (Cassells.)

Matter and Motion. -/10 (Chambers.)
Botany for Beginners, by Maxwell Masters.
Physiology for Schools, by Mrs. Bray. 1/- (Longmans.)
The House I live in. Edited by T. C. Girtin. (Parker & Son.)
A Popular Manual of Physiology by H. Lawson. 2/6 (Hardwicke.)

B.—APPLIED.

The Philosophy of Common Things. 1st & 2nd Series. 1/- each. (Religious Tract Society.)
Lessons on Houses, Furniture, Food, and Clothing, being the 2nd vol. of the "Instructor." 2/- (Parker.)
Outlines of Technical Knowledge by D. M. Smith. 1/- 12mo. (Kent.)
The Chemistry of Common Things, by S. M'Adam. 1/6 (Nelson & Sons.)
Lankester's Lectures on Food. (Hardwicke.)
A Handybook on Food and Diet, by C. H. Cameron. 1/- Post 8vo. (Baillière & Co.)
A Handybook on Health, by C. H. Cameron. 1/- 12mo. (Cassells.)
The Hygiene of Air and Water, by W. Proctor. 2/6 Post 8vo. (Hardwicke.)
Provident Knowledge Papers, a series of Penny Pamphlets by G. C. T. Bartley, containing advice and information to the Industrious Classes on the following and analogous subjects: Annuities, Life Insurance, Penny and Post Office Savings Banks, Pawnbroking, Domestic Service, Thrift. Published by the Provident Knowledge Society, 112 Brompton Road, S.W.

SECTION 3.—THE THREE UPPER GRADES OF SCIENTIFIC INSTRUCTION.

Whilst every Candidate, even for a Minor Degree, will be required to possess Grade 1 of the range of Elementary and Applied Knowledge described as General Scientific Foundation, and whilst most Candidates for a Major Degree will be expected to master the whole of Grade 2, Superior Candidates, *i.e.* those aspiring to a Diploma of Excellence, will similarly be prescribed a Grade 3.[1]

There is nevertheless a considerable difference between this and the preceding Grade; not in the substance, but in the style of the instruction, which being addressed to a more advanced class of Students, may assume a proportionately higher tone. The second Grade, though a considerable step higher than the first, still involves the difficult task of teaching Physics with scarcely any Mathematics, Chemistry without anything beyond the comprehension of the uneducated, and Science in general with a minimum of hard words; whereas in Grade 3 the conventional phraseology of Science may be freely used within reasonable bounds.[2]

---

[1] There will be no 4th Grade for the collective range, though many of its component parts may be carried to a 4th or even a 5th Grade.

[2] I say "within reasonable bounds," because I am supported with very good authority in asserting that in recent scientific and especially chemical phraseology, the bounds of reason have been frequently overstepped. The celebrated French Chemist Dumas complains of the complexity and uncouthness of modern chemical nomenclature, and admits that with all his assiduity, he finds it impossible to keep pace with the innovations, new theories, and verbal alterations now flooding his favourite Science.

The consequence is that whereas it will be one of the very first duties of the Central Technical University to provide standard Text Books for the two lower Grades of General Scientific Foundation, the same urgency will not exist as regards Grade 3, on account of the difference in the class of Students, owing to which the scientific literature of the day can to a considerable extent be made available. It will however be necessary, particularly as regards the Applied Division, to adopt the plan of forming a detailed Grade 3 Syllabus of the whole range of Subjects, and of referring for each to the most appropriate source. This SYLLABUS OF REFERENCE will be valuable not only to the few Students whose Curricula prescribe Grade 3 for the whole of their Foundation, but to the much larger number who will be recommended to carry a portion of it up to that level. In showing Candidates the exact instruction that suits their purpose, it will keep them from mixing in their present studies what had better be left for future ones. Many a Student loses his love of learning through taking up by himself comprehensive Books that are partly within his intellectual grasp, and partly above it; whereas he might imbibe knowledge from them with delight, in consecutive readings on an expanding scale under judicious guidance.

The following list was originally intended to contain only a few Books carefully selected to represent the Grade 3 level of studies, but so great has been the number of publications useful in various ways which have passed under consideration in attempting to make that selection, that I have been induced to include as many as are likely to be found acceptable in preparing the proposed Syllabus of Reference, or perhaps convenient in ultimately compiling the Grade 3 Course.

Time has however failed for thoroughly sifting so vast an amount of intricate literature, and moreover in many cases different levels of study occur in the different parts of the same book, and more especially in the different books of the same series. I have accordingly been induced to unite publications of unascertained level in a Supplementary List, corrected to the latest date, and which will be found with similar ones in the Appendix at the end of this volume. Altogether the Lists given in the present Chapter should be accepted as they are offered, simply in a prodromous character. I should indeed be thankful for any additions or emendations that might fit them for rendering useful service by the time they are wanted by the Instructional Board of the proposed University.[1]

[LIST 3.]
SUGGESTIVE EXAMPLES
OF BOOKS MORE OR LESS DESERVING OF CONSIDERATION IN FRAMING A SYLLABUS OF REFERENCE FOR

GRADE 3 STUDIES.[2]

PHYSICS.
GENERAL.
    Lessons in Elementary Physics, by Balfour Stewart. 18mo.[3] 4/6 (Macmillan's Elementary Scientific Class Books.)

---

(1) I take this opportunity of acknowledging the advantage I have derived from G. C. T. Bartley's Catalogue of Modern Works on Science and Technology (Chapman & Hall). I understand that a new edition will shortly appear.

(2) Some of the Series indicated for Grade 2 include Books available for the present Grade.

(3) As the sizes of the Books will be mostly given, it may be well to remark that the technical terms commonly used for this purpose, are far from expressing fixed Standards, on account of the various dimensions of

A Manual of Natural Philosophy, by Comstock, edited by Hoblyn. 12mo. 3/6 (Blackie.)

Elements of Experimental and Natural Philosophy, by Jabez Hogg. (Bohn's Scientific Library.) 5/- (Bell & Daldy.)

An Elementary Handbook of Physics, by Wm. Rossiter. 5/- (Blackwood & Sons.)

The Cambridge Course of Elementary Natural Philosophy, by Snowball & Lund. Crown 8vo. 5/- (Macmillan & Co.)

Natural Philosophy for General Readers, &c., by Prof. Ganot, edited by E. Atkinson. Cr. 8vo. 7/6 (Longmans.)

SPECIAL.

Handbook of Natural Philosophy, by D. Lardner. In 5 vols. viz.:—Mechanics; Hydrostatics and Pneumatics; Heat; Optics; Electricity, Magnetism and Acoustics. Small 8vo. 5/- each vol. (Lockwood.)

Elementary Treatise on Natural Philosophy, by A. Privat Deschanel. Translated by J. D. Everett. 4 Parts, viz.:— Mechanics, Hydrostatics, and Pneumatics; Heat; Electricity and Magnetism; Sound and Light. Medium 8vo. 4/6 each part. (Blackie & Sons.)

Exercises on Mechanics and Natural Philosophy, by Thomas Tate. 12mo. 2/- (Longmans.)

Mechanical Philosophy, by W. B. Carpenter. 12mo. 5/ (Bell & Daldy.)

An Elementary Class Book on Sound, by G. C. Foster. (Rivington.)

Notes of a Course of Nine Lectures on Light, by John Tyndall. Crown 8vo. Cloth, 1/6 (Longmans.)

An Introduction to the Science of Heat, by T. A. Orme. Foolscap 8vo. 3/6 (Groombridge & Sons.)

An Elementary Treatise on Steam, by J. Perry. 4/6. (Macmillan.)

Notes of a Course of Seven Lectures on Electrical Phenomena, &c., by John Tyndall. Crown 8vo. Cloth, 1/6 (Longmans.)

the paper employed. I became particularly aware of this in bringing out my 'Handbook of Economic Literature' in 1862, and a friend specially versed in the technicalities of Bibliography, prepared for me the following list of approximate standards measured by height of page :—

4to (quarto), 11¼ inches.
8vo (octavo), 8½.
12mo (twelvemo or duodecimo), 7¼.
18mo (eighteenmo or octodecimo), 5¾.

Electricity, by R. M. Ferguson. (Chambers' Educational Course.) 3/6.

An Elementary Class Book on Electricity, by G. C. Foster. (Rivington.)

N.B. Chemical Physics are also well given in the introductory portions of several of the Works on Chemistry hereafter mentioned.

CHEMISTRY.

GENERAL. (Including Inorganic and Organic.)
> Elementary Chemistry, by Rev. H. M. Hart. Small 8vo. 3/6 (Cassell & Co.)
> Principles of Chemistry, by J. A. Stockhardt. Small 8vo. 5/– Bohn's Scientific Library. (Bell & Daldy.)
> Lessons in Elementary Chemistry, by Henry Roscoe. 18mo. 4/6 (Macmillan's Elementary Scientific Class Books.)
> Notes for Students in Chemistry, by A. J. Bernays. 3/6 12mo. (Churchills.)
> Outlines of Chemistry, by W. Odling. 7/6 Cr. 8vo. (Longmans.)

INORGANIC.
> Inorganic Chemistry, by George Wilson, edited by G. F. Madan. Foolscap 8vo. 4/– (Chambers' Educational Course.)
> Chemistry for Schools, by C. H. Gill. Small 8vo. 4/6 (Walton.)
> Introduction to the Study of Inorganic Chemistry, by W. A. Miller. Small 8vo. 3/6 (Longmans' Text Books of Science.)
> Inorganic Chemistry, the Non-Metals, by T. E. Thorpe. 2/6 (Collins' Advanced Science Series.)

ORGANIC.
> Organic Chemistry, by Hy. E. Armstrong. Small 8vo. 3/6 (Longmans' Text Books of Science.)

PRACTICAL.
> Practical Chemistry, by Stevenson Macadam. Foolscap 8vo. 2/6 (Chambers' Educational Course.)
> Practical Chemistry, including Analysis, by E. Bowman and C. L. Bloxam. Foolscap 8vo. 6/6 (Churchills.)
> Laboratory Teaching, by C. L. Bloxam. Crown 8vo. 5/6 (Churchills.)

NATURAL HISTORY.

A.—THE MINERAL KINGDOM.

MINERALOGY.

Rudiments of Mineralogy, by Alex. Ramsay, Jun. (Weale's Series.) Demy 12mo. 3/- (Strahan & Co.)
Introductory Text Book of Geology, by David Page. Crown 8vo. 2/- (Blackwood & Sons.)
Rudimentary Treatise on Geology, by Ralph Tate. Weale's Series. (Lockwood.)
    Part I. Physical Geology. 2/
    „ II. Historical Geology, reviewing the chief formations.
Elementary Geology, by J. C. Ward. 12mo. 6/ (Trübner.)

B.—THE VEGETABLE KINGDOM.

A Manual of Structural Botany, by M. C. Cook. Bound. 1/6 (Hardwicke.)
Lessons in Elementary Botany, by D. Oliver. 18mo. 4/6 (Macmillan's Elementary Scientific Class Books.)
Botany, by Edward Smith. (Orr's Circle of Sciences.) 8vo.
School Botany, by John Lindley. Demy 8vo. 5/6 (Bradbury & Evans.)
Rudiments of Botany, by Prof. Henfrey. 3/6 (Van Voorst.)
Vegetable Physiology and Systematic Botany, by W. B. Carpenter. Bohn's Scientific Library. 8vo. 6/- (Bell & Daldy.)
Profitable Plants, by T. C. Archer. 8vo. 5/- (Routledge.)
Systematic and Economic Botany, by J. H. Balfour. Post 8vo. 2/6 (Collins' Advanced Science Series.)
The Vegetable Kingdom and its Products, with an account of their uses, by Robert Hogg. Small 8vo. 7/6 (Kent & Co.)
N.B.—No Floras are here named, because minute and time-taking Botanical Diagnosis, whether cultivated for its own sake, and as a fascinating pastime, or for adding new data to the registers of " *habitat,*" or with the less frequent intent of throwing light on points of theory, would scarcely be indulged in to any notable extent by Artisan Students, without a sacrifice of more necessary studies.

C.—THE ANIMAL KINGDOM.

Zoology, by R. Patterson. 8vo. 6/- (Longmans.)
An Introductory Text Book of Zoology, by H. Alleyne Nicholson. Small 8vo. 3/6 (Blackwood.)
Natural History of the Animal Kingdom, by W. S. Dallas. 8vo. 8/6 (Griffin.)
Lessons in Elementary Physiology, by Thomas H. Huxley. 18mo. 4/6 (Macmillan.)
Animal Physiology for Schools, by D. Lardner. 12mo. 3/6 (Walton.)
Animal Physiology, by W. B. Carpenter. Bohn's Scientific Library. Small 8vo. 6/ (Bell & Daldy.)

DOMESTIC AND SANITARY ECONOMY.

A Manual of Domestic Economy, by W. B. Tegetmeier. 12mo. 1/6 (Groombridge & Sons.)
Household Economy, by M. Gordon. Small 8vo. 2/- (Gordon.)

Every-day Chemistry, by Alfred Sibson. Small 8vo. 2/6 (Routledge & Co.)
The Hygiene of Air and Water, by W. Procter. Post 8vo. 2/6 (Hardwicke.)
The Science of Home Life, by A. Bernays. 6/ (Allen.)
The Chemistry of Common Life, by J. F. W. Johnston. 2 vols. Crown 8vo. 11/6 (Blackwood & Sons.)
Adulterations of Food, by R. J. Atcherley. 2/6 (Isbister.)

Health; a Handbook for Households and Schools, by D. E. Smith. (Isbister.)
Principles of Physiology applied to Health, by A. Combe, edited by J. Coxe. Cr. 8vo. 4/6 (Simpkin.)
Public Health; Introduction to Sanitary Science, by W. A. Guy. 12mo. 2/6 (Renshaw.)

[LIST 4.]

SUGGESTIVE EXAMPLES

OF WORKS AVAILABLE FOR

GRADE 4 STUDIES.

As before stated (see note 1 to page 203) there will be no regular Course of General Scientific Foundation above the Third Grade; but in many instances Candidates for Diplomas of Excellence will be called upon to push a Grade higher certain portions of their studies, and it is for their convenience that the present List has been framed.

PHYSICS.

GENERAL.
    Elements of Natural Philosophy, by Charles Brooke. Foolscap 8vo. 12/6 (Churchills.)
    An Elementary Treatise on Physics, by Prof. Ganot. Atkinson's Translation. Post 8vo. 15/- (Longmans.)
    A Cyclopædia of the Physical Sciences, by Prof. Nichol. 8vo. 21/ (Griffin.)

SPECIAL.
    Mechanics for Beginners, by Isaac Todhunter. 18mo. 4/6 (Macmillan.)
    Elementary Statics and Dynamics, by Rev. Harvey Goodwin. 2 vols. Foolscap 8vo. 6/- (Deighton.)
    An Elementary Treatise on Mechanics, by S. Parkinson. Cr. 8vo. 9/ (Macmillan.)
    A Treatise on Statics, by S. Earnshaw. 8vo. 10/ (Deighton.)

A Treatise on Motion, by S. Earnshaw. 8vo. 14/– (Deighton.)

Elementary Hydrostatics, by J. B. Phear. Crown 8vo. 5/6 (Macmillan.)

The Elements of Hydrostatics and Hydrodynamics, by W. H. Miller. 8vo. 6/ (Deighton.)

Eight Lectures on Sound, by John Tyndall. Cr. 8vo. 9/ (Longmans.)

A Treatise on Acoustics, by W. J. Donkin. Crown 8vo. 7/6 (Macmillan.)

On Sound and Atmospheric Vibrations, by G. B. Airy. Crown 8vo. 9/– (Macmillan.)

A Treatise on Optics, by Sir David Brewster. In the Cabinet Cyclopædia. 8vo. 5/– (Longmans.)

Six Lectures on Light, by John Tyndall. 16mo. (Longmans.)

A Treatise on Optics, by S. Parkinson. Post 8vo. 10/6 (Macmillan.)

The Undulatory Theory of Optics, by G. B. Airy. Crown 8vo. 6/6 (Macmillan.)

The Spectroscope and its Applications, by J. Norman Lockyer. Vol. I. of "Nature Series." Crown 8vo. 3/6 (Macmillan.)

An Elementary Treatise on Heat, by Balfour Stewart. Foolscap 8vo. 7/6 (Macmillan.)

Heat; a Mode of Motion, by John Tyndall. Cr. 8vo. 10/6 (Longmans.)

A Manual of Electricity, by H. M. Noad. 8vo. 24/ (Lockwood.)

A Treatise on Electricity, by A. de la Rive. Walker's Translation. 3 vols. 8vo. 73/– (Longmans.)

Rudimentary Magnetism, by Sir W. Snow Harris. Edited by H. M. Noad. Demy 12mo. 4/6 Weale's Series. (Lockwood.)

A Treatise on Magnetism, by G. B. Airy. Crown 8vo. 9/6 (Macmillan.)

Chemical Physics, being Part I. of the Elements of Chemistry, by W. A. Miller; revised by Macleod. 8vo. 15/– (Longmans.)

Elements of Natural Philosophy, by Sir W. Thompson and P. G. Tait. Part I. 8vo. 9/– (Macmillan.)

Elements of Physical Manipulation, by Pickering. Royal 8vo. Part I. 10/6 (Macmillan.)

CHEMISTRY.

GENERAL.

Chemistry for Students, by A. W. Williamson. Extra fcap. 8vo. 8/6 (Macmillan.)
Chemistry, by Brande and Taylor. 12/6 (Hardwicke.)
A Manual of Chemistry, by George Fownes. Foolscap 8vo. 15/– (Churchills.)
Inorganic and Organic Chemistry, by C. L. Bloxam. 8vo. 16/ (Churchills.)
The Elements of Chemistry, by Prof. Graham. 2 vols. 8vo. 40/ (Baillière.)

INORGANIC.

First Principles of Modern Chemistry, by U. J. Kay-Shuttleworth. Crown 8vo. 4/6 (Churchills.)
Introduction to Inorganic Chemistry, by W. G. Valentin. 5/6 (Churchills.)
First Steps in Chemistry, by R. Galloway. 12mo. 6/6 (Churchills.)
Second Steps in Chemistry, by R. Galloway. 10/ (Churchills.)
Lecture Notes for Chemical Students, by E. Frankland. 8vo. 4/ (Van Voorst.)
Inorganic Chemistry, being Part II. of the Elements of Chemistry, by W. A. Miller; revised by H. Macleod. 8vo. 21/– (Longmans.)

ORGANIC.

Organic Chemistry, being Part III. of the Elements of Chemistry, by W. A. Miller. 8vo. 24/ (Longmans.)
A Manual of the Carbon Compounds, by C. Schorlemmer. 8vo. 14/ (Macmillan.)

PRACTICAL.

Chemical Manipulation, by Michael Faraday. 9/–
A Handbook of Chemical Manipulation, by Greville Williams. 15/.
Exercises in Practical Chemistry, by A. G. V. Harcourt and H. G. Madan.
    Series 1. Qualitative. Cr. 8vo. 7/6.
    Series 2. Quantitative. (Macmillan.)
Qualitative Analysis, by R. Galloway. Post 8vo. 8/ (Churchills.)
A Course of Practical Chemistry, by W. Odling. Cr. 8vo. 7/6 (Longmans.)
Chemical Analysis, by Dr. C. R. Fresenius.
    Vol. I. Qualitative Analysis. 8vo. 12/6
        II. Quantitative ditto. 8vo. 18/–
    A. Vacher's Edition. (Churchills.)

NATURAL HISTORY.

A.—MINERAL KINGDOM.

MINERALOGY.
    Elementary Course of Mineralogy and Geology, by D. Ansted. 8vo. 12/ (Van Voorst.)
    Elements of Mineralogy, by James Nicoll. 12mo. 5/- (Longmans.)
    A Manual of Mineralogy, by J. D. Dana. Post 8vo. 7/6 (Trübner.)
    Elementary Mineralogy, by Brooke and Miller. 8vo. 18/- (Longmans.)
    Crystallography, by Mitchell. 3/ (Orr's Circle of the Sciences.)

GEOLOGY.
    Advanced Text Book of Geology, by David Page. 8vo. 7/6 (Blackwood & Sons.)
    The Student's Elements of Geology, by Sir Charles Lyell. Post 8vo. 9/- (Murray.)
    Student's Manual of Geology, by J. B. Jukes. 8vo. 12/6 (Longmans.)
    A Manual of Geology, by J. D. Dana. 8vo. 25/ (Trübner.)
    The Principles of Geology, by Sir Charles Lyell. 2 vols. 8vo. 32/- (Murray.)

B.—THE VEGETABLE KINGDOM.

The Elements of Botany, by John Lindley. Demy 8vo. 9/- (Bradbury & Evans.)
A Manual of Botany, by Robert Bentley. Foolscap 8vo. 12/6 (Churchills.)
A Manual of Botany, by J. H. Balfour. 8vo. 12/6 (Black.)
An Elementary Course of Botany, by Henfrey and Masters. 8vo. 12/6 (Van Voorst.)
The Vegetable Kingdom, by John Lindley. One thick volume. 8vo. 25/- (Bradbury & Evans.)
Medical and Economic Botany, by John Lindley. 12mo. 5/ (Bradbury & Evans.)
The Useful Plants of Great Britain, by Johnson. 8vo. 12/ (Hardwicke.)
Domestic Botany, by John Smith. Post 8vo. 16/ (Keene.)

C.—THE ANIMAL KINGDOM.

ZOOLOGY.
An Introduction to the Classification of Animals, by Thomas H. Huxley. 8vo. 6/- (Churchills.)
Manual of Zoology, by Milne-Edwards. Knox's edition. 8/6 (Renshaw.)
A Manual of Zoology, by H. Alleyne Nicholson. Crown 8vo. 12/6 (Blackwood.)
Forms of Animal Life, by G. Rolleston. 8vo. 16/- (Macmillan.)
General Outline of the Organization of the Animal Kingdom, by T. Rymer Jones. 8vo. 31/6 (Van Voorst.)

ANATOMY AND PHYSIOLOGY.
Handbook of Physiology, by W. S. Kirkes. Small 8vo. 12/ (Walton.)
A Manual of Physiology, by W. B. Carpenter. Foolscap 8vo. 12/6 (Churchill.)
The Principles of Human Physiology, by W. B. Carpenter. Power's edition. 8vo. 28/- (Churchill.)
Outlines of Physiology, by John Marshall. 2 vols. Crown 8vo. 32/- (Longmans.)
Atlas of Anatomy and Physiology, by Turner and Goodsir. 25/ (Johnston.)

DOMESTIC AND SANITARY ECONOMY.

Domestic Economy, by M. Donovan. 2 vols. 12mo. 7/- (Longmans.)
An Encyclopedia of Domestic Economy, by Webster and Parkes. One thick vol., 8vo. 31/6 (Longmans & Co.)
A Manual of Practical Hygiene, by E. A. Parkes. 8vo. 16/- (Churchill.)
Lectures on Public Health, by E. D. Mapother. 12mo. 6/ (Hardwicke.)
Food, by H. Letheby. 5/- (Baillière.)
Water Analysis, by Wanklyn and Chapman. Post 8vo. 5/ (Trübner.)
Traité d'Hygiène publique et privée, par Michel Lévy. 2 vols. 20 fr. (Baillière.)
Traité d'Hygiène Générale, par Ad. Motard. 2 vols. 16 fr. (Baillière.)

[LIST 5.]
GRADE 5 STUDIES.

The examinations of this Grade will be on nearly the same level of difficulty as those for the Doctor of Science Degree at the University of London.

The Books that might be indicated as suited for studies of this highest Grade, are mostly those named for the previous one; a difference being made by the omission in the one case of certain portions of comparatively unnecessary or superlatively difficult matter, which must be retained in the other case. The following are, however, a few Works of a decidedly 5th Grade character. They will serve to indicate that in Studies of this level, which will be chiefly encouraged among Students in training for Professorships, the History of Science should be added to Theory and Research. It might be argued that this is not strictly within the province of the proposed University; but in training Professors of Applied Science for Provincial and Colonial Institutions, it is essential to qualify them for leading on deserving Pupils to the higher paths of Scientific Research, as well as for exercising an enlightened and high-minded influence in the development of local industrial resources.

On Matter and Force, by H. Bence Jones. Foolscap 8vo. 5/- (Churchills.)
The Correlation of Physical Forces, by W. R. Grove. 8vo. 10/6 (Longmans.)
The Connexion of the Physical Sciences, by M. Somerville. Post 8vo. 9/- (Murray.)

Molecular and Microscopic Science, by M. Somerville. Post 8vo. 21/- (Murray.)

The Elements of Molecular Mechanics, by Joseph Bayma. Demy 8vo. 10/6 (Macmillan.)

Natural Philosophy, by Sir Wm. Thompson and P. G. Tait. 8vo. 4 vols. (Macmillan.) Vol. I. 25/-.

Elementary Treatise on the Wave Theory of Light. 8vo. 10/6 (Longmans.)

Lectures on Polarized Light, by Baden Powell.

A Sketch of Thermo-Dynamics, by P. G. Tait. 8vo. 5/ (Edmonston & Douglas.)

Sur la Chaleur, par E. Peclet. 3me édition. 8vo. 3 vols. 38/- (Paris, 1860-61.)

On Heat, by Clausius. 8vo. 15/- (Van Voorst.)

A Treatise on Electricity and Magnetism, by J. C. Maxwell, 2 vols. 8vo. 31/6 (Macmillan.)

A History of Chemical Theory, by Ad. Wurtz. Crown 8vo. 6/- (Macmillan.)

Principles of Chemical Philosophy, by J. Cooke. Post 8vo. 12/- (Macmillan.)

The Chemistry of Vegetable and Animal Physiology, by J. G. Mulder; edited by Johnston. 8vo. 30/- (Blackwood & Sons.)

Physiological Chemistry, by Lehmann, translated by Day. 8vo. 21/- (Harrison.)

A Cyclopædia of Quantitative Chemical Analysis, by F. H. Storer. Part I. Royal 8vo. 7/6 (Spon.)

Spectrum Analysis, by Hy. E. Roscoe. Royal 8vo. 21/- (Macmillan.)

Spectrum Analysis in its Applications to Terrestrial Substances, by Dr. H. Schellen, edited by Wm. Huggins. 8vo. 28/ (Longmans.)

A General System of Botany, by E. Le Maout and J. Decaisne. Royal 8vo. 52/6 (Longmans.)

A Handbook of Zoology, by Van der Hoovens. 2 vols. 8vo. 120/

The Philosophy of the Inductive Sciences, by W. Whewell. 30/

A History of the Inductive Sciences, by W. Whewell. 3 vols. Crown 8vo. (Bell & Daldy.)

SECTION 4.—ATTAINMENTS NOT INCLUDED IN THE FOREGOING SECTIONS.

Special Scientific Preparation.[1]

Without trenching on the detailed account of the instructional requirements of Artisans reserved for the next chapter, I may mention that they will be divided into two kinds, *preparatory* and *technical*.[2] The former will mainly consist of,—1stly, a certain Grade of General Scientific Foundation; 2ndly, any Special Scientific Preparation for which there may be occasion; and 3rdly, any Artistic Attainments required. The first of these points has been already dealt with; the two others demand a few explanations.

Under SPECIAL SCIENTIFIC PREPARATION will be included,—in the first place the pushing forward of any particular department or departments of either the Elementary or the Applied Division of the Standard Courses, a Grade higher than the rest. Then in many instances certain offsets will be grafted on the main branches of knowledge, as MECHANISM on the elementary study of the Mechanical Powers, HYDRAULICS on Hydrostatics, STEAM POWER, including the Machinery for applying it, on the study of Heat, CHROMATICS, or the rationale of Primary, Secondary and Tertiary Colours, combined with the Laws of Taste in relation thereto, on the study of Light, METALLURGY on Inorganic Chemistry, and so on. Experience will best decide to what extent these and analogous branches

---

(1) For available Works see List 6, p. 224.
(2) The Preparatory Knowledge here adverted to will be independent of Reading, Writing and Arithmetic, respecting which see the Article on Preliminary Attainments at the beginning of Chapter V.

of SUPPLEMENTARY KNOWLEDGE of which only a few general features can be included in the Applied Division of the Standard Courses may be satisfactorily taught from existing Works, or should as regards Grade 2 be brought out in distinct publications, or embodied in the Technical Cyclopædias, and especially in the Student's Edition thereof (see Sect. 6 of this Chapter). I mention the latter plan, because knowledge applicable to divergent Industries should, as much as possible, adapt itself to the particular character and requirements of each.

Mathematical Preparation.[1]

Allusion has been made to Studies which it will be desirable that Artisans should carry on, even where necessity does not actually compel, as subsidiary to those included in the Standard Courses, and MATHEMATICS have been particularly named. They are highly conducive to mental acumen, and afford valuable assistance in the attainment of various branches of Physical and Industrial Science. Among this description of SUPPLEMENTARY STUDIES, Mensuration will most frequently occur. It will be named in the suggestive outlines of Trade Curricula, to designate all those branches of Mathematical Study which converge to a direct bearing on the Measurement of Areas and Contents, and which include the necessary portions of Advanced Arithmetic, Algebra, Geometry, Trigonometry, &c. Grade 1 of Mensuration will probably consist of a selection of the most easy and most frequently needed Chapters of Todhunter's "Mensuration for Beginners," whilst Grade 2 will embrace the whole

---

[1] For available Works see List 7, p. 224.

of that valuable little Book. Grade 3 will rise somewhat higher, calling in more freely the assistance of the above-named branches of Mathematical Study. At Grade 4 the specialty of Mensuration will lose itself in a general course of Mathematics.

Any Student who has a certain Grade of Mensuration assigned to him in his Trade Curriculum, will be welcome to try for a higher distinct Mathematical Certificate.

Artistic Abilities.[1]

Various special considerations attach to this important element of industrial success, which will be best treated apart from the scientific attainments, and on different principles of gradational progress. Among those considerations are the following :—

— Without ignoring certain openings for improvement which exist in the system of instruction of the Art Department, I may safely say that it is so well organized, and presents such valuable facilities for the establishment of affiliated Schools of Art, that every Town should have one of these, and that every young Artisan should attend who can do so without neglecting more urgent duties.[2] His selection of a style of

---

(1) For Works available for Artistic Studies see List 8, p. 225.

(2) Respecting openings for improvement, I am told among other things that the Masters at some of the Schools of Art, actuated perhaps by an impression that red tape is the best proof of a connection with Government, affect to ignore the attainments already possessed by Art Workmen who wish to rise higher, and not only make them begin again at the rudiments, but do not allow them to pass to higher classes when their ability to do so is proved. I could name a case in which a young man, whilst he was thus kept in a low grade, was actually offered by the same Master, artistic employment involving the possession of a high grade of abilities.

Art should be guided by the character of his occupation, as far as there is any connection between the two, and thus far his Artistic Attainments will form an integral portion of his Technical Training. Beyond this he will be at liberty to follow the turn of his inclination, which may possibly be the voice of genius, though he should be cautious in trusting to the pleasing idea; for the country is more in want of Artistic Workmen, than of third-rate Artists. Subject to this consideration, every encouragement and facility will be offered to Students for obtaining, in addition to University Trade Certificates, the Art Certificates or Prizes granted under the Government Art Department. Further remarks on Art Studies will be found in the part of the next Chapter dealing with Art Industries.

— Viewing Art from a purely *industrial* point of view, and independently of the scale of merit recognized in the regular Art Schools, we perceive that there is a certain artistic cleverness of the hand and eye, which should be possessed by almost every Artisan, and which he might acquire partly in the advanced period of his Schooling, say between 12 and 14 years of age, and partly during his Apprenticeship. The determining of the extent to which this cleverness might reasonably be expected to be developed in skilled Workmen, must for the present be left an open question, because it must necessarily be for many years to come, a compromise between what ought to be obtainable, and what in the present state of popular education, it is possible to obtain. Nevertheless the following progression of Artistic Ability considered merely in reference to Industrial pursuits, beginning with the lowest grade of handicrafts in which work is done according to a Drawing or Pattern, may facilitate to some extent the devising of the artistic portion of the respective Curricula. The Degrees do not correspond with the Grades which have been adopted in reference to Scientific Knowledge, and in order to avoid confusion, they will be called STAGES. It will be seen that their import will vary according to the nature and requirements of the various trades.

1. Ability to understand and work from a Drawn Pattern or Design, and also to copy the same. The style will be governed by the requirements of the Trade in view, and on these will much depend the degree of skill involved.[1]

2. Ability to *make* Working Drawings, correct but not elaborate, and simply sufficient for the ordinary requirements of the Trade in question.

3. Ability to represent with proper shading, in free hand or mechanical Drawing, artistically done, and either in perspective, isometrically, or in elevations and sections, any complex article, ornament, or piece of machinery which there may be probable occasion to produce or to use.

4. Ability to originate as well as to execute ornamental designs.

5. Ability to execute articles of Art-workmanship in which the artistic element is susceptible of assuming a high character; *e.g.* certain branches of House Decoration, Mosaic and Inlaying, Wood-carving, Wood-engraving, Chasing, Filigree, Repoussé Work, Cameo-engraving, Seal-engraving and Die-sinking, Copper-plate and Steel-engraving, Etching and Aquatinta, Lithography, and many more. To determine

---

[1] Wherever necessary "Modelling" must throughout these stages be understood in addition to or instead of "Drawing."

standards of ability with a view to establishing a certain degree of fixity in the Examinations of Trades and Occupations like these, in which numberless gradations ascend from humble Handiwork to high Art, will be a task requiring special study and much thought, but for which valuable indications may be collected from the records of the Prize Exhibitions of Art Workmanship held at the Society of Arts.

Artistic Taste.

Apart from the ability to wield artistically the pencil, the brush, or the modelling tool, and more important perhaps for those whose business it is to direct the productive abilities of others, is a sound knowledge of the RATIONALE OF THE BEAUTIFUL AND APPROPRIATE in form, colour and subject, or in other words, of those natural and conventional rules which constitute the THEORY OF TASTE. It is highly desirable that in the furtherance of this important branch of instruction, the Central Technical University should have the hearty co-operation of Provincial and Colonial Institutions, including the Technical Colleges. Collections of Works of Art, far less numerous than many of the existing ones, but made in strict accordance with the principles of selection indicated in Chapter II. Section 8, G, should by the respective Art Instructors, be made the subject of critical discourses throwing the light of a refined judgment on the respective merits and demerits of Works of Art of various kinds. The place of all others where everything will be done to facilitate the acquisition of this refined judgment, in the exact sense in which it should be possessed by those who are to be the pilots of our Art Industries, will be among the Art Collections specially devised for the purpose in the Museum of the Central Technical University, and of which moreover the *catalogue raisonné*, suitably illustrated, will do all that can be done in print, for giving to the Æsthetics of High Art a practical Industrial bearing.

Several of the analyses of Trades collected in the next Chapter, will show the need for special instruction in what it has been proposed to call CHROMATICS (p. 191) consisting in a knowledge of the phenomena of the Solar Spectrum, of primary, secondary and tertiary colours, of their harmonies and contrasts, and of the principles which should regulate their application to industrial purposes. This and other analogous branches of Preparatory knowledge, of which the treatment should vary in some measure according to the Industry in view, will independently of special publications existing or to be brought out, be included as far as required in the Introductory Portion of the Technological Handbooks and Manuals described further on in Sect. 6 of this Chapter.

[LIST 6.]
EXAMPLES OF WORKS
AVAILABLE FOR SPECIAL SCIENTIFIC PREPARATION.

Practical Mechanics, by Imray. 12mo. 1/6 (Griffin.)
Mechanics and Mechanism, by R. S. Burn. 8vo. 2/ (Ward.)
The Elements of Mechanism, by T. M. Goodeve. 3/6 (Longmans' Text Books of Science.)
Principles of Mechanism, by R. Willis. 8vo. 18/ (Longmans.)
Descriptive Mechanism, by C. P. B. Shelley. 3/6 (Longmans' Text Books of Science.)
Construction and Working of Machinery, with plates, by C. D. Abel. 4to. 9/ (Lockwood.)
The Elements of Practical Hydraulics, by Samuel Downing. 8/– (Longmans.)
A Treatise on Hydro-Mechanics, by W. H. Besant. 8vo. 10/6 (Deighton.)

Metals, by C. L. Bloxam. 3/6 (Longmans' Text Books of Science.)
Useful Metals and their Alloys, by Dr. A. Ure. 7/6 (Houlston.)
A Manual of Metallurgy, by J. A. Phillips. Cr. 8vo. 12/6 (Griffin.)

[LIST 7.]
EXAMPLES OF WORKS
AVAILABLE FOR MATHEMATICAL PREPARATION.

Mensuration for the Million, by C. Hoare. 12mo. 1/ (Wilson.)
Technical Arithmetic and Mensuration, by C. W. Merrifield. 3/6 (Longmans' Text Books of Science.)
Mensuration for Beginners, by I. Todhunter. 2/6 (Macmillan.)
An Elementary Treatise on Mensuration, by B. T. Moore. 12mo. 5/ (Deighton.)
Geometry, by I. Todhunter. 3/6 (Macmillan.)
Geometry, by H. W. Watson. 3/6 (Longmans' Text Books of Science.)
An Elementary Course of Mathematics, by Rev. H. Goodwin, edited by P. T. Main. 8vo. 16/ (Deighton.)

[LIST 8.]
EXAMPLES OF WORKS
AVAILABLE FOR ARTISTIC STUDIES.

Drawing Copies, Models, &c., in every style of Art, supplied under the sanction of the Science and Art Department. (See the Art Directory, Appendix C; and the Catalogue of Messrs. Chapman & Hall.)

Hermes' "Drawing School" and other Series, forming a Comprehensive and Systematic Collection. (Published at Berlin, and to be had of A. N. Myers, 15 Berners St., Oxford St., W.)

Society of Arts' Drawing Book, by J. Bell. Oblong 4to. 3/6 (Bell & Daldy.)

Elementary Drawing, by E. A. Davidson. 8vo. 3/– (Art Department.)

A Handbook of Practical Perspective, by Jewett. 12mo. 1/6 (Cornish.)

Perspective, by G. Pyne. 2/– Weale's Series (Lockwood.)

School Perspective, by J. R. Dicksee. 8vo. 5/– (Strahan.)

Linear Perspective for the use of Schools of Art, by R. Burchett. Post 8vo. 7/– (Chapman & Hall.)

The Elements of Geometrical Drawing, by T. Bradley. Folio. Parts I. & II., each 16/– (Chapman & Hall.)

Principles of Decorative Art. 1/– (Chapman & Hall.)

Decorative Art, by Crace. (Chapman & Hall.)

On Design, by Richard Redgrave. (Chapman & Hall.)

The Principles of Decorative Design, by Christopher Dresser. 7/6 (Cassell & Co.)

The Art of Decorative Design, by Christopher Dresser. 7/6 (Cassell & Co.)

Analysis of Ornament, by R. N. Wornum. 8/– (Chapman & Hall.)

The Encyclopædia of Ornament, by Henry Shaw. Royal 8vo. 21/– (Pickering.)

The Munich Collection of Ornamental Designs. 15/– (Myers.)

Grammar of Ornament, by Owen Jones. £5.

Dessin Industriel, by Aimengaud aîné. (Paris, 1860.)

Dessin Industriel, by Aimengaud jeune. (Paris, 1860.)

Manual and Catechism of Colour, by Richard Redgrave. –/9 (Chapman & Hall.)

A Grammar of Colouring, by G. Field. 12mo. 2/– Weale's Series. (Lockwood.)

Lessons on Colour. 36 coloured discs. 2/6 (Myers.)

Colour, by Professor Church. 2/6 (Cassell & Co.)

A Manual of the Science of Colour, by Wm. Benson. 12mo. 2/6 (Chapman & Hall.)

An Alphabet of Colour. 4to. 3/– S. K. M. (Chapman & Hall.)

Rudiments of Colours and of Colouring, by G. Field. 12mo. 4/6 (Strahan.)

Laws of Contrast of Colour and their Application to the Arts, by M. E. Chevreul; translated by J. Spanton. Foolscap 8vo. 5/– (Routledge.)

Colour as a Means of Art, by F. Howard. 12mo. 6/– (Bohn.)

Ornamental Designs in Colour, approved by Art Department. 15/– (Chapman & Hall.)

Principles of the Science of Colour, by Wm. Benson. 4to. 15/– (Chapman & Hall.)

Principles of Beauty in Colouring Systematized, by D. R. Hay. 18mo. (Blackwood.) 1845.

A Nomenclature of Colours, &c, by D. R. Hay. 228 examples. 63/

SECTION 5.—TEXT BOOKS FOR THE STANDARD COURSES.

It has already been explained that as regards Grade 3 of General Scientific Foundation, both the Elementary and the Applied Knowledge may for a time be obtained from the Scientific Literature of the Day, the appropriate Books or parts of Books being indicated by a SYLLABUS OF REFERENCE prepared for the purpose; but that it is of urgent necessity that the Central Technical University should bring out two STANDARD COURSES to serve respectively as Text Books for the first and second Grade of General Scientific Foundation. Each of them will form in its appointed degree of comprehensiveness, and in its appropriate style, a single-minded, connected and progressive discourse, divided into Lectures for delivery on the Binary System. It has been shown in Section 2 of this Chapter, that various serial and detached publications exist from which, under judicious guidance, Artisan Students may possibly cull information of the kind they want; but that no satisfactory trial could thus be made of the proposed plan of Examinations and Certificates. To say nothing of expense and inconvenience to the Students, it would be impossible to make *excerpta* from Books of varying calibre, and still more varying style, dovetail and harmonize with each other; and it is indispensable that the Central Technical University should, at least as regards the two lower levels, have absolute control over its Standard Text Books, not only for thoroughly moulding them at the first to the requirements of a new system, and impressing them with the stamp of its own authority, but also for modifying them subsequently as experience and the progress of the times may demand. It will be on the Board of Instruction (see p. 92) appointed in the manner to be shown in Chapter VI. (sec. 2), that will devolve the duty of providing for, and supervising the production of all necessary educational requisites, and it is accordingly for their consideration that are chiefly intended the following suggestive remarks :[1]—

a.—For Manuals of earnest study in advanced Science, the prevailing style is, naturally enough, what a literary man would call no style at all. Information split up into dry bits has the advantage of being sorted without trouble into strictly methodical order; but the painstaking plan of threading nearly the same information in smoothly-running, well-connected sentences, pleasant to read and more pleasant to hear delivered by a competent Reader, produces a kind of instruction far more palatable and attractive to those whose mind has not through habit acquired a taste for concentrated thought.

b.—The foregoing especially applies to the two lower Grades, and more particularly to the first Grade of the range of Knowledge which has been selected to serve the double purpose of affording a Scientific Foundation to the Artisan, and of supplying the Science of Daily Life to the general mass of the Industrial Population.

[1] It has already been explained that the Text Books issued by the Central Technical University, will be closely adhered to in all Examinations held under its authority, and of course by the Professors charged with preparing Students for them.—It must also be remembered that the Text Books will, as a rule, be submitted to a triennial revision, and that in the intervals, important novelties will be made known by means of Supplements.

It evidently must be made attractive as well as easy, if in either case it is to constitute a voluntary evening's supplement to a day of toil.[1]

c.—The Text Books in question should be printed in bold and clear type, so as to afford comfortable evening reading to those who are not much accustomed to read, or whose eyes have had a good deal to do in the course of the day; and they should, by an abundance of carefully executed Woodcuts, present to those who are confined to home study, as good a compensation as can be had, for the absence of the more impressive paraphernalia of a Lecture Table.

d.—Nevertheless the most valuable utilization of the said Text Books will probably be as Lecture Texts; that is to say, they will be made to afford by the joint action of a Reader and a Demonstrator, in the manner described as the BINARY SYSTEM in the previous Chapter, as near an approximation as possible to an oral Address by a first-rate Popular Professor.

e.—Whilst therefore these Text Books will be enlivened with every attraction of colloquial oratory compatible with brevity and consonant with the nature of the subject, they must necessarily conform to every requirement of the said system, as regards the crosses in the Text addressed to the Reader, the list of articles and code of instructions for each Lecture addressed to the Demonstrator, and so on.

f.—Without prejudicially disturbing the methodical distribution of the subjects, or interfering with the due apportionment of respective amounts of instruction according to their relative deserts, the Discourse should

---

[1] Various suggestions for winning the attention of uneducated audiences to Scientific Lectures, will be given further on under the heading "Science Popularized."

be divided into Lectures occupying each as a rule about one hour in the delivery, and in no case less than three-quarters of an hour, or more than an hour and a quarter. It is far better to vary in length, within these limits, than to omit valuable, or insert irrelevant matter. Of course proper allowance must be made for the *display* of Specimens or Diagrams, and the *performance* of Experiments; the one an affair of only a few moments when the joint delivery is well performed, the other varying considerably in time according to the nature of the experiments.

g.—It would be convenient if there were a *break* in each Lecture, not far from half-way, to facilitate its being divided into two Lessons, each of which, making allowance for various circumstances inherent to the altered mode of tuition, would form about the right quantity for an ordinary Evening Class. This break is however only mentioned as a matter of convenience, and not as obligatory; for competent Teachers, aided by indications which will be supplied to them, will have no great difficulty in cutting up the Text according to the requirements of their respective Classes in Institute or School.

h.—It will probably be expedient to publish both Grade 1 and Grade 2 Course in Parts of one or two Lectures each, for the convenience of delivery, as well as in Volumes for home study.[1] It may further be found desirable to bring out Grade 2 in Sections embracing the Chief Departments of Study, as for

---

(1) A printed specimen which I have had prepared with a view to the publication in parts of my Popular Course, intended to follow that of the present work, is in Quarto with double columns of Pica. This arrangement is very suitable for a Lecturing Desk, and affords at the same time, for the benefit of the home Reader, a great variety of space for Woodcuts.

example in Division 1, Inorganic Chemistry, Organic Chemistry, Physiology, &c.; and in the 2nd Division, Dwellings, Fabrics and Clothing, Food, Warming Materials and Appliances, and so on; which topics will occur in the Curricula of certain "Trades Examinations," and in some instances may form the theme of special Examinations for "Subject Certificates."

i.—Possibly by means of matter of an advanced character distinguished from the ordinary text by a rather smaller type, or given in Notes and Appendices, it may be possible to make an edition of the Grade 2 Course serve, to a certain extent, the purpose of a Grade 3 Course.

k.—Should it be found expedient to bring out, particularly as regards Grade 2, special Text Books for the branches of Supplementary Knowledge named in the previous Section (p. 217), the same rules will mostly apply as have been suggested in respect of the Standard Course for that Grade.

l.—Though, as stated elsewhere, much may be done for supplying local Institutes with Illustrations for their Educational Courses, by the circulation of Portable Collections under the charge of Official Demonstrators empowered to act as Organizers of local Instruction, it will be unquestionably desirable that by degrees every Industrial Institute should acquire its own stock of Illustrations for at least the whole range of Grade 1, and the Chemistry of Grade 2.[1] As however it cannot be denied that in the expense for Illustrations will probably lie the greatest obstacle to be

---

[1] A variety of suggestions for accomplishing the foregoing and other analogous purposes in provincial localities, are collected in a Memorandum intended to form part of the Introduction to my Popular Course. In the mean time a few indications of available Diagrams and Apparatus are given in List 10, p. 252.

overcome by small Institutes and Schools, notwithstanding the assistance proposed to be extended to them from the Museum Department of the Central University, as well as through Government Grants, the Authors of the Text Books for the Lower Grades and for Grade 1 in particular, must carefully avoid involving expensive Illustrations where cheaper ones can be made to serve the purpose nearly as well. In many cases too, it will be desirable to indicate contrivances and makeshifts which may be resorted to with a slight alteration of the Text, where it prescribes things that cannot everywhere be had.

m.—As the possibility of uniting cheapness with good quality in publications of any kind, and especially in illustrated works, depends very materially on the MULTIPLICATION OF COPIES, I cannot lay too much stress on the ECONOMY that may in this as well as in other essentials of Instruction, be made to result from the coincidence of what the Artisan requires as a General Scientific Foundation to his special studies, with what every man requires as the guiding Science of Daily Life. It will be one of the duties of those who may be charged with preparing the Text Books of the lower Grades, to turn this coincidence to the best account, by writing so that they may not only be understood, but read with pleasure by adults desirous of retrieving the deficient education of their youth, and that these familiar discourses, when well delivered as Lectures on the Binary Plan, may everywhere be listened to with a satisfaction increasing in earnestness as the tale proceeds. This more particularly applies to Grade 1, and it may not be amiss to quote here almost textually a few suggestions concerning Lectures of this Grade, contained in Sect. 2 of " Science for the People."

Science Popularized.

1. When you intend to treat an untutored audience with a course of scientific instruction, you must do your utmost to make the heaven-born muse of Science leave for a time the clouds among which she is wont to recline in rapturous self-contemplation, casting now and then a look of pity and contempt on all below. You must absolutely induce her to come down to this nether world, put on homely garb, and enter a working family's dwelling, to ventilate it and make it wholesome and comfortable, to inspect the furniture and wardrobe, the kitchen utensils and the contents of the larder, nay actually to light the fire and cook a model meal, not forgetting the care of the young ones and of the sick person in the next room. Now if you show the Working Classes that Science can do all this to perfection, and be a saving instead of an expense, you may be sure that they will give her a hearty welcome.

2. I have found no difficulty in making Working People clearly see, that in order to understand what Science has to say about the concerns of Daily Life, they must first take the trouble to make themselves acquainted with a few indispensable scientific facts and expressions. They readily undertake this, provided they be explicitly assured (and of course every pledge given must be conscientiously redeemed) that all unnecessary technicalities will be avoided, and all difficulties smoothed down as much as possible; that there will be abundance of specimens, models, diagrams, experiments and other devices for making the senses act as helpmates to the memory, and that through all these contrivances, the *preparatory* and more strictly

scientific part of their studies, will be rendered as entertaining as the subsequent *economic* part, in which the several departments of Domestic and Sanitary Economy will be successively reviewed.

3. One of the chief difficulties to be encountered in selecting the subjects of these two necessary divisions of any methodical course of Practical Bionomy, is the overwhelming abundance of the matter as compared with the limits beyond which common sense forbids us to reckon on the regular weekly attendance of a large popular audience. This difficulty can only be overcome by great pains and discrimination bestowed on the selection of fundamental facts and typical illustrations, by studying the manner in which they can be best and most closely fitted together, and by endeavouring in repeated revisions to condense the matter of many pages into a few, without squeezing out the juice of the subject, and rendering it dry and unpalatable.

4. The necessity of dividing the subject matter into tolerably equal Lectures, so that each of these may take from an hour to an hour and a quarter in the delivery, cannot without considerable trouble be kept from interfering with the natural divisions, and renders more arduous the always difficult task of allotting to each portion and sub-portion of a subject, an amount of attention proportionate to its relative importance. The best remedy is to reserve certain items of elementary Science, especially of Chemistry, for the second division of the Course; allotting them to the several departments of Household Economy to which they specially apply, where they will afford a welcome variety.

5. One can scarcely be too fastidious as to the accuracy of every statement, or too patient and persevering in the confronting of different Authors on

points on which one has not a direct and personal certainty. The discrepancies between various *standard* books on the same subject, and even sometimes between one part and another of a work of high authority, are only known to those who have subjected these matters to scrupulous research.

6. In proportion as Practical Bionomy surpasses many other branches of applied Science in the actuality of its bearings, the grievous multiplicity of the errors, abuses, and frauds on which it throws a detective light, and the radical nature of the reforms which it might seem to suggest, so also is proportionate prudence required in inveighing against existing prejudices, and in running foul of existing interests. The Economic Teacher of the Working Classes should not so much tell them what to buy and where to get it, as give them the insight which will enable them to judge for themselves; not so much hurl offensive epithets, however well deserved, against those who live by adulteration and fraud, as sow knowledge that will make their present dealings unprofitable, and at the same time, open their way to better ones.

7. Moderation, and a discreet adaptation of precepts to circumstances, are nowhere more necessary than in matters of Hygiene. Exaggerated remedies generate reaction, or replace one evil by another. Systems that work admirably in the combination of circumstances which some countries afford, prove lame and unprofitable elsewhere, and contrivances that have been marvels of success under the management of the inventors or of their intelligent friends, may come to grief in the hands of ignorance and prejudice.

8. Unity of purpose throughout the Course, and the occasional connecting of one department with another

by means of mutual references, should bind the heterogeneous subjects of which the Science of Common Life is composed, into a harmonious whole, pervaded from its leading features down to its merest details, with a methodical spirit of forethought and classification. It is a great mistake to deal lightly with considerations of this kind, in writing for the uneducated. METHOD assists both the intelligence and the memory, and the less cultivated and clear the mind to be taught, the clearer should be the teaching.[1] Great care should be taken to explain all hard words the first time that they are used, or so to construct the phrase as to make their sense self-evident; also (and this applies particularly to the Chemical Departments), to avoid mixing the names of substances not yet described, in the account of those in hand. There should be, as much as possible, a gradual and logical progression from the simple to the complex, from the easy to the difficult; a development of information, step by step, in that connected sequence which makes little children remember so well the Nursery Tale of the " House that Jack Built."

9. Rigidity of principle in the selection and arrangement of the subject matter, does not by any means imply rigidity of style. The character of the language which suits a Scientific Discourse, varies immensely, according to the subject and the purpose. It should be solemn where the object is to raise the mind to a true conception of the power and beneficence of an allwise-Creator; it should be sedate and didactic when useful physiological facts and advice can thus be more appro-

[1] " Chercher constamment la clarté par la simplicité du discours, la justesse de l'expression, l'enchaînement logique, la succession graduelle et bien calculée des notions et des pensées, telle est, selon nous, la poétique à suivre dans l'exposition familière des faits scientifiques."—' La Terre avant le Déluge,' par Louis Figuier, Préface, page 10.

priately imparted; but the general tone to adopt in teaching Working Men the practical difference between Knowledge and Ignorance, is decidedly a cheerful and colloquial one, and in many cases ridicule will be found the best weapon to use against an absurd contempt of the Laws of Nature, and a silly subserviency to those of Fashion. But what as much as anything will secure for wholesome truths a ready and sincere acceptance, is their being offered in a spirit so plainly fraught with sincere Christian benevolence, as to be above all suspicion of an interested motive.

Science for Primary Schools.

It has been shown that the greater part of the range of Knowledge embraced by the Grade 1 Syllabus in the first Section of this Chapter, is intended to be gradually included by progressive selections in the Elementary Schooling of the Children of the People.[1] Till a special School edition can be provided, the ordinary one in a lecturing or colloquial form may supply the required materials, but at all events the manner of using them with children demands in many respects to be the subject of special instructions.[2] Not only will much discrimination be required by a Schoolmaster teaching the *elements* of the Natural Sciences as sketched out in the first division of the above Syllabus, but considerable caution will be necessary in dealing with their *applications* to the actual concerns of Daily Life as indicated in the second Division. Much injury has been done to the cause of Hygiene

[1] See Chapter I., Section 2.
[2] Of the nature of these I have endeavoured to convey an idea by a kind of Questionary previously mentioned, which may be inspected at the Economic Museum with various other Manuscripts on kindred subjects.

by exaggerating its precepts, and by overlooking the impediments which too often render a strict compliance with them next to impossible for the Working Classes. This particularly applies to the construction and ventilation of Dwellings, and the selection and preparation of Food. Even unfounded prejudices and pernicious habits must be conquered rather by quietly inducing a state of mind that will naturally lead to better things, than by exciting the children to dissatisfaction and open antagonism with what they see at home. Particular discretion is required in the question of Temperance versus Total Abstinence, which is one on which considerable disagreement exists even among the well-educated.

Everywhere to a certain extent, but especially in rural districts, the Schoolmaster is or should be one of the oracles of the parish. No doubt can be entertained as to the vast amount of beneficial influence which, when trained to a sound and methodical knowledge of Science applied to Daily Life, he may be able to exercise either as a friendly adviser, or in conjunction with the Clergyman and the Doctor, as a supporter of public measures of a sanitary character, or as a promoter of Evening classes, Mutual Instruction Societies, and other means for pursuing in Youth and Manhood the knowledge initiated at an early age. For purposes like these the Schoolmaster will find convenient reminders in the portions of the Course relating to Hygiene, Industrial Pathology, Household Management, Providence, and other matters evidently not intended for children.

SECTION 6.—TECHNICAL TEXT BOOKS, AND ILLUSTRATIONS.

Four years ago I complained in "Science for the People," that no adequate steps had been taken to supply the various Industries with repertories of Technical Information calculated to secure their success, and promote their progressive advancement. I adverted to the high price which mostly debarred Working Men from the possession of the few thoroughly good and well illustrated existing works, and to the impossibility of bringing down those prices to meet limited means, without a certain amount of pecuniary sacrifice which an ordinary publisher could not be expected to incur, but which would most decidedly be money well spent on the part of the public purse. As regards minor Trade Manuals, I mentioned that Messrs. Houlston and Wright, and a few other Publishers, had done very creditably as much as could be expected from unassisted speculation, having brought out over 38 little detached volumes mostly ranging from one shilling to half a crown; but that these could scarcely be compared to the *Manuels Roret*, published systematically at Paris, as the result of the united labours of an association of Savants and Technologists.

Notwithstanding the appeal of the London Workmen through the medium of the Club and Institute Union, no essential change has supervened as regards strictly Technological Literature, since the above was written, and no powerful Body or influential Association has been induced to meet on public-spirited principles the multiplied wants of the Industrial Community. Under these circumstances I feel that I cannot do better than endeavour to give some idea of what our Artisans are likely to require under the proposed organization of Technical Studies, leaving for those on whom may devolve, under the authority of the University, the duty of seeing that those requirements are properly provided for, to take note of all available resources, and to utilize them as far as practicable.

For the simple reason that one can seldom form a correct estimate of the value of an article without having some idea how it is made, the scientific review of the Things of Daily Life forming the second Division of an Artisan's General Scientific Foundation, will naturally give him, even in Grade 1, a certain insight into all the ordinary operations of Technical Industry, showing him how they support each other, and how his own pursuit, whatever it may be, is directly or indirectly connected with a number of others. In Grade 2 the account given of the various Departments of Domestic Technology, will be not only extended but so much improved by being based on more advanced scientific knowledge, that in many cases these departments may, as already stated, be picked out from the course to form part of Technical Curricula, or to constitute the theme of special examinations for Subject Certificates. Thus for example the Department of Textile Materials and Fabrics, including a *general* insight into Spinning, Weaving, Dyeing, and other processes, will form a very appropriate basis to the technical knowledge of Minor Students whose *special* occupation is connected with any portion of this range. But in the study of all industrial pursuits of any importance, generalities must be followed up by a close examination of *technical details*, and it is accordingly to the consideration of the best means for supplying STANDARD REPERTORIES OF TECHNICAL INFORMATION, that our attention will now be turned.

Technical Cyclopædias.

a.—It is proposed that two TECHNICAL CYCLOPÆDIAS should be prepared,[1] the one more compendious and easy, to be called the JUNIOR CYCLOPÆDIA, the other more exhaustive, and assuming in the Student a higher standard of preliminary knowledge, to be called the ADVANCED CYCLOPÆDIA. Their general range will be nearly identical, both being intended to include, more or less, the Technical Knowledge appertaining to every notable Industrial or Commercial Trade or Occupation, whether or not it be of a nature and magnitude to claim special Examinations and Certificates.[2]

b.—It will probably be found expedient to adopt in both Cyclopædias, as in the Standard Courses described in the previous section, a discursive or colloquial form of instruction, which will not only afford agreeable reading to Home Students, but will be convenient to the Professors, who in their Lectures and Lessons will

[1] In some instances it may be feasible to purchase the copyright of an existing Manual, and to use portions, or adapt the whole for one or the other Cyclopædia.
[2] It will be seen in the following Chapter, that in respect of certain occupations, the University will have occasion to offer Major Certificates and no Minor ones. In a number of cases there will not be sufficient scope for either, but even then the well-digested information supplied by the Cyclopædias will be in itself a very acceptable boon.

be expected to adhere in substance and level of difficulty to the printed matter, though they need not use its exact wording.

c.—The simple and easy as well as compendious style of the first or JUNIOR CYCLOPÆDIA, will render it suitable for *Minor* Students whose knowledge will not be supposed to rise above the level of Grade 2.

d.—The ADVANCED CYCLOPÆDIA will similarly be adapted for *Major* Students, and as they will as a rule be expected to have pushed forward to the third Grade all chemical or other preparatory scientific knowledge involved in their respective Trades, use will be made of the accepted phraseology of modern Scientific Books.

e.—This same Advanced Cyclopædia can be made to serve also the purpose of *Superior* Candidates. With this view, matters of a nature to be included in their studies only, will be given in Notes or Appendices, and further supplemented wherever necessary by references to special Works.

f.—Both Cyclopædias, besides being published in uniform and sightly Volumes, forming an EDITION FOR REFERENCE, that should be on the Library Shelves of every Industrial Institute, will also be published as a STUDENT'S EDITION, in as many PARTS as they contain distinct Trades or Occupations. These Parts, of which the Junior Series is proposed to be called HANDBOOKS, and the advanced Series MANUALS, may perhaps be of uniform size throughout, but will necessarily be of greatly varying thickness; for no attempt will be made to obtain uniformity in this respect by squeezing some subjects and dilating others.

g.—It is the STUDENT'S EDITION that will mainly be used for purposes of Study and Examination, and accordingly in each of the Handbooks and Manuals, the Text will be cut up into a number of Sections and Articles or Paragraphs, with such indications attached, as to be conveniently referred to, singly or collectively, in the Curricula of Studies and the Examination Questionaries. See Section 7 of this Chapter.

h.—Information required identically or nearly so by two or more Industries, need only be given in one place in the "Editions for Reference," and simply referred to from other places; but it must necessarily be repeated in each of the respective Handbooks or Manuals. Independently of convenience, it can thus be more perfectly adapted to the particular requirements of the respective Trades. For the same reason any Preparatory Knowledge susceptible of being specialized in its application, will also be included in INTRODUCTORY PORTIONS of the said Handbooks and Manuals. This particularly applies to the Subjects which have been described as Offsets of the Standard Course, such as Mechanism, Metallurgy, Mensuration, Chromatics, &c.— Indications will also be supplied of accessory attainments that may be cultivated with advantage, and full advice given as to the facilities for study, and openings for reward, afforded by the Central Technical University and its affiliations.

i.—Great pains will be bestowed on the Woodcuts which will abound in the two Cyclopædias. Economy will be practised by using for both, those which come within the scope of the Junior one; but the best ally of economic production is *extensive sale,* and this will soon be realized if the Central Technical University is known to be favoured with the influence and co-operation of the highest educational, scientific, and artistic authorities, and to be prepared, should it prove necessary, to make any pecuniary sacrifice demanded by the true interests of our manufacturing and commercial Industries.

Scientific and Technical Illustrations.

In describing the Museum Department (Chapter II. Section 8), mention has been made of the direct production of SETS of EDUCATIONAL ILLUSTRATIONS of every kind by clever Students under proper guidance (p. 113); but independently of this, it will be one of the duties of the University to organize and direct the publication through appropriate channels, of EDUCATIONAL PRINTS, DIAGRAMS and the like, of more than ordinary scientific correctness, methodical completeness, and artistic merit, and all this at, or if necessary under, cost price.[1]

I have alluded in "Science for the People" to the difficulty which I have experienced in obtaining suitable Diagrams for my Lectures, though I use many foreign as well as English ones. Certain sets have been published deserving of much praise, but they are far from embracing in appropriate style, and on a proper scale, all the departments of popular knowledge; and as for the cheaper productions (though still scarcely so cheap as I should wish to see them), some of their errors and short-comings call for the interference of supervisors well up in science and art, and having higher purposes in view than that of speculating on public ignorance and the prevailing want of artistic culture.

[1] Suggestive examples are given in List 10, p. 252.

[LIST 9.]
EXAMPLES OF WORKS
AVAILABLE FOR THE COMPILATION OF THE PROPOSED TECHNICAL CYCLOPÆDIAS.

N.B.—Books relating to the Trades and Professions included in Category A (see Chapter I., Sect. 6), such as Mining, Agriculture, Navigation, Civil Engineering, Architecture, and Pharmacy, are not included in this List, unless they seem likely to supply useful introductory information for Trades belonging to Category B. Books rich in Technological details are repeated from previous Lists.

General Reference.

The Encyclopædia Britannica. 21 vols. £25 4s.
The English Cyclopædia. 27 vols. Quarto. £16 12s. (Bradbury, Agnew & Co.)
The National Cyclopædia. 12 vols. £3. (Routledge.)
Popular Cyclopædia. 7 vols. 8vo. 140/- (Blackie.)
The Encyclopædia of Universal Knowledge. 10 vols. 8vo. 90/- (Chambers.)
Ure's Dictionary of Arts, Manufactures, and Mines. Re-edited by R. Hunt. 3 vols. £4 14s. 6d. (Longmans.)
Encyclopædia of Civil Engineering, by E. Cresy. £2 2s. (Longmans.)
A Dictionary of Science, Literature, and Art. Re-edited by W. T. Brande and G. W. Cox. 3 vols. £3 3s. (Longmans.)
A Dictionary of Science, by G. F. Rodwell. 8vo. 18/- Haydn Series. (Moxon.)
The Technical Educator. Royal 8vo. 4 vols. 6/- each vol. (Cassells.)
The Cyclopædia of Useful Arts, Mechanical and Chemical. Edited by Charles Tomlinson. 3 vols. £3 15s. (Virtue.)
The Useful Arts and Manufactures, by C. Tomlinson. 2 vols. 10/- (S. P. C. K.)
Illustrations of Trades, by C. Tomlinson. 4/- (S. P. C. K.)
Book of Trades, a Circle of the Arts and Manufacturers, by J. Wylde. 3/6 (Houlston.)
The Gallery of Useful Arts, by Charles Knight. (Knight & Co.)

Lardner's Cabinet Cyclopædia. (Longmans.)
Cyclopædic Science Simplified, by J. H. Pepper. Cr. 8vo. 9/- (Warne.)
M'Culloch's Dictionary of Commerce, &c. Edited by H. G. Reid. One thick volume. £3 3s. (Longmans.)
A Commercial Dictionary of Trade Products, Manufacturing Terms, &c., by P. L. Simmonds. 7/6 (Routledge & Sons.)
A Cyclopædia of Commerce, by Waterston; edited by Simmonds. 8vo. 12/- (Bohn.)
Philosophy of Manufactures, by Dr. Ure; edited by Simmonds. 12mo. 7/6 Bohn's Library. (Bell & Daldy.)
History of Inventions, by Beckman. 2 vols. 12mo. 3/6 each. Bohn's Library. (Bell & Daldy.)
Waste Products and Undeveloped Substances, by P. L. Simmonds. 9/- (Hardwicke.)
A Course of Technical, Industrial, and Trade Education, by Dr. J. Yeats, comprising the Raw Materials of Commerce, Skilled Labour applied to Production, Growth of Trade, and Modern Commerce. Crown 8vo. 5/- each. (Virtue & Co.)
Serial Classifications of Trade Products, by M. Bernardin, Curator (Conservateur) of the Industrial and Commercial Museum of the Maison de Melle, Lez-Gand, Belgique.

Scientific Information.

N.B.—Are here included Scientific Works likely to supply either direct Technical Information concerning Industrial Processes, or Preparatory Information suited for the Introductory Portions of the various Handbooks and Manuals. Some of the works are repeated from Lists 3, 4, 5 and 6, to which reference may be made for further additions.

A Handbook of Natural Philosophy, by D. Lardner. 5 vols. 8vo. 5/- each vol. (Lockwood.)
Elements of Natural Philosophy, by C. Brooke. 8vo. 12/6 (Churchills.)
An Elementary Treatise on Physics, by Prof. Ganot; edited by Atkinson. Post 8vo. 15/- (Longmans.)
An Elementary Treatise on Natural Philosophy, by A. P. Deschanel; edited by J. D. Everett. 4 parts. 8vo. 4/6 each part. (Blackie & Sons.)
Theoretical and Applied Mechanics, by Richard Wormell. Foolscap 8vo. 4/- (Groombridge & Sons.)

Elements of Mechanism, by T. M. Goodeve. Small 8vo. 3/6 (Longmans.)
Elementary Introduction to Practical Mechanics, by John Twisden. Cr. 8vo. 10/6 (Longmans.)
A Manual of Applied Mechanics, by Rankine. Post 8vo. 12/6 (Griffin.)
Machinery and Millwork, by Rankine. 12/6 (Griffin.)
Practical Hydraulics, by Thos. Box. Post 8vo. 5/- (Spon.)

Artificial Lighting. (Orr's Circle of the Sciences.)
Chemistry of Artificial Light, by Scoffern. 12mo. 1/6 (Griffin.)
Chemistry of Light, Heat, &c., by Scoffern. 12mo. 3/- (Griffin.)
A Practical Treatise on Heat, by T. Box. Cr. 8vo. 8/6
Ventilation, Warming, &c., by D. B. Reid. (Longmans.)

The Elements of Chemistry, by Prof. Graham. 2 vols. 40/- (Baillière.)
Inorganic and Organic Chemistry, by C. L. Bloxam. 8vo. 16/- (Churchills.)
Chemistry, by Brande & Taylor. 12/6 (Hardwicke.)
A Dictionary of Chemistry and the Allied Branches of other Sciences. Edited by Henry Watts. 5 vols. £7 3s. Supplements to ditto, £1 11s. 6d. each.
A Cyclopædia of Chemistry, including the Applications of the Science to the Arts, by R. Thompson. (Griffin.)
Chemistry applied to Arts and Manufactures, by Dr. Muspratt. 2 vols. 8vo. 67/- (Mackenzie.)
Chemical Technology, by Knapp, Ronald, and Richardson. £1 16s.
A Commercial Handbook of Chemical Analysis, by A. Normandy. Edited by H. M. Noad. (Lockwood.)
A Supplement to the Pharmacopœia, by Th. Redwood. 22/- (Longmans.)
Précis de Chimie Industrielle à l'usage des Ecoles Industrielles, des Fabricants et des Agriculteurs; par A. Payen. 2 vols. (Paris, 1851.)
Traité de Chimie Générale, contenant les Applications à l'Industrie, à l'Agriculture, &c.; par Pelouze et Fremy.
Traité de Chimie appliquée aux Arts et Manufactures; par Dumas.

Traité de Chimie Technique appliquée aux Arts et à l'Industrie, à la Pharmacie et à l'Agriculture ; par Barruel. 3 vols.
Inorganic Chemistry, being Part II. of the Elements of Chemistry, by W. A. Miller. 8vo. 21/- (Longmans.)
Useful Metals and their Alloys, by Dr. Ure. 7/6 (Houlston.)
Elements of Metallurgy, by J. A. Phillips. 12/- (Griffin.)
A Practical Treatise on Metallurgy, by Kerl, adapted by W. Crookes and E. Röhrig. 3 vols. £4 19s. (Longmans.)
Metallurgy, by J. Percy. 5 vols. 8vo. (Murray.)
The Metallurgy of Iron, by H. Bauermann. 4/6
Electro-Metallurgy, by A. Watt. Weale's Series. 12mo. 2/- (Lockwood.)
Galvano-Plastic Manipulations, by A. Roseleur. 8vo. 30/- (Philadelphia.)
Organic Chemistry, by Hy. Armstrong. 8vo. 3/6 (Longmans' Text Books of Science.)
Organic Chemistry, being Part III. of the Elements of Chemistry, by W. A. Miller. 8vo. 24/- (Longmans.)
Cantor Lectures on Fermentation, by A. W. Williamson. 1870. (Society of Arts.)
The Chemistry of Food and Diet, by Bronner and Scoffern. 12mo. 1/6 (Griffin.)

Applications of Geology to Arts and Manufactures, by D. T. Ansted. 12mo. 4/- (Allen.)
The Vegetable Kingdom, by R. Hogg. 8vo. 7/6 (Kent.)
Systematic and Economic Botany, by J. H. Balfour. Post 8vo. 2/6 (Collins' Advanced Science Series.)
Medical and Economic Botany, by John Lindley. 12mo. 5/- (Bradbury & Evans.)
Profitable Plants, by T. C. Archer. 8vo. 5/- (Routledge.)
A Cyclopædia of Agriculture. Edited by J. C. Morton. 2 vols. Large 8vo. £3 15s. (Blackie & Sons.)
Guide to the Collection of Animal Products at the East London Museum. 1d. (S. K. M.)

Technology.

The INDUSTRIAL LIBRARY, a series of cheap Trade Manuals, from 1/- to 2/6 each. (Houlston & Sons.)
Manuals analogous to the foregoing published by H. Elliott, New Oxford Street.

SECT. 6.] *LIST 9.* 249

ENCYCLOPÉDIE RORET, a comprehensive Series of Illustrated Trade Manuals, specially prepared by a Committee of Savants, and published at Roret's Library, Rue Hautefeuille, No. 12 Paris, and to be had at Hachette's, King William Street, London.

N.B. —The order of the following Books is conformable to that of the Trades' Analyses in the next Chapter, of which the explanation is given at p. 269.

A Manual of Dyeing, by James Napier. 7/6.
A Handbook of Dyeing and Calico Printing, by Wm. Crookes. (Longmans.)
Manual of Colours and Dye Wares, by Slater. Post 8vo. 7/6 (Lockwood.)
Aniline and its Derivatives, by M. Reimann. 10/6. (Longmans.)
Complete Art of Brewing, by D. Booth. 8vo. 1/- S. D. U. K. (Simpkin.)
A Practical Treatise on Brewing, by W. Black. 10/6 (Longmans.)
Breweries and Maltings. Construction and Machinery, by G. Scamell. Royal 8vo. 15/- (Fullarton.)
On the Manufacture of Beet Root Sugar, by Wm. Crookes. 8vo. 8/6 (Longmans.)
Works on Building, by Dobson. Weale's Series (Virtue.)
Practical Rules for the Builder and Young Student, by G. Pyne. 4to. 7/6 (Lockwood.)
Science of Building, by Tarn. 8vo. 8/6 (Lockwood.)
Examples of Building Construction, by Hy. Laxton. 4 parts. each 50/-.
Limes, Cements, &c., by G. R. Burnell. 12mo. 1/6 (Virtue.)
Building Construction, including Stone, Brick, Slate, Timber and Iron-work. Junior and Advanced Handbooks, by R. S. Burn. 8vo. 2/- and 2/6 (Collins.)
Brick and Tile-making. 12mo. 3/- Weale's Series. (Virtue.)
Masonry and Stone Cutting. 12mo. 2/6. Weale's Series. (Virtue.)
Various Works on the Application of Art to the Trades of Construction, by Ellis A. Davidson. (Cassells.)
Brief Chapters on British Carpentry, by Thos. Morris. 8vo. 6/6 (Simpkin.)
Carpentry and Joinery, by Robinson and Tredgold. Weale's Series. 12mo. 1/6. Plates to do. 4to. 4/6 (Lockwood.)
Elementary Principles of Carpentry, by T. Tredgold; revised by J. T. Hurst. Post 8vo. 18/- (Spon.)
The Elementary Principles of Carpentry, by T. Tredgold; edited by P. Barlow. One large vol. 4to. £2 2s. (Lockwood.)

The Carpenter and Joiner's Assistant, by James Newlands. 4to. £3. (Blackie & Son.)
The Carpenter, Joiner, &c., by J. Riddle. Folio. 45/- (Simpkin.) Supplement to do. 22/- (Simpkin.)
A Manual of Carpentry and Joinery, by S. T. Aveling. 12mo. 2/6 (Warne.)
The Cabinet-maker's Design Book, by King. 4to. 21/- (Bohn.)
Practices of Hand-turning, by Campin. Cr. 8vo. 6/-
Lathes and Turning, by Northcott. 8vo. 18/- (Longmans.)
Patterns for Turning, by H. W. Elphinstone. 4to. 15/- (Murray.)
Turning and Mechanical Manipulation, by Holtzapfell. 3 vols. 8vo. 50/-
Geometric Turning, by H. S. Savory. 8vo. 21/- (Longmans.)
A Practical Guide to House Decorating, by W. Sutherland. 12mo. 2/6 (Heywood.)
The Grainer and Decorator. (Brodie & Middleton.)
Sign Writing and Glass Embossing, by Jas. Callingham. Post 8vo. 5/- (Simpkin.)
Textile Fabrics, by Dr. Rock. 8vo. 31/6 (Chapman & Hall.)
The Cotton Manufacturer's Assistant, by E. D. Foley. 12mo. 1/6 (Heywood.)
The Embroiderer's Book of Design, by F. De la Motte. Ob. 8vo. 2/6 (Lockwood.)
An Encyclopædia of Domestic Economy, by Webster and Parkes. 31/6 (Longmans.)
The Economical Housekeeper, by J. H. Walsh. 12mo. 15/- (Routledge.)
Domestic Economy, by M. Donovan. 2 vols. 12mo. 7/ (Longmans.)
On Food. 1/- 18mo. (Cassells.)
Common Salt. 1/6 (S. P. C. K.)
Rudimentary Treatise on Coal and Coal Mining, by W. W. Smyth. 12mo. 4/6 (Lockwood.)
Gas Works, by Hughes. 3/- Weale's Series.
Moulder's and Founder's Pocket Guide, by Overman. 4/6
Modern Workshop Practice, by Winton. 3/- (Lockwood.)
The Operative Mechanic, by Nicholson. 8vo. 10/6 (Bohn.)
Works for Engineers, Assistants and Mechanics, by Templeton.
Illustrated Handbook of Machinery and Ironwork, by Appleby. 8vo. 12/6 (Spon.)
Cyclopædia of Machine and Hand Tools, by Rankine. In 25 parts, each 2/ (Mackenzie.)

Treatise on Mills and Millwork, by Sir Wm. Fairbairn. 2 vols. £1 12s.
The Miller, Millwright and Engineer, by Pallett. 12/-
The Engineer's Guide Book, by Ainsley. Demy 8vo. 6/- (Philip.)
The Engineer's Handbook, explaining the Principles which should guide the young Engineer in the Construction of Machinery, by C. S. Lowndes. 5/- (Longmans.)
The Engineer and Machinist's Assistant. 2 vols. Imperial quarto. £4 4s. (Blackie & Son.)
Useful Information for Engineers, by Sir Wm. Fairbairn. 3 vols. Cr. 8vo. 31/6 (Longmans.)
A Practical Treatise on Mechanical Engineering, by Campin. 8vo. 12/- (Lockwood.)
The Mechanician and Constructor for Engineers, by Cameron Knight. 52/6
Theory of the Steam Engine, by De Pambour. 12/-
The Steam Engine, by Dr. Lardner. (Weale's Series.) (Strahan.)
A Treatise on the Steam Engine, by J. Bourne. £2 2s. (Longmans.)
Manual of the Steam Engine and other Prime Movers, by Rankine. Post 8vo. 12/6 (Griffin.)
The Engine Room, by Ainsley. 12mo. 3/6 (Philip.)
The Marine Steam Engine, by T. J. Main and T. Brown. 12/6 (Longmans.)
The High Pressure Steam Engine, by E. Alban; edited by Dr. Pole. 8vo. 16/6 (Lockwood.)
The Locomotive Engine, by De Pambour. 18/-
History and Progress of the Telegraph, by R. Sabine. Weale's Series. 12mo. 3/- (Lockwood.)
Handbook of the Telegraph, by R. Bond. Weale's Series. 12mo. 1/6 (Lockwood.)
A Handbook of Practical Telegraphy, by R. S. Culley. 8vo. 16/- (Longmans.)
Optics, Theoretical and Practical, by E. Nugent. Post 8vo. 5/- (Strahan.)
Photographic Optics, by Van Monckhoven. Cr. 8vo. 7/6 (Hardwicke.)
Principles and Practice of Photography, by J. Hughes. 12mo. 1/- (Simpkin.)
A Manual of Photographic Chemistry, by T. F. Hardwick. Foolscap 8vo. 7/6 (Churchills.)
Private Book of Useful Alloys for Goldsmiths, by J. E. Collins. 16mo. 3/6 (Hotten.)

[LIST 10.]
EXAMPLES OF ILLUSTRATIONS

OF VARIOUS KINDS AVAILABLE FOR LECTURES ON SCIENTIFIC AND INDUSTRIAL SUBJECTS, OR WHICH MAY SERVE TO SUGGEST THE PRODUCTION OF APPROPRIATE SERIES.(1)

Diagrams, &c.

Diagrams of the Working Men's Educational Union (28 Paternoster Row, E.C.). They are of large dimensions, and coarsely but effectively printed on calico, and grouped in sets including the following :—

		£	s.	d.
The Mechanical Powers .. 3 diagrams, per set	0	9	0
The Steam Engine .. .. 9 do. do.	1	7	0
Geology .. .. 15 do. do.	2	5	0
Coal Mining .. .. .. 10 do. do.	1	10	0
Physiology in relation to Health .. .. 10 do. do.	1	10	0
The Human Eye and Telescope .. .. 6 do. do.	0	18	0
Physiology of the Foot .. 5 do. do.	0	15	0

Diagrams of various sizes illustrating the several branches of Science. Published by Reynolds, 174 Strand. The execution is good, and many of the larger ones, sold at 5/- and 10/- each, are available for Technical Training.

Diagrams of Machinery, somewhat similar to those of the foregoing publisher, are issued at Paris under official authority, and may be had of Hachette, King William Street, Strand, who also supplies various sheets of Trade Illustrations for Primary Instruction.

The Edinburgh and London firms of Keith Johnston & Co., and Chambers & Co., publish well-executed Diagrams in Physiology, and Science generally. Many objects being mostly united on one sheet, these diagrams are better adapted for Class teaching than for Lecture purposes.

---

(1) The chief motive for placing here the present List instead of inserting it at the end of Sect. 5 is that it includes materials for facilitating Technical as well as Scientific Instruction.

Day's Physiological Diagrams, life size, in illustration of Marshall's Physiological Work, occupy the first rank in productions of this kind. Price for the series of eleven sheets, mounted on rollers and varnished, £11 11s.

Among the Educational Diagrams published under the authority of the Science and Art Department by Messrs. Chapman & Hall, those most deserving of notice in connection with the proposed System of popular Instruction, are perhaps Henslow's Botanical and Patterson's Zoological Illustrations.(1)

Among the Illustrations published by the Society for Promoting Christian Knowledge, Great Queen Street, London, may be mentioned those of Structural Botany, Poisonous Plants, and Useful Animals.

Barfoot & Co., 295 Strand, successors to Darton & Hodge, publish a variety of cheap Diagrams for Primary Schools, many of which illustrate Economic and Industrial Subjects.

Among the publications by Messrs. Groombridge and Sons, Paternoster Row, are Tegetmeier's Diagrams and Tablets for Schools.

Among the Educational Repositories well deserving of a visit may be mentioned that of A. N. Myers, 15 Berners Street, Oxford Street, which is rendered particularly interesting by an influx of educational prints, models, &c., from various parts of the Continent, and especially from Germany.

A series of 20 Wall sheets illustrating Natural History, with specimens, by A. Boucard. (Murby.)

The Abbé Moigno, well known for his exertions in furtherance of Popular Instruction, has used with advantage the Magic Lantern with its latest improvements. A comprehensive Catalogue of Slides, prepared for the purpose and sold by MM. Molteni, 44 rue du Château d'Eau, Paris, may be consulted at the Twickenham Economic Museum.(2) Here also may be inspected examples of the prints prepared by M. Charles Buls, the enlightened Secretary of the Belgian Educational League, and which are distributed freely to the audiences at his Lectures on Decorative Art as an appropriate means of ensuring lasting impressions.

(1) A series of Botanical Illustrations by Prof. Oliver has just been brought out by the same Publishers.

(2) Photographic and other Educational Slides are also extensively prepared by Messrs. Childe and Doubell, Wholesale Manufacturers, 3 Edith Grove, West Brompton, S.W.

Apparatus and Chemicals.

It may be convenient to the organizers of Provincial Institutes to have the following addresses of Firms mostly known to me through satisfactory personal experience, and whose Catalogues may be obtained by Post.

Elliott Brothers, Mathematical Instrument Makers, 449 West Strand.

Negretti and Zambra, Holborn Viaduct, E.C., New Encyclopædic Catalogue of Philosophic Apparatus with 1100 Engravings. Royal 8vo. 5/6.

J. J. Griffin and Sons, Chemical and Philosophical Instrument Makers, 22 Garrick St., W.C. Beside special Catalogues, this Firm publishes Illustrated Volumes on "Scientific Handicraft."

M. Jackson, 65 Barbican, E.C. Manufacturer and Importer of Scientific Apparatus and Chemicals. Among the Catalogues of this Firm, is one describing 38 distinct sets of cheap educational Apparatus, mostly chemical.

Townson and Mercer, late Jackson and Townson, Wholesale and Export Dealers in Scientific Apparatus and Chemicals. Laboratory, 89 Bishopsgate St. E.C.

Poths, Haas, and Semple, 23 Bevis Marks, St. Mary Axe, E.C. Importers of Chemical Appliances, &c.

Hopkin and Williams, 16 Cross St., Hatton Garden, E.C. Manufacturers of Chemicals of every description, including the most recent products.

Henry Barnes, 38 Long Acre, W.C., and Messrs. Skilbeck Brothers, 202 Upper Thames St., E.C. Supply all Chemicals connected with the Dyeing Trade.

C. W. Noble, 13 James Street, Covent Garden, W.C. Varnish and Colour Manufacturer, dealer in all varieties of Gums and Resins.

James R. Gregory, Mineralogist and Geologist, 15 Russell St., Covent Garden, W.C. Supplies Educational sets.

Mottershead and Co., 1 Market Place, and St. Mary's Gate, Manchester. Chemical Laboratory Furnishers.

James Woolley, Sons, and Co., 69 Market St., Manchester. Chemical Laboratory Furnishers.

[The above two Firms deliver all orders exceeding £2 free to any Railway Station in England.]

A Catalogue of Apparatus for Instruction in Geology, Physiology, &c., has been issued under the authority of the Science Department, by Messrs. Chapman & Hall.

Various devices for the production of cheap but serviceable home-made Apparatus for Scientific Lectures may be inspected at the Twickenham Economic Museum.

SECTION 7.—THE EXAMINATIONS.

Before proceeding to consider the essential features of the proposed System of Technical Examinations, let us dispose of one or two questions which stand in the way.

Mention has been made of an impression which formerly prevailed as to the difficulty of testing that manual dexterity, and those practical abilities, which constitute so important a portion of an Artisan's industrial worth. Now independently of the answer afforded by the Society of Arts' special examinations in Art Workmanship, by the competitions for the Whitworth Scholarships, and by the trials of skill recently organized by the City Companies of Turners and Paper Stainers, the number of Handicrafts to which a combination of intellectual and manual Examinations may be applied, is shown to be very comprehensive by the experience of the old German Guilds, adverted to in my "Letters on Nassau." Yet even they had seldom at their disposal any facilities comparable to those proposed to be afforded by the *Ateliers* of the Central Technical University, and of the affiliated Provincial Universities. — Nevertheless there are undoubtedly many Industrial Employments in which the skill of the Workman can only be manifested amid the resources of a regular Factory; but here an expedient suggests itself, of which the eligibility is guaranteed by its having been adopted in Major Donnelly's scheme of

Technical Examinations, now being applied by the Society of Arts to the Industries represented at the International Exhibitions. It consists in requiring the Artisan to produce a Certificate of efficiency signed by his Employers.[1]

A much more serious obstacle is the difficulty which a considerable proportion of the Industrial Population experience in expressing even ordinary thoughts and incidents in writing; so that it would in many cases be hopeless to expect them to convey to paper in the time and under the restrictions of an Examination, a legible, intelligible and not altogether uncouth exposition of their knowledge. This difficulty which will not be satisfactorily overcome till Primary Education connects mental training with the management of the pen, varies according to circumstances independent of scientific or technical attainments, of which it often falsifies the appreciation, and I fear that among the lower range of Candidates, it will in numerous instances prove so serious, that recourse must, at all events for some years to come, be had at the discretion of the Examiners, to *oral* instead of *written* tests. This is one of those points on which the proposed Scheme must for a time be worked tentatively, before it can settle into a definite groove.

The deficiency of educational culture which in extreme cases will altogether preclude the idea of demanding from Working Class Candidates written expositions of their knowledge, would in far more numerous instances, engender diffidence and disinclination to enter the lists, or raise perplexity and confusion in the way of their success, if they could not be afforded

[1] See the Programme of Technological Examinations issued by the Society of Arts.

from the beginning of their studies, a clear perception of what their Examinations are to be. That this may be done both safely and effectively by the "open-handed Method" described in the preceding Chapter, and which is equally applicable to *Competential* and to *Competitive* Examinations, has I believe been sufficiently established; nor do I indeed see any other means likely to be found so satisfactory for securing, if not in the Superior, at all events in the Minor and Major Examinations of the proposed System, that uniformity and permanency of standards without which its verdicts would be fallacious, and its Certificates would soon fall to the ground. — Assuming then the adoption of the method in question, and recapitulating in connection with it the measures concerning Curricula, Text Books, and the like, proposed to be carried out by the Board of Instruction of the Central Technical University, we see that Board entrusted with a line of operations which may be summarised as follows :—

a.—To prepare in detail a thoroughly well-considered plan, or CURRICULUM, of the Course of Studies to be prescribed for each Description and Grade of Students.[1]

b. — To supervise the preparation by competent Authors and Artists, and the publication with appropriate Illustrations at, or in some cases under cost price, of the Text Books which have been described as necessary for imparting the scientific, technical, or other attainments included in the Curricula; such publications to be conspicuous for unity of purpose and mutual adaptation, and at the same time to be in every respect suited to the prevailing educational status of the

---

[1] Suggestive indications for facilitating this task will be found in the next Chapter.

Students for whom they are designed, as well as to the convenience of the respective Professors.

c.—To arrange for periodical revisions of the Curricula and Text Books, the latter receiving any required Supplements, and being re-edited from time to time, say every three years.

d.—Besides the *General* Programme or Synopsis of Examinations in the yearly Calendar, to publish a *Special* Programme for each Examination or group of Examinations, giving in detail the CURRICULUM or SYLLABUS of the knowledge required, with indications of the Text Books and other facilities available, and moreover including a QUESTIONARY, or full enumeration of the Questions liable to be asked.

e.—The preparation of the Questionaries, and their use by the Examiners, to be *mutatis mutandis*, as described in Chapter III., Section 5.

f.—It is desirable to carry out to the fullest extent in a spirit of courtesy towards Students of every Degree, the practice of publishing freely, and without any unnecessary delay, the number of Marks obtained in each Subject by every successful Candidate, and of supplying similar data to unsuccessful Candidates if desired.

The subject of MATRICULATION EXAMINATIONS to precede the admission of Artisans as regular Students at the Central or Provincial Technical Universities, is one of those which must be dealt with tentatively, and according to circumstances.

The Time required for the Examinations of various kinds and degrees may, like many other points of practical detail, be left for the present an open question. It would however be probably not far from the mark to assume, firstly, that as at the London University, each Examination Day will have two sittings of three hours each, say from 10 to 1, and from 3 to 6; secondly, that Subject Examinations will mostly involve only two sittings; thirdly, that Minor Trade Examinations will probably occupy on an average three to four sittings, and corresponding Major Examinations six to ten sittings; fourthly, that Tests of Manual Skill must be reckoned apart from these computations.[1]

It would be premature to attempt determining to what extent Minor Examinations may be held by delegation at District Colleges, and Major and Superior ones similarly at Provincial or Colonial Universities, or on the other hand how far the system of centralization adopted by the Society of Arts and at South Kensington, may be found applicable. One thing however will be essential in all Examinations held by or under the authority of the Central Technical University, namely, that concerted action should everywhere prevail, and that in all matters relating to the Valuation of Attainments, and Awarding of Certificates, the Examiners themselves should heartily co-operate in establishing and maintaining DEFINITENESS OF PRINCIPLE, and UNIFORMITY OF STANDARDS.

[1] The more advanced Students are, the more comprehensive will be their knowledge of each Subject; but in order to prevent too much time being taken up with Major Examinations, two measures may be resorted to:—The one already mentioned consists in the use of Blank or Skeleton Forms for being filled up, a plan which, in passing, I may suggest to be susceptible of useful adoption in ordinary Chemical, Botanical, and other Examinations; the other is to increase the number of Questions in each Subject given in the Questionary, without increasing the number asked at the Examination, and thus proportionately to diminish the range of knowledge embraced by each Question. Moreover it must be remembered that Major Candidates need not be examined again in a Subject in which their Minor Certificate shows that they have already satisfactorily passed.

CHAPTER V.

ANALYSES OF TYPICAL OCCUPATIONS.

SECTION 1.—INTRODUCTORY CONSIDERATIONS.

THOSE Industrial Pursuits which enjoy or deserve the benefit of special educational Institutions, and which it has been agreed to designate collectively as Category A, have been sufficiently adverted to in Chapter I., Sect. 7, and we may now proceed at once to devote our attention to the less favoured Categories which come more immediately within the scope of the proposed University ; namely :—

Category B. — Normal Pursuits deserving regular complex Curricula, Composite Examinations, and Corresponding Certificates which being delivered for proficiency in a given Trade or Industry will be called TRADE CERTIFICATES (p. 42).

Category C.—Abnormal pursuits not deserving regular Curricula, but for the benefit of which the study of Detached Subjects will be recommended, and SUBJECT CERTIFICATES will be awarded (p. 42).

It will suit our purpose to take first the latter Category, for which a few remarks will suffice ; but before attempting to deal with either, let us take note of a few considerations which apply to both.

Preliminary Attainments.

Under this title are included Reading, Writing, and Arithmetic, which will not be mentioned either in the sketches of Trades Curricula, or among the Detached Subjects recommended for distinct study, inasmuch as they can be sufficiently compassed by the following general rules:—

1stly. Even Minor Candidates will ultimately be expected to be able to read well enough for improving from books what they learn through oral teaching, and to write well enough for undergoing an Examination in the ordinary way. In the meantime, some allowance may (as mentioned in the preceding Chapter) be necessary for meeting existing ignorance; but nothing had better be determined till necessity occurs, in order that full advantage may be derived from the rapid improvement that may be expected from the working of the Education Act.

2ndly. Major Candidates may without exception be expected not only to read perfectly well for purposes of study, and to write sufficiently well for purposes of Examination, but also to understand Arithmetic sufficiently for the keeping of such accounts as their respective Trades may require.

3rdly. Candidates at Superior Examinations will be expected not merely to write a fair hand, but to add a certain refinement of style to grammatical accuracy in the expression of their ideas. The knowledge of Arithmetic and Book-keeping which they should possess, will depend on the nature of the position to which they aspire.

4thly. Under "Preliminary Attainments" are not

included the higher or more technical kinds of Commercial and Financial Accounts, respecting which due provision will be made.

Studies for Subject Certificates.

The imparting of Instruction, the holding of Examinations, and the awarding of Prizes in DETACHED SUBJECTS to Students of CATEGORY C, will constitute only a secondary and auxiliary department of the operations of the proposed System, and one of which the development must to a considerable extent be guided by circumstances. The working details should however be devised, as much as possible, in conformity with those adopted in respect of Category B Students, and at the same time attention should be paid to the following considerations :—

1stly. Detached Subjects should not be taken up at the commencement of operations, unless they either are of sufficient importance to warrant the expectation of a large concourse of Candidates, or form part of Courses of Instruction subserving other purposes at the same time. As regards the latter point, everything will be done both in the arrangement of the Text Books, and in the organization of the classes, for enabling Students whose Pursuits are ultimately divergent, to learn together what they want in common.

2ndly. So great is the importance of inducing the industrial classes to establish their Knowledge, whatever it may be, on a broad and sound scientific basis, that it will be advisable to require that every Candidate for a Subject Examination, should first pass an Examination in the Standard Course of General Scientific Foundation, unless he can show by the possession of a

Trade Certificate or otherwise, that he has already passed one.[1] By this precaution if the Student takes up only one Subject, it will not be too isolated in his mind, and if he takes up two or more Subjects, there will always be a common basis to connect as well as support them.—It will probably be convenient to have Major and Minor Subject Examinations, and only to require of Minor Candidates that they should possess a Grade 1 of General Scientific Foundation, whilst a Major Candidate will be required to possess a Grade 2, either of the whole, or at all events of the first Division.

The following are a few examples selected to illustrate the sort of easy SUBJECTS for which it may be expedient to offer Minor Certificates.[2] As a *general rule*, if a given Subject cannot well be understood without the previous knowledge of some other one, that Preparatory Subject must be taken first, in addition to the indispensable Grade 1 of G. S. F.[3]

a.—Grade 2 of Chemistry, or of any other department of the Elementary Division of the Standard Course, suited for being useful by itself, or for serving as preparation to any of the subjects indicated under the following headings :—

b.—A Subject, Group of Subjects or Department, selected from the

(1) The rule of an obligatory General Scientific Foundation, which will be seen to prevail equally in reference to Category B Students, is on a small scale a measure analogous to that which has been adopted with such good results by the London University, in instituting its Matriculation Examination.

(2) It must always be remembered that the selection is to be strictly governed by the nature of the calling in view, or to quote the advice given by the Bishop of Exeter in distributing prizes to the Evening Classes at Bristol on Oct. 3, 1872, the Students should "choose some "study which in some way or other would be bound up with their own "occupation, and that they would constantly come across in the work "they had to do."

(3) These initials will be frequently used for convenience sake in lieu of the lengthy expression "General Scientific Foundation."

Applied Division of Grade 2 of the Standard Course. The *general rule* given above, will frequently apply to subjects of this description. Thus for example, any Department of an essentially chemical character, must be preceded by a Grade 2 of Inorganic or Organic Chemistry, or of both.

c.—A portion selected either from the Junior or from the Advanced Cyclopædia, and embracing the history of a single material (say SILK), or the account of some Technical Process, useful to many, though not of sufficient magnitude to constitute a trade by itself. The above rule respecting Introductory or Preparatory Subjects, must not be forgotten; many subjects will succeed best if preceded by a double introduction, supplied by *b* as well as *a*.

d.—MECHANISM, as taught by some popular Work suited to serve as Text Book.[1] This Subject and others of an analogous kind, should be preceded by Grade 2 of the part of Mechanical Physics relating to the Mechanical Powers.

e.—CHROMATICS, as defined in the previous Chapter, Sections 2 and 4. To be preceded by Grade 2 of the part of Chemical Physics relating to Light.

f.—METALLURGY. To be preceded by Grade 2 of Inorganic Chemistry, of which it must adopt the level.[2]

g.—COMMERCIAL ARITHMETIC AND BOOK-KEEPING.

h.—Grade 2 of MENSURATION, as explained in Section 4 of the preceding Chapter, taking as Text Book Todhunter's 'Mensuration for Beginners.'

i.—An analogous Grade of either Geometry, Algebra, or Trigonometry, taking as Text Books appropriate portions of the respective Handbooks " for Beginners," by Todhunter.

Most of the foregoing Subjects may be rendered suitable for Major Students, by simply raising the Foundation and all additional knowledge of a preparatory character, one Grade, and taking the Technical Knowledge from the " Advanced," instead of from the " Junior Cyclopædia."

Many other Subjects will suggest themselves, either

---

(1) The small book by Thomas Tate, entitled the 'Elements of Mechanism,' appears to be suitable for a Minor Candidate, whilst a Major Candidate would take the work of the same title by T. M. Goodeve, M.A.

(2) Candidates for higher Grades would be referred to the School of Mines.

through analogy or through the existence of publications suited for being adopted as Text Books. To find such Books will be difficult in the Minor, but much easier in the Major Standard, which will correspond to an ordinary level of educational Works. Applying to the most appropriate among these, the proposed system of Examinations, under due precautions against hollow or unsupported knowledge, the field of operations may if deemed desirable, be considerably extended.

THE ARTS OF DESIGN, besides being included to a limited extent in the majority of Trade Examinations, will receive every encouragement as Detached Subjects, but always as much as possible in connection with the South Kensington System.

MUSIC may perhaps be best promoted by fostering a kind of autonomic development of its own.

Studies for Trade Certificates.

It must be borne in mind that Composite Examinations of three degrees, with corresponding Certificates of Competency, are contemplated for Trades and Occupations of Category B, viz. :—

A MINOR EXAMINATION, which the young Artisan, say emerging from his Apprenticeship, should pass in order to obtain a Certificate of Competency for a Journeymanship, or MINOR CERTIFICATE.

A MAJOR EXAMINATION which the Journeyman or ordinary Workman should pass in order to obtain a Certificate of Competency for a Mastership, or MAJOR CERTIFICATE, showing his fitness either to set up Trade for himself, or to take charge of a Department in a large business or a factory, as Foreman, Leading Man, or under any other equivalent title.(1)

A SUPERIOR EXAMINATION, the corresponding Certificate to which will be called a DIPLOMA OF EXCELLENCE, and denote qualifications of a high character, such as those required for a position of Director, Manager, Clerk of Works, or the like.

Of course the foregoing three degrees will only occur in *normal* or fully and regularly developed Trades. As we proceed we shall find the normal examples few, the variations manifold, and the educational difficulties to be overcome proportionately great. I believe however that the elasticity of the proposed system, and the variety of its resources, will be found equal to every obstacle that is fairly scanned and advisedly encountered.

Purpose of the Proposed Analysis.

The want of pains-taking inquiry into the working details of the numberless individualities which crowd the vast field of Technical Industry, and the easy confidence with which those who never tried, can speak of dissipating in all directions the darkness of ignorance

---

(1) It will be found as we proceed, that though a given Certificate for a given Trade must represent an unvarying level of Knowledge, whether that Certificate may have been granted at the Central University, or at any Provincial or Colonial Examining College, yet there may be a great difference between the amount of studies implied by a Certificate in one Trade or Pursuit, as compared with a corresponding Certificate in another Trade. Thus for instance a Builder's Major Certificate will represent considerably more attainments than the Major Certificate of a Mason, a Bricklayer, or a Carpenter, seeing that it involves, if not much manual cleverness at these various trades, at least a sound knowledge of them all. Again, the Minor Certificate in a Trade like that of the Watchmaker, demanding considerable abilities, and generally practised by an intelligent class of Men, will involve nearly as much Science as the Major Certificate in an unaspiring Trade like that of a Baker.

by the light of Science, have led to far more sanguine anticipations of the achievements of a Central University than are likely ever to be realized, to say nothing of realizing them at the first onset. Prudent beginnings will best secure permanent success. The first operations must necessarily be more or less tentative, and as the tide of public opinion will be much influenced by their results, a careful selection must determine, even in Category B, the Standard Trades that most deserve an honourable preference for the inauguration of the System. But at the same time we must faithfully observe the principle of thoughtfully planning from the beginning, the scheme of future developments, in order that each of these in its due season may spontaneously take its place in the symmetrical growth of the System. For this purpose the infinitely diversified and more or less urgent demands for intellectual furtherance on the part of the various Industrial and Commercial Occupations, must be scrutinized and classified, and a proper provision appointed for each. To accomplish this definitively must be reserved for the Instructional Board of the proposed Central Technical University, but any data tending to break the ground for so laborious a task, will doubtless prove acceptable, and may serve in the meantime to inspire confidence in the practical working of the System. It is for supplying such data that the present Chapter will be mainly devoted to a review of the Industrial Range from the standpoint of the proposed University. Such however has been the rapid progress of civilization within the present century, and such in consequence the luxuriant growth of our wants, and of the efforts to meet them, that to attempt anything like an exhaustive review, even of those Industries which are of a comparatively permanent and well-defined character, would be tedious and confusing. The Analyses of a few Representative Occupations, or groups of Occupations, will produce clearer impressions, and afford better guidance for a judicious selection of those to which, as above stated, we should first offer a helping hand. As the educational machinery of the Central Technical University becomes perfected, and public confidence in its management becomes confirmed, the sphere of its operations will acquire spontaneous extension.

The plan will be adopted of interweaving with the *special* remarks suggested by the requirements of each Trade or Group of Trades, such *general* observations as are prompted by the occasion. They will thus be better understood, and seem less formal than if collected by themselves.

Wherever practicable, the Article will be concluded with a condensed SUMMARY showing approximately at a glance the leading attainments and sources of instruction proposed for the respective Candidates for a Minor Certificate, for a Major Certificate, and where there is occasion, for a Diploma of Excellence.

The outlines of Courses of Studies thus given, are but suggestive and provisional, forming at the best only an imperfect sketch to be completed in detail by the official Curricula or Syllabuses, whilst the degree of comprehensiveness and standard of difficulty of the respective Examinations, will be further defined and made patent by the permanent Examination Questionaries of which the plan has been explained. All selections and apportionments of matter can be carried into effect with comparative ease, through the use of these Syllabuses and Questionaries, which will serve to the Students as guiding threads through the Labyrinth of Knowledge.

Order of Trades and Occupations.

After various trials, I have been induced to adopt the plan of taking first a few highly scientific Industries, forming convenient types for comparison, and favouring the discussion of various difficulties to be overcome; and then to pursue the order suggested by the Syllabus of the Second or Applied Division of General Scientific Foundation given in Chapter IV., Section 1.

Order in which the Attainments will be Considered.

The order of the remarks concerning the various instructional requirements of each Trade, will as far as practicable be that indicated by the following six Questions, which the persons entrusted with preparing the Official Curricula will be supposed to ask themselves.

Preparatory Knowledge.

a.—What Grades of GENERAL SCIENTIFIC FOUNDATION, ought the Students of different degrees in the Trade under consideration, to possess?

As previously stated, no Certificate of Competency in any Trade or Subject will be granted to a Candidate, even for a Minor Certificate, who does not show proof at the Examination, of satisfactorily possessing at least the minimum of General Scientific Knowledge described as Grade 1 of this Collective Range. Major Candidates will mostly be required to possess Grade 2, and Candidates for a Diploma of Excellence, Grade 3.

b.—What Special Scientific or Mathematical Preparation is required for understanding the Technology of the Trade under consideration?

It will be found that in many cases this object will be best met by pushing forward one Grade beyond the rest of the General Scientific Foundation, either the whole or a part of Chemistry, or some branch of Physics, or some Department of the Second Division which is specially devised for supplying links between Facts and Applications, Principles and Practice. In some cases however, Supplementary Scientific Subjects, such as Mechanism or Metallurgy, will be indicated, and more frequently Mathematical ones, such as Mensuration or Advanced Arithmetic.

c.—What Artistic Abilities are required by the Trade under consideration?

As far as Drawing is concerned, the answer to this question may be reduced to a general rule, which except as regards Occupations in which artistic attainments have a prominent part to perform, may almost dispense with adverting to them in the Trades' Analyses, viz. :—Every Minor Candidate whose Trade requires that he should work from a drawn Pattern or Design, should sufficiently understand Drawings to do so,[1] and every Major Candidate of such Trade should be competent to draw a Pattern suited for being worked from. —Respecting Candidates for a Diploma of Excellence, it is not easy to frame a rule. There are for instance, Artistic Industries in which a person competent through superior mental abilities and refined taste to direct others, need not be so clever at perform-

---

(1) See No. 1 of the five stages of artistic ability described in Section 4 of the preceding Chapter.

(2) See No. 2 of the stages above referred to.

ing certain parts of the manual work, as some of his subordinates.

It must be borne in mind that independently of the amount of Art inherent to the several Trades, it is proposed to favour any further artistic proficiency of a practical tendency, and especially to foster the development of inventive genius, taking advantage of the facilities offered for that purpose by the system of Art Instruction centred at South Kensington.

ARTISTIC TASTE, apart from Artistic Abilities, will not unfrequently be prescribed. Its importance and the best means for its promotion have been sufficiently adverted to in Section 4 of the preceding Chapter, to which reference may also be made respecting CHROMATICS, or the LAWS OF COLOUR, that will be occasionally mentioned in the Analyses of Trades whose connection with Art rests rather on Harmony of Colour than on Elegance of Design.

Technical Knowledge.

d.—What useful insight into the RESOURCES of the Trade under consideration, can be appropriately conveyed in a scientific review of the Shop, Atelier, or Factory, embracing its Construction, its special adaptations of Motive Power, its Appliances and Tools of every kind, and especially the Materials employed?

It has often been urged as an objection to any comprehensive plan of Technical Instruction, that many Industries can only be carried on with an amount of extensive as well as expensive paraphernalia, which it would be neither practicable nor desirable to show in actual operation in an educational Institution. But it must be remembered that our future Professors, in describing the appliances and processes proper to each Manufacturing Art, will not have to deal with pupils altogether ignorant thereof. Even the Minor Students may be expected to have passed through an ordinary Apprenticeship, and to be competent to understand Models and Diagrams; and indeed, what they will chiefly want to know, will not be how things are done, but the rationale of so doing them, in order that as Journeymen they may do justice to the means placed at their disposal, and as Masters they may improve on them.

e.—What practical information can be imparted in a scientific, or in some cases in an artistic review of the PROCESSES or Operations, and PRODUCTS or Results of the Trade under consideration, or of the Articles dealt in, if it is a Commercial Trade?

Though Appliances and Processes are here separated under distinct Headings, they will frequently be united in the Suggestive Curricula. In fact many cases will occur in which the Technological Professor will find it preferable to unite in the same discourse, the description of a particular part of an Establishment, and the explanation of the particular Process carried on there.

Here may be mentioned as a rule embracing this and the preceding question, that a Candidate for a Major Certificate will be expected to possess a sound knowledge of all the *available* resources and processes of his Trade, and more or less of each branch of his Trade if it should be a complex one. The Minor Candidate on the contrary, need not generally speaking go beyond the *ordinary* resources and processes of his Trade, and if it should be a complex one, he may be allowed to take a Certificate of Competency in one main branch only. A knowledge of the other branches may be left to be acquired, tested, and certified subsequently, for which encouragement will be afforded by ADDITIONAL or SUPPLEMENTARY CERTIFICATES.[1]

f. — What INCIDENTAL INFORMATION and ADVICE should be given in reference to the Trade under consideration ?

Under this heading are intended to be included the following and kindred topics :—The History and prospects of the Trade, and its variations under local influences. Substitutes and makeshifts available for emergencies.—The Hygiene of the Trade, and considerations connected with it.

Recapitulation.

Preparatory Knowledge.[2]

a.—General Scientific Foundation.
b.—Further Scientific Preparation.
c.—Artistic Preparation.

Technical Knowledge.

d.—Review of Plant and Materials.
e.—Review of Processes.
f.—Incidental information and advice.

The terms JUNIOR, ADVANCED, and FULL TECHNOLOGY, will be used for designating the Technical Knowledge concerning any given Trade respectively

[1] Additional Certificates will also be awarded to Tradesmen who add to their regular calling other occupations of a kindred nature.

[2] The division into Preparatory and Technical must not be taken in too strict a sense. Artistic, and sometimes other Studies, classed as Preparatory, may have to be pursued during the whole of the Technical Studies, and on the other hand, under Technical Knowledge will be included by way of Introduction, various Subjects of a preparatory nature susceptible of being with advantage specially adapted, and as it were moulded to the Industry which requires them.

required by Students of the Minor, Major, and Superior Degrees.

Sources of Information.

a.[1]—For Grades 1 and 2 of General Scientific Foundation, the Books used will be the respective STANDARD COURSES brought out by the University in a lecture form, and in Parts well illustrated for the convenience of Students.—For Grade 3, till a Standard Course can be prepared, use will be made of the Books or parts of Books to be indicated by the SYLLABUS OF REFERENCE described in the third Section of Chapter IV. (p. 204).[2]

b.—The foregoing sources will be utilized as far as available. — In the case of Mechanism, Metallurgy, Mensuration, &c., recourse may be had to such Books as those of Lists 6 and 7, p. 224.—The Technical Handbooks and Manuals forming respectively the Student's Edition of the Junior and Advanced Cyclopædia, will as explained, include in their Introductory Portion, all Preparatory Knowledge of which the mode of treatment should vary according to the Industry to which it is applied.

c.—Freehand and Mechanical Drawing, Modelling, and the like, will be mostly acquired at Art Classes connected with South Kensington, unless Artistic should be so blended with Technical performance, as to render the Technical Class or the Atelier a more suitable source of instruction. Artistic Taste in form, colour, or subject, will be obtained from such publications as those given in List 8, p. 225.

(1) See above, the heads of information referred to under this, and each of the following letters.

(2) Books available for the preparation of the Syllabus are suggested by List 3, p. 205; Books for Grade 4 by List 4, p. 210; and Books for Grade 5 by List 5, p. 215.

For d, e, and f, that is to say, for the whole of the Technical Knowledge, the respective Handbooks and Manuals will as a rule be the official Text Books, viz.: — for Minor Students, the Handbooks; for Major Students, the Manuals, with the omission of certain Notes, Appendices, and other Reserved Matter; for Superior Students, the Manuals to their full extent, supplemented in many instances by references to Special Works of an exhaustive character.

SECTION 2.—SELECT EXAMPLES OF CHEMICAL TRADES.

THE DYER.

Few of the familiar Trades combine with Manual Labour, so obvious a necessity for extensive and multifarious Chemical Knowledge, as that of the Dyer; hence the choice commonly made of it to exemplify a Scientific Trade; nor could a more convenient starting-point be selected for our Analytical Review. Following the Order of Questions indicated in the foregoing Section, we arrive at the following conclusions.[1]

a.[2]—Grade 1 of the "General Scientific Foundation," Elementary and Applied, may suffice for the Journeyman Dyer, and Grade 2 for the Master Dyer; but we have further to consider that the Dyeing Trade is one of those in which the positions of Managing

[1] Remarks suggested by one Trade or Group of Trades will hold good for all similar cases. Hence the amount of these remarks will be found to decrease very considerably as we proceed.

[2] The division of the subject into distinct headings is fully carried out in the present instance, in order to give a clear idea of the proposed routine of analysis. It will subsequently be dropped.

Directors or Foremen of large Works, generally demand, and sometimes recompense scientific attainments of a very high order. This is accordingly a case for instituting a "Diploma of Excellence," the Candidates for which may reasonably be expected to master Grade 3 of that Science of Daily Life, which besides benefiting themselves, will so much strengthen their hands for benefiting every one in their employ.

b.—Respecting "Further Scientific Preparation," it must be remembered that Preparatory Knowledge of a nature to vary according to its application, will be mostly associated with the Technical Knowledge in the Cyclopædias. This applies to the Physical Knowledge of the properties of Steam as a Heating Agent, and also to the portion of the Study of Light bearing on Chromatics. The case is different with Chemistry, of which a Special preparatory Study will be necessary; the Minor Candidate pushing forward this part of his Foundation to Grade 2, and the Major Candidate to Grade 3.[1] They should do the same with the Department of Applied Knowledge (Second Division of the

---

[1] The following is an Abstract of the evidence supplied some time since by a very intelligent London Dyer:—

"The Apprentices, as a rule, have no preliminary scientific training; they learn by experimenting on the goods of the Customer, much to the disgust of the latter. It is not an unusual occurrence for Apprentices, and even Journeymen, to mix Acids and Alkalies together, in ignorance of their neutralizing properties.—The waste of drugs by carelessness and ignorance, is enormous." As an example of the way in which this occurs, may be cited the leaving out of stoppers of jars containing Ammonia or Hydrochloric Acid, through ignorance of their being aqueous solutions of Gases.

Striking instances of the results of ignorance of Chemistry in the Dyeing Trade, are afforded by the statement of the Dewsbury Chamber of Commerce, that through this cause, Woollen Fabrics are ofttimes spoilt in the dyeing, and by the fact related by Mr. Bennoch, of large quantities of Silk sent by our Manufacturers to the Continent at a heavy expense, to be dyed and then brought back again.

Standard Course), relating to Textile Materials and Fabrics, unless it should be found preferable to include all that a Dyer should know of these matters in the Handbook and Manual of the Trade.—The Candidate for a Diploma of Excellence should possess Grade 4 of Chemistry.

c.—As regards " Artistic Preparation," nothing need be included in the training of the Dyer, beyond a certain knowledge of Chromatics, slight in the Minor, rather more considerable in the Major Degree, and which should be comprised in the Handbook and Manual of the Trade.[1] We must indeed bear in mind that our Technical Instruction, in order to become popular among Artisans, and to disarm prejudice among their Employers, must strictly avoid consuming any unnecessary amount of time and expense, and steer clear of everything that might be construed into ambition or pretence, keeping to a safe channel, straight and short. Accordingly we should trouble as little as possible the Candidate for a Certificate of Journeyman Dyer with obligatory artistic attainments. The Master Dyer himself might be let off with a general insight into the applications of the Theory of Colour to his Trade, and as Pattern Designing is a distinct business, even the Director or Foreman of Dye Works might obtain a Diploma of Excellence, without any further artistic training than might be necessary to keep him from giving any grievous offence to the intelligent eye, in his attempts to please the eye of fashion.[2]

---

[1] See the heading "Artistic Taste" Chapter IV., Section 4. Among the many branches of Technological Nomenclature, concerning which it will be the duty of the proposed University to establish and maintain a clear and methodical understanding, aided by a classified collection of Standard Patterns, is the nomenclature of COLOURS.

[2] The blocks prepared for carrying out a beautiful design in well

d.—Having disposed of the three headings of "Preparatory Knowledge," we proceed to consider the first heading of "Technical Knowledge," viz.: the "Review of Plant and Materials," or in other words of the Dye-House and its contents. As the Junior and Advanced Cyclopædias (Artisan's edition), or in other words as the official Handbook and Manual are to be respectively the Text Books for the Minor and Major Studies in this as in other Trades, it is to the persons charged with preparing those important repertories of information for the guidance of the instructors, as well as for the use of the instructed, that the following and all similar pieces of advice must be considered to be addressed.

The more we enter into the task of Technical Instruction, the more we become convinced of the necessity of proceeding with cautious discrimination in any endeavour to effect real and durable improvement.— Let us take first the case of the intelligent young Artisan who having completed his Apprenticeship, and possessing a tolerably good Scientific Foundation, comes to us to get his knowledge up to the required standard, and then to pass his examination and take his Certificate of Competency as Journeyman Dyer.— Nothing can be better calculated to enhance what may be called his *working value*, than to review in succession, under the light of Science, the contrivances and appliances of a Dyeing Establishment, together with the Fabrics commonly dyed, and the Drugs commonly used in dyeing them. We may to a certain extent make him acquainted with appliances and materials *not* com-

assorted shades of the same colour, once fell into the hands of an ignorant Calico Printer who, to please the gaudy taste of his customers, spoilt the whole by using different colours instead of different shades.

monly used, but which he may come across in some of the larger or more advanced establishments; but we must remember that it is from the *Master* that improvement in these matters, as well as in processes, should chiefly emanate, and that *he* would probably not think the *working value* of a Journeyman increased by a dissatisfied and reforming spirit, however well founded it might be on scientific principles. Accordingly our anxiety will be to make our Candidate understand the nature, purpose, and intelligent use of things as they are, rather than as they ought to be, and to teach him how to work well and be contented even under difficulties.—Not so with the advanced Journeyman who comes up to us that we may raise his knowledge to the Major Standard, and give him a Certificate of Mastership. As Master, his duty, not only towards himself and those who may depend on him, but also towards his Country, will be to strive according to his means, that his establishment may do honour to him; which if each individual Dyer were to do, the aggregate Trade could not fail to be an honour to our National Industry.[1]

All this applies, *à fortiori*, to the higher votaries of the Dyeing Art, who may aspire to Diplomas of Excellence, and nothing in the way of instructional advantages will be spared at the University, in order to enhance their future influence among the industrial community.

[1] According to the source above quoted, a "Model Dye-house" is somewhat sneered at in the Trade. Some pin their faith on make-shift contrivances, and glory in the ingenuity with which they succeed under disadvantages. But the most meritorious way of conquering disadvantages is to make them cease to exist, and the best way to save a Model Dye-house from being a conspicuous object of remark, is to make it succeed and induce imitation, till efficiency and order become the general rule, and working *anyhow* the exception.

e.—The "Review of Processes" adds to the foregoing considerations another motive for cautious instructions. I allude to the jealousy with which communities or firms guard their "Trade Secrets,"—secrets sometimes scarcely worth guarding, but at other times really forming the main stay of individual or local prosperity, and with which in the latter case, it must not be supposed for a moment that our scientific propaganda is intended to interfere.

f.—The Dyer's Trade is suggestive as regards "Incidental information and Advice," no less than as regards the previous headings. The History of the origin and progress of the Art of Dyeing, from the days of Joseph's Coat of many Colours to the days of the glory of Coal Tar, may be made instructive as well as interesting, but only for Students sufficiently advanced to understand the technical terms necessarily involved. For those of the Upper Grades it should be treated as introductory matter, but the reverse for Beginners. Both should be taught to argue the prospects of the future from the experience of the past, and to deduce from the History of Dyeing, a conviction that the permanent success of this important Industrial Art, depends on developing to the fullest extent the tutelary influence of Science over Labour.

It is particularly when Industry labours under difficulties, that the helping hand of Science is most welcome; and it is in this sense that must be interpreted the insertion among the subjects for consideration, of "substitutes and makeshifts available for emergencies."

As regards "Hygiene," there are few Occupations that have not in this direction some weak points claiming scientific surveillance.—Among those which occasionally give the Dyer opportunities for exercising his intelligence and his knowledge, are the following:—

A Warm Atmosphere laden with Steam. Transition from it to Cold Air. Inflammability of Benzole, Spirits of Turpentine, &c.—Poisonous nature of some of the Articles used, of which the injurious effects may assume the form of Blood-poisoning when they are absorbed through a wound or an abrasion; of Absorption through the skin, which might produce serious consequences if the hands and arms were immersed for some time in certain solutions; of inhalation of deleterious vapours, or of injury through incautious handling, as of Sulphuric Acid and other acrid compounds; to say nothing of errors through sheer ignorance, which it may be hoped will soon be anecdotes of the past.— Among the noxious Articles that might be mentioned, are several Metallic Compounds that are becoming less used than formerly, but are amply made up for as regards danger, by Aniline and its products.[1] To the chance of being poisoned, is added that of being blown up; for though Picric Acid is fortunately not used in the form of Picrate of Potassa, of warlike fame, there is a Morone Dye, (Potassic Isopurpurate) so explosive, that to prevent accident one must constantly keep it in a moist condition.

All this is said, not that it should be supposed for a moment that a Dye-house is necessarily a dangerous place; (an intelligent Dyer has been heard to say that the greatest annoyance he found there was simply the smell of Safflower)—but in order to show what a difference it may make whether the Artisan who works there is neglectful and ignorant, or well trained and a Chemist.

[1] In the preparation of one of the most beautiful of these, Magenta, it is unfortunately found economical to adopt the Arsenic Acid process.

Dyeing forms, with the operations which prepare and complete its results, a compact group round which generally arrange themselves several minor processes requiring similar knowledge and similar conveniences, such as Cleaning, Scouring, and Calendering. We are here reminded of the plan mentioned in the preceding Section, of allowing a Minor Candidate to obtain at first a Certificate in one main branch of a Trade, and to try subsequently for Supplementary Certificates; but the present is not so strong a case as will occur in the Leather-producing Trades.

Summary.

Explanatory Remarks.

A few explanatory remarks on this Summary, may serve at the same time to explain many others on the same principle which will occur in the sequel of this Chapter.

—The main object is to sum up the results arrived at in the preceding analytical enquiry as to the instructional requirements of the Trade in question; thus supplying approximate and suggestive outlines which may in some measure facilitate a fair and systematic apportionment of studies, and the ultimate elaboration of the detailed official Examination Syllabuses.

—Some of the Summaries will be more precise and complete, and others more vague and imperfect, and I have found it best not to attempt strict uniformity of plan,[1] but the Sources of Information will mostly be

---

[1] Variety of form and expression may be rather acceptable than otherwise to the persons charged with preparing the definitive Schedules, Syllabuses, &c., who will be able to select what seems to answer best.

indicated, and the order of succession of the various attainments, as far as they occur, will generally be as follows :—

1. A whole Course of G. S. F. (General Scientific Foundation) of a given Grade.
2. One or more further portions of Preparatory Knowledge, selected from the Elementary or Applied Division of the Course next above the preceding.
3. After these branches of knowledge for which the Standard Courses are to serve as Text Books,[1] " Supplementary Studies " will occasionally be indicated, such as Mechanism, or Metallurgy, for which special Text Books are to be assigned by the Examination Syllabus.
4. More frequently some branch of " Mathematical Studies " will occur, such as Mensuration.
5. In many cases a certain stage of Artistic Ability will be noted, the Source of Instruction being indicated by the title of ART CLASS, or the more comprehensive one of ART STUDIES.[2]
6. Those branches of Preparatory Knowledge which demand a special adaptation to certain Industries, will as explained, be reckoned as Technical Knowledge, and included in the Introductory Portion of the respective Handbooks and Manuals, whilst

---

[1] As previously explained the 3rd Grade Course will, till ready, be replaced by existing publications, or parts thereof, indicated to the Student by a " Syllabus of Reference."

[2] It must be remembered that though the foregoing are all reckoned as Preparatory Attainments, the acquisition of some, as for instance of mathematical and especially of artistic Abilities, may have to be continued throughout the duration of the Technical Studies.

the Main Text of these will supply the Student with the regular Technology.[1]

7. Though as a rule tests of Manual Skill or Practical Ability will be required in all Trade Examinations in which they may be found feasible, they will only be noticed in the Summaries in cases where their mention seems particularly called for. The Source of instruction will not be indicated. It should be the place of Apprenticeship, but may have to be sought in the atelier of a Technical College.

Minor Examination.

GRADE 1 COURSE. Whole range of G. S. F. (General Scientific Foundation) as explained in Chapter IV. Section 1.

GRADE 2 COURSE. Chemistry, Inorganic and Organic.[2]

HANDBOOK. (Introductory Portion.) Leading features of the following Subjects considered from the point of view of the Dyer :—Production and Properties of Steam as a Heating Agent, &c. Rationale of Colours and Chromatic Taste. Review of Materials and Fabrics commonly dyed.

HANDBOOK (Main Text). Review of Plant, Materials,

---

[1] It must be borne in mind in carrying out the proposed Scheme of Technical Instruction, that Preparatory Knowledge of a nature to be included as such in the Standard Courses, can thus be made to serve for the joint teaching of a large number of Students, and that there will be to some extent a corresponding saving of Power at the Examinations. On the contrary what is taught in a special form at a Technical Class, can only be addressed to a comparatively small number, but then it will be exactly adapted to their wants, and they will always have it at their convenience in the Handbook or Manual intended to be their permanent *vade mecum*.

[2] Possibly it may be found desirable that the Student should borrow from the Applied Division of this Course, the general insight afforded there into the subject of Fabrics and Clothing.

SECT. 2.] THE DYER. 285

and Processes ordinarily employed. Select data concerning the History, Resources, and Hygiene of the Trade.

Major Examination.

GRADE 2 COURSE. Whole range.
GRADE 3. Chemistry, Inorganic and Organic.
MANUAL. The Knowledge, both introductory and technical, indicated above as being offered to Minor Students by the Handbook, will be so developed and extended in the Manual, as to embrace the scientific rationale and practical management of the Trade in all its branches, including select information concerning foreign improvements, new resources, and hygienic measures.

Superior Examination.

GRADE 3. General Scientific Foundation, as supplied by the Books indicated by the "Syllabus of Reference."
GRADE 4. Chemistry.—The special Examination Syllabus will indicate the Book or Books most particularly eligible.
MANUAL (duly supplemented). Introductory and Technical Knowledge on the most comprehensive scale, including an acquaintance with subsidiary and kindred Industries, and a large amount of hygienic, historical, and commercial information.

THE TANNER AND THE CURRIER.(1)

We here come across a more striking example of a difficulty slightly touched upon in speaking of the Dyer, and which will be of frequent recurrence; that of accommodating the proposed system of Examinations and Certificates to the varying forms, sometimes simple and sometimes singularly mixed, which Trades exhibit under local influences, or according to personal convenience. Great is the diversity of processes and many are the stages through which Hides and Skins of different kinds become endless varieties of Leather, and it is not surprizing that among the occupations involved, there should be some which under more or less definite names, claim to be considered as distinct trades. In such cases it would be premature to decide exactly what collateral occupations should be united with the main one in planning the Examinations and Certificates, and

---

(1) It will be convenient to include under this title, the whole of the leather-producing processes briefly described as follows by Miller at p. 818 of his 'Elements of Chemistry,' Part III. 4th edition:—

"Four principal processes are in use for the preparation of leather. They consist of—

"1. *Tanning.*—This is employed for the thicker kinds of leather; it is essentially a process for combining the astringent principle of plants with the hide.

"2. *Sumaching.*—This operation is similar in its results to tanning, but is less laborious and tedious; it is performed upon the thinner leathers or *skins*, which are often subsequently dyed.

"3. *Tawing.*—This process is followed in preparing white and black kid, principally for the glover: in this operation alum and common salt are worked with some oily matter into the skin.

"4. *Shamoying.*—This process is used in the preparation of wash-leather. It consists essentially in combining some suitable fatty matter with the texture of the softer part of the hide."

The term "currying" properly applies to the finishing stage of tanning; but it may be also considered as including all operations in which the mechanical manipulation of the skin, plays as important a part as its chemical treatment.

what should be taken separately; but a few points may at once be established:—

1stly. Provided there be no material difference in the Preparatory Knowledge required by the various branches of a complex Trade, it will be easy by an appropriate arrangement of the Technical Knowledge in the Handbook, to carry out to any extent found desirable, the plan of allowing Minor Candidates to take first a Certificate in only one main branch of a Trade, and to follow it up by Supplementary Certificates in other branches. 2ndly. As a rule, Major Candidates will be required to embrace at once in their studies, all the branches that belong to a common industrial stem, and have the same Preparatory Knowledge, but the production of Leather may possibly be found to claim exemption from this rule, by the variety of its processes and products. 3rdly. As regards Candidates for Diplomas of Excellence, they should not only be thoroughly acquainted with all the branches of their own Trade or Profession, but possess a good knowledge of all those with which it is directly connected.

Proceeding to the educational analysis of the Leather-producing Trades, we notice that as regards "General Scientific Foundation," the requirements are the same as in the Dyeing Trades, but that respecting "Further Scientific Preparation," the knowledge involved is less comprehensive. There is nothing equivalent to the facts concerning Decomposed Light which we have seen the Dyer borrowing from Physics, in order to graft on them notions of Chromatic Taste. Nor do we find on examining the various forms of Tanning or Sumaching, Tawing, Chamoying, and the like, that they involve a knowledge of Chemistry comparable for depth and breadth to that involved in the numberless details of the modern Dyeing Trade. Accordingly, if found desirable for economy of time or similar reasons, something might be subtracted from the chemical studies; otherwise I should prefer seeing them kept up to the following standards, which may be considered as the normal ones for all Trades in which Chemistry has a leading part to perform; viz.—for Minor Students Grade 2; for Major Students Grade 3; for Superior Students (aspiring to a Diploma of Excellence) Grade 4.

It might be well to pay special attention to the Zoology of the Mammalia, as an introduction to the study of Hides and Skins.

No "Artistic Preparation" is required; nor is there in the "Review of Plant and Materials," or that of "processes," anything to detain us. As regards "Accessory Knowledge," the History of Tanning is perhaps less instructive than that of Dyeing, but enlivened with what may be called its ethnological aspects, it is not without interest for advanced Students. These will find matter for useful research in the peculiarly prepared articles supplied by certain countries,[1] and still more so in various more or less new materials, both for being tanned and for tanning, for obtaining which commercial facilities now exist.[2] As regards "Hygiene," the Tanning Trade is one of those few which may be made a subject of special congratulation; at least as far as the reputed healthful influence of the Oak Bark Tanneries extends.

The Trade in Leather, whether wholesale, or assuming the forms of Leather-cutting and Grindery, belongs

---

(1) *E.g.* Russia Leather, Shagreen, &c.
(2) Among the materials for being tanned, the introduction of the Porpoise Hide is very suggestive of further steps in this direction.

to a Group of Commercial Trades, remarks concerning which will be found under the heading "Trades connected with Fabrics and Clothing."

Summary.

Minor Examination.

GRADE 1 COURSE of G. S. F.

GRADE 2 COURSE. The whole of Chemistry if found conveniently practicable; if not, the Examination Syllabus will indicate the portions to be selected as particularly bearing on the nature, preparation, and staining of Skins.—The portion of Zoology relating to Mammalia.

HANDBOOK. The contents to be so arranged that the Students may have the option either of embracing all the branches of Leather-producing Industry, or of taking at first only one, the Certificate being made out accordingly.

Major Examination.

GRADE 2 COURSE of G. S. F.[1]

GRADE 3. Select portions of Chemistry to be indicated by the Examination Syllabus.

MANUAL. In consideration of the peculiarly divided character of the Leather-producing Trades, the Student may, against the ordinary rule of the Major Degree, be allowed to take out a Certificate in a single branch, but he will be required to show a good general knowledge of all the others.

---

[1] The outlines of Zoology here included will probably suffice.

Superior Examination.[1]

GRADE 3 of G. S. F.

GRADE 4. Any Chemical Works or parts thereof found specially appropriate.

MANUAL (duly supplemented). To include the introductory and special Technology of all the Leather Industries, thoroughly embracing all interesting and useful information concerning available resources, improved processes, former or foreign practices, &c.

CONSIDERATIONS ON MANUFACTURING INDUSTRIES.

It is not often a wise policy to look for difficulties, but our present course is exceptional in that respect. It is by confronting the proposed system with the various educational obstacles which are unfortunately so numerous in the Industrial line, that we shall best prove its capabilities. The Dyeing Trade was selected first as pre-eminent for the amount of scientific knowledge it involves. The Trade or rather Group of Trades engaged in the production of Leather, has served to illustrate the difficult task of satisfying the lax and arbitrary requirements of occupations sometimes united, and sometimes carried on separately. The next obstacle to be considered is the Division of Labour in Industries commonly conducted in large establishments on what may be conveniently called the FACTORY PRINCIPLE. Here the work is divided into more or less separate Stages or Departments, each assigned to a special set of Workmen. Each Department may be essentially dif-

---

(3) Candidates for Diplomas of Excellence are likely to be fewer in the Tanning than in the Dyeing Trade.

ferent in character from the rest : one being mechanical, another chemical, and a third artistic. Each of them may in a large concern have its Apprentices, its Workmen of different degrees of cleverness, and its special Foreman; the whole of the Departments being under the collective supervision of the Heads of the Firm, or of their Manager. In an Industry thus organized, there might be no positions exactly equivalent to a Journeymanship or a Mastership; but there would generally be no great practical difficulty in determining for most of the well defined Departments, certain standards of attainments respectively deserving a Minor and a Major Certificate; nor would there be, under the proposed plan of publishing explicit Examination Syllabuses, as well as of providing appropriate and special Text Books, any ambiguity as to the standard which any given certificate is intended to mark, or any uncertainty as to its real value. If however the question of practicability is not likely to stop us, we must all the more bear in mind the question of expediency. Attention has already been drawn to the particular care which must be exercised in avoiding to raise prejudice by indiscriminate zeal. Many Industries are split up into small Handicrafts, or are composed of a string of factory operations, each of which is neither likely nor intended to have any intrinsic meaning for the individual engaged. His pair of hands admirably perform their task *mechanically*, and he would neither work better, nor please his employer so well, if he tried to be scientific.

Passing to the other extremity of the scale, we find on the contrary the Diploma of Excellence particularly at home among large and complex Industries, being specially designed to recompense that comprehensive and thorough knowledge which without necessarily involving cleverness at every manual performance, gives an ability to supervise, and a right to command.

The system of division of occupations is perhaps pushed farthest in some of the mechanical Manufactures, but even among the chemical Industries towards which our attention naturally directs itself first, on account of their calling out for Science more loudly than the others, examples are not wanting of establishments in which many distinct occupations collectively constitute a single Industry. Among the considerations which an inspection of them suggests, are the following:—

1.—Rough manual labour requiring no special technical training, does not come within the compass of our Analysis. We may however comfort ourselves with the reflection that the proposed System will benefit even the common labourer, by including the Rudiments of the Science of Daily Life in his Primary Schooling, and by multiplying facilities for their development in all parts of the country.

2.—All persons preparing themselves for positions in manufacturing establishments, corresponding approximately to the position of a Journeyman in a normal Working Trade, will, if the Industry in question should be sufficiently important and well-defined, be offered the opportunity of Special Studies, Examinations and Certificates; and if not, they will be directed to try for appropriate Subject Certificates.—An analogous rule will prevail respecting positions corresponding to the Major Degree.

3.—In most manufacturing Industries, the higher the position, the more comprehensive are the attributes. Hence not unfrequently in a manufacture where the studies of the Journeymen or Workmen have nothing worthy of attention, there may be Foremen of Departments deserving of a Major Certificate, and *à fortiori* the Director of the whole establishment may be expected to possess a range of knowledge well worthy of a Diploma of Excellence.

4.—In manufactures in which Chemistry is the leading principle, that Science should be specially studied even by those employees whose department is not a chemical one. This will add to the zeal and *esprit de corps* which should prevail throughout the establishment, and above all, will facilitate mutual kind offices, such as temporary help in cases of illness, of holidays, &c.[1]

5.—In many cases the nature of a Manufacture will suggest pushing the study of certain *parts* of Chemistry to a Grade beyond the rest. Thus a Brewer, Distiller, or Vinegar Manufacturer, will give a preferential lift to his Organic Chemistry, or simply to the Chemistry of the Fermentations. A Glass or Porcelain Manufacturer will on the contrary push forward his Inorganic Chemistry.

6.—Attention has, in the introductory portion of this Chapter, been drawn to the fact that the magnitude of the paraphernalia employed in most of the great manufacturing Industries, need not be considered an impediment to their being made the subject of academic instruction, any more than the inconvenient or noxious nature of the processes involved; inasmuch as even the lowest grade of Students admitted to the technical

---

[1] The amount of Chemistry here adverted to will generally be that contained in the Grade of G. S. F. next above the Grade of which the respective Employees will be expected to study the whole range. What is here said of Chemistry would of course apply more or less to any Science or Art constituting the leading principle of any particular Industry.

Courses will be expected to have had the advantage of a practical Apprenticeship, or of some equivalent, and will have no difficulty in understanding Models or Diagrams of Objects more or less analogous to familiar ones.

7.—Hygienic considerations should always have their full share of attention in every rational system of Technical Training, and this attention should be the more anxiously bestowed on Chemical Manufacturers, because their dangers are often of an insidious character, hurtful to the ignorant, but easily avoidable through scientific knowledge. Thus even in the Brewer's Trade, which on the whole is by no means an unhealthy or dangerous one, mischief sometimes occurs through inhalation of the Carbonic Acid Gas engendered in the process of Fermentation. Of course even the dangers most obvious to the scientific eye, must not be omitted on that account in addressing those unaccustomed to make a scientific use of their understanding. Continuing to borrow examples from the Brewer's Trade, we see accidents happen in moving ponderous barrels, which a trained mind would foresee, and cleverness in the application of the Mechanical Powers might easily prevent. As for the occurrence of Brewer's Men falling into Vats of almost boiling Wort, it must be ascribed to the callousness produced by the constant presence of danger, but this callousness itself is precisely an evil which it will be an essential point of Technical Training to counteract, by inducing the Artisan to cultivate habits of thoughtful caution, and making him always have his wits about him.[1] Speaking of Hygiene in

---

(1) There are a few branches of Manufacture, such as those of Gunpowder, Gun Cotton, Nitro-Glycerine, Dynamite and the like, respecting which remarks of this kind could not be too emphatically expressed; but

connection with a Brewery, one is strongly reminded of the controversial question whether or not, considered from a Hygienic point of view, a Brewery should exist at all; but controversial questions should be avoided as much as reasonably possible in an Institution that aspires to be national. Its influence should be fully exerted to prevent adulteration and fraud from having any part in the production of Beer, Spirits or any other article, but as long as that production is one of the great Industries of the Country, our task will be to teach rather than to discuss.

THE BREWER.

Taking then the Brewery without any thoughts of the Beer-shop, we may derive from it very convenient illustrations of some of the points which we have been considering.—Thus to give an actual example of the variety of departments in a single business, we find in a certain provincial Town of the East of England, the Staff of a Brewing Establishment numbering about 24 hands, divided as follows :—

a.—Management and Business Department.
b.—Malting Department.
c.—Brewing Department.
d.—Engine Department.
e.—Cooperage Department.
f.—Conveyance Department.
g.—Supernumeraries or Men-of-all-work.

though unequivocally belonging to the Chemical Industries, they are not likely to claim much special attention from the Central Technical University, the Establishments devoted to them being limited in number, and conducted on the plan of subdivided labour to such an extent, that scientific responsibility rests only with a very few. Of course it weighs on these the more heavily.—For further remarks on the subject of Industrial Hygiene or rather of Industrial Pathology, see the conclusion of Section 6 of this Chapter.

Among these the most important department as regards actual products, is that marked *c.*—which is entrusted to a BREWER, who may be considered as the pivot of the whole establishment, and is little interfered with.[1] Towards the success of this Brewing Department are concentrated not only the regular functions of each of the others, but also what may be called their extra-functional assistance. Thus the Engine Driver and Stoker, besides attending to their Engine, look after certain things which it is the duty of the said Engine to perform, the head Cooper attends to certain matters besides actual Cooperage, and indeed the stiffness of attributes which elsewhere is so frequent a cause of friction, is in some measure merged by each Member of this little Community, in one common purpose,—the production of Beer that may be creditable to the Brewery; corroborating what has been said above of the desirableness of promoting solidarity of interest through solidarity of knowledge.

It is however obvious that to diffuse knowledge throughout such an Establishment is no easy task, and that if we can find means of overcoming the educational difficulties presented by its complex organization, those same means will be likely to help us over many similar obstacles, not only in Breweries more or less differently constituted, but in other kinds of Factories.

To determine what should be required of a Brewer, Manager of a Brewery or Professor of the Brewing Art aspiring to a Diploma of Excellence, is the least difficult portion of our task, and it will suit our purpose to take first this highest Degree of Requirements, descending thence to the lower ones.

---

[1] In the present instance he has charge of and is assisted by two Pupils.

For a Diploma of Excellence.

GRADE 3 of General Scientific Foundation. The Second or Applied Division of this Range will include a tolerably comprehensive review of the amylaceous and saccharine articles which in various parts of the world are or have been commonly submitted to Fermentation, and of the Drinks thus produced, directly or by Distillation.

GRADE 4 of the part of Organic Chemistry treating of the Fermentations.

A sufficient knowledge of APPLIED MECHANICS in general, and of the STEAM ENGINE in particular, for judging of the construction, condition and capabilities of a Brewery Engine, as well as of the manner in which the Engine Driver performs his duties.

A thorough TECHNICAL KNOWLEDGE not only of the Plant, Materials, Processes, and business management of Malting, Brewing and allied Industries, as commonly carried on in this country, including all that relates to Excise Duties, but also of all resources and contrivances to which recourse may be had in a spirit of improvement, or for conquering temporary difficulties.

A thorough knowledge of the HYGIENE of the Brewery as regards both its inmates and the public, and an acquaintance with the commercial organization of the Liquor Traffic in general, and the regulations bearing thereon, together with such information of a historical, statistical, or economic nature, as may best raise the rather common-place subject of Beer, to a distinguished position in a cultivated mind.

For a Major Certificate.

GRADE 2 may suffice for the general range of Scientific Foundation, but the Student should master GRADE 3 of Chemistry, or at least of all parts involved in his Trade. He will be abundantly recompensed for the trouble bestowed on its advanced phraseology, by the advantages which symbolic notations present for understanding the mutations involved in the saccharine, alcoholic, and acetic Fermentations.

The TECHNICAL KNOWLEDGE, including an insight into the machinery used, will resemble that indicated above for a Diploma of Excellence, but be mostly limited to such Appliances, Materials, Processes and Products, and such matters of business management and Excise interference, as commonly occur in, or are connected with, the ordinary manufacture of Malt Liquors in this Country.—The HYGIENE of the Establishment will be included, together with a selection of the most important among the points mentioned above in connection with it.

For Subject Certificates.

Descending to the subordinate staff of the Brewery in question, we find little occasion for regular Curricula of Studies, but favourable scope for Certificates in appropriate Subjects. Besides the ordinary resource of purely scientific subjects, the technical one of " Malting" without Brewing might be picked out from the Junior Cyclopædia for Minor Candidates, and from the Advanced Cyclopædia for Major Candidates, and supported in the one case by a second, and in the other by a third Grade of Chemistry, with a sufficient knowledge of Mensuration to check any errors of the Exciseman in the measurement of the *Couches*.[1]

Far be it from our purpose to engender a *Cacoethes docendi*, or to endeavour to make Philosophers of Draymen, but at the same time it is satisfactory to find that the means at our disposal will as far as found expedient, enable us to supply educational doses for all intellectual temperaments and requirements : compound ones for those whose normal Pursuit involves a comprehensive range of attainments, simpler ones for those whose Occupation can be benefited by the study of one or more detached Subjects, and for all a wholesome infusion of General Scientific Knowledge.

MISCELLANEOUS CHEMICAL INDUSTRIES.

After applying the experience acquired in the Brewery to other establishments in which FERMENTATION is the leading principle, such as the Distillery, the Manufactory of British Wines, of Vinegar, &c., let us allow our mind to wander from the production of Acetic Acid to that of the Inorganic Acids, and we soon find ourselves in a crowd of scientific Manufactures, claiming for all employed in them a certain general chemical foundation, and for the upper members of their staff a considerable amount of special chemical knowledge. Several of them illustrate in a peculiarly striking manner, the power of modern Science, and are eminently calculated to encourage the spirit of studious emulation. Such are for instance the manufacture of Oxalic Acid from Sawdust, of common

---

[1] See also the Article on "Studies for Subject Certificates" at the commencement of this Chapter.

Soda, of Sal Ammoniac, of Alum, of Ultramarine and of Alizarine. Fraught with encouragement are also the Palm Oil Products, and those distilled from Bog Head Coal, and above all the numberless products, as varied as they are important, yielded by the common Coal Tar of the Gas Works; from the fulminating Picrate that will blow up a Man of War, to the choice Dye that is the charm of a Lady's Dress, or the fragrant mock-essence that perhaps perfumes her Handkerchief. There is something unmistakeably pregnant with discovery, profit, and renown, in these strange mutations of Organic Matter, which would seem miracles if the perfection of speculative analysis had not appointed for each phenomenon its rationale, and marked beforehand for each new product its proper place. Then again in the Sugar Refinery, how instructive are the Vacuum Process and the more recent Hydro-Extractor which so neatly illustrate the joint action of various departments of Science, and how worthy of note is the substitution of inodorous means of purification, for the offensively impure ones formerly used, showing as it does the Manufacturer in the very satisfactory light of one who thinks of his neighbours as well as of himself.

Uniting the commercial with the scientific element, the mixed Trade of the COLOUR and VARNISH MAKER may be named as one of those in which a good knowledge of Chemistry should be allied with practical notions of Economic Botany, and Commercial Geography, in order that the strange ignorance which prevails as to the origin and nature of many common Drugs, may be gradually dispelled, and a correct nomenclature may be established, permanently attaching standard names to standard types; so as not only to put an end to the present ridiculous blunders, but in some measure to prevent fraudulent substitution from happening under colour of a mistake.(1)

GLASS MANUFACTURE is essentially chemical, but in its higher departments it allies with that character a considerable amount of Artistic Design and Decorative Taste. By a few easy transitions, coloured glass leads us to Enamel, and Enamel to Porcelain, the manufacture of which blends Mineralogy, Chemistry, Mechanics, and Art, in one of the happiest combinations to be found in the whole range of the Fictile Industries.—Descending from Porcelain and China to Earthenware, and down to the lower grades of Pottery, we find at last Science and Art merged almost entirely in a mere mechanical labour.

Outlines for Summaries.

If we add to the intrinsic divergencies of Chemical Industries, the diversity arising from their various scales and forms of organization from the small Business to the potent Company, it becomes obvious that the endeavour to mark out for them special Curricula, would be as tedious as it would be premature. It will suffice for the present to add to the general indications already supplied in speaking of the Brewery, a suggestive Formulary of an elastic character for normal Candidates, whilst those belonging to Category C are referred to the article on "Studies for Subject Certificates" at the beginning of this Chapter.

---

(1) As a specimen of a common misnomer, may be mentioned the term *Gums* used in speaking of Resins. Thus one constantly hears of Gum Copal, Gum Dammar, &c., and the confusion has reached the extent of calling GAMBOGE *Gum-booge*.

Minor Examination.

(It has been explained that in many Industries there will be no Examination of this Degree.)

GRADE 1 COURSE of G. S. F.

GRADE 2 COURSE. Inorganic or Organic Chemistry, or both, according to the nature of the Industry in question.—Any other department of the Elementary Division of the Course that may be similarly required.—Any portion or portions of the Applied Division of the Course that may serve to give a general view of the Industry, or a preparatory insight into its rationale.

HANDBOOK. The Junior Technology of the Trade in question. The Introductory Portion to include as usual any Preparatory Knowledge required to be in a state of special adaptation.

MATHEMATICAL STUDIES. In the case of any Chemical Industry involving Mensuration, Advanced Arithmetic, or any other branch of Mathematical Attainments, there will be little difficulty in determining the grade of proficiency required, and assigning the appropriate Text Books.

ART STUDIES. The attainments to be derived from this source will for the most part resolve themselves into the following alternatives:—Either the Workman will only require that stage of Art which will enable him to understand and work from a Drawing or Model,—or he will require distinct artistic ability, so that he will be reckoned as an Art Workman, and be thus relieved of the necessity of knowing the same amount of Science as his companions in other Departments of the Factory.

In certain departments of manufacture, as for instance, in the Glass and Ceramic Industries, an amount of manual dexterity is involved equivalent to Art Workmanship, and which it will be best to consider as a distinct Handicraft.

Major Examination.

(Approximately corresponding in Industries conducted on the Factory System to the requirements of a First Rate Workman, Leading Man, or Foreman.)

GRADE 2 COURSE of G. S. F.

GRADE 3. Chemistry and any other department of " Elementary or Applied Knowledge " specially required.

MANUAL. Advanced Technology of the Industry in question, including any Introductory Matter required.—Hygienic Advice to be freely given as regards processes unhealthy for those employed, or tending to the annoyance of the neighbourhood.

MATHEMATICAL STUDIES. Requirements in this direction will more frequently occur in the present than in the Minor Degree, but can be met with equal ease.

ART STUDIES. The above remarks concerning Minor Students are to some extent here applicable. As for the Artistic Taste which in the Ceramic and other kindred Industries may be required by many

Supervisors of work who need not themselves possess artistic abilities, it will, according to circumstances, be acquired either in the Art Classes or in the Technical Ateliers.

What has been said above concerning manual ability, acquires additional strength in this Advanced Degree.

Superior Examination.

GRADE 3 of G. S. F.

GRADE 4 of Inorganic or Organic Chemistry, or of both; as also of any other branch of preparatory Science specially required.

MANUAL (duly supplemented). The complete Technology of the Industry in question, embracing in a comprehensive manner everything that may tend to make a Chemical Manufacture commercially successful and nationally creditable.

MATHEMATICAL STUDIES. Any attainments that may possibly be required in this direction will be indicated in the respective Examination Syllabuses.

ART STUDIES. In those few Chemical Industries in which Art is concerned, the Directors of large establishments, without being artists themselves, should possess that training of the mind and eye which may enable them to appreciate Artists, and to see that all things susceptible of it, be carried out artistically.

SECTION 3.—TRADES CONNECTED WITH THE CONSTRUCTION, DECORATION, AND FURNISHING OF DWELLINGS.

THE BUILDER.

Having illustrated in some measure by means of a few convenient Types, the intended mode of assigning to the various Trades their respective educational requirements, we will now proceed, as explained at p. 269, to review seriatim the leading Industrial Occupations which minister to the necessities and comforts of Daily Life, taking first those without which there would be no House to live in.

Architecture has two constituent Elements, the *Artistic* and the *Economic* Element, which ought to harmonize, but which are too often made to seem antagonistic, one being developed at the expense of the other. Looking at what is done abroad, one feels fully entitled to hope that if not both, at least the Artistic Element will ere long be accommodated with a regular and complete system of Instruction of a national character, notwithstanding the adverse influence of the Members of the Profession, who might see in any Institution for that purpose, whether in connection with South Kensington or otherwise, an encroachment on the advantages which they derive from their Apprenticed Pupils. As regards the Economic Element through which a Building is made to suit the purpose for which it is required, and an equitable compromise is effected between convenience and space, pretensions and means, it is less likely that these utilitarian considerations should receive due educational attention, unless they be taken up by the proposed Central Technical University, where accordingly Instruction in Economic Architecture will be abundantly provided in the Museum, in the Class Room, and through the Editorial Department. Everything will be done in order that this intellectual *pabulum*, which is equally required by the Architect and the Builder, may prove equally palatable to both, so that being ostensibly provided for the latter, it may also attract the former.

There are few Occupations through which a more open path leads the intelligent and persevering Workman to a high position in the social scale, or in which Knowledge more readily effaces the barrier between a Trade and a Profession, than that of the BUILDER, which taken in its full comprehensiveness, unites an extraordinary variety of interesting departments of study. This, coupled with the consideration of the vast number of subordinate hands it employs, and the tangible importance of its results, sufficiently explains the attention paid to it in continental countries, where either distinct Colleges, or distinct Courses of Instruction, are provided for its Students. Assuming that the latter measure will be preferred in this country, and proceeding accordingly to ascertain what Instruction should be provided, we find that as regards General Scientific Foundation, the Trade of the Builder does not, like the Industries we have hitherto dealt with, require a predominant knowledge of any particular part, but rather a substantial knowledge of the whole range; inasmuch as the Technical Knowledge which it involves, is characterized by a comprehensiveness which few Trades surpass. It embraces if not necessarily manual cleverness, at least ability to command and supervise in such matters as the following:—

The attainments described further on as within the compass of the Stone Mason.

All that relates to the different varieties of Bricks and Tiles, including materials, manufacture, and modes of use for various purposes.

A sound theoretical and practical knowledge of Mortars, Cements, Stuccoes, and the like.

A good acquaintance with Roofing and Paving Materials.

A thorough knowledge of the business of the House Carpenter, including the materials he uses, and the best way of using them.

A considerable insight into the Trades of the Painter and Decorator, Glazier, Plumber, and Sanitary Engineer, General Ironmonger, and in short into everything for making an abode healthy, comfortable, and if required, elegant.

To all this Technical Knowledge must be added proficiency in the working out of Quantities and Estimates, and the like, and though we may put down Drawing at a very moderate standard in order to keep the Builder from interfering with the Architect, we find before us on summing up, such a total as to lead us to the following conclusions:—

There is no scope for a Minor Certificate.

The Candidate for a Major Certificate must add to a broad and sound groundwork of Preparatory Knowledge, the comprehensive range of Technical Knowledge indicated above, which must be supplied by the Official Manual.[1]

---

[1] Such portions as are identical with or merely an abridgment of what is required by the Stone Mason, the Bricklayer, the Carpenter, &c., need

His Certificate will have a significancy above that of the Major Certificate of any single one of the several Trades embraced in it, though at the same time the knowledge of them will be rather mental and supervising than manual and self-sufficient.

Hygienic considerations should be fully gone into, not only as regards the health of those for whose use Dwellings are erected, but also as regards the safety of the Men employed in erecting them.

The Diploma of Excellence should denote not only earnest studies, but also good natural abilities, and prove a distinction equally honourable and remunerative. As regards Preparatory Knowledge the increase above the Major Degree will indeed be less notable than in most Industries, unless a tendency to amalgamate the attributes of the Architect with those of the Builder, should induce a special development of artistic attainments; but the expansion of the Technical Knowledge will be very considerable. Firstly, it will embrace more effectively the various Trades blended for a common purpose. Secondly, it will not only add stately mansions to the ordinary dwellings for Town and Country, but will also include the construction of Churches and public Buildings of every description. Thirdly, it will search other countries for time-honoured types and new resources. Fourthly, it will discuss for the benefit of the less wealthy classes, the means of combining cheapness with comfort and durability, and scan for general enlightenment the union of prosaic convenience with artistic aspirations.

probably not be repeated under the heading "Builder" in the edition of the Advanced Cyclopædia printed as a set of volumes for reference, but this information must be given in full, and exactly as the Builder wants it, in the Manual printed separately for his use, and which is to be the Text Book for his Technical Studies (p. 242).

Summary.

Major Examination.

GRADE 2 COURSE of G. S. F.

GRADE 3. Statics and Dynamics. — Inorganic Chemistry.—The Departments of Applied Knowledge bearing on Dwellings Improvement.[1]

MATHEMATICAL STUDIES. Mensuration, &c., sufficient for working out estimates and the like.[2]

ART CLASS. Stage 3 of the appropriate style of Drawing.

MANUAL. The Advanced Technology of the Trade, well leavened with considerations of Health, Comfort, and Economy.[3]

Superior Examination.

GRADE 3 of G. S. F. will supply most of the Elementary and Applied Science required. Additions will much depend on the Books available.

MATHEMATICAL STUDIES. It should be shown by the Examination Syllabus that the Studies assigned

---

(1) See in Chapter III., Section 2, the Synopsis of the late Twickenham Economic Museum.—The Subject of Dwellings Improvement considered from an economic and sanitary point of view, and in special reference to the requirements of the Industrial Population, is one of those on which a methodical and well-digested Book of Instruction on a Grade 3 level is very desirable. Till this can be forthcoming, selections from the best existing publications must be pointed to in detail by the Syllabus of Reference.

(2) Some recent work like Nesbit's 'Treatise on Practical Mensuration,' or parts thereof, might serve as Text Book.

(3) The construction of Dwellings, and other ordinary work, will be chiefly had in view. It will have to be decided to what extent other more difficult or more aspiring branches of the Trade should be included here, or reserved for the Superior Degree.

to Major Candidates are to be considerably extended without rising much in difficulty.

ART STUDIES. The Artistic Attainments will considerably depend on the degree of approximation to the functions of the Architect.

MANUAL (abundantly supplemented). The full Technology of the Trade, thoroughly scientific, and including on a scale much more expanded than that of the Major Degree, its more arduous and aspiring undertakings.

THE MASON.(1)

Having furnished our ideal type of a Builder with every attainment that may fit him for command, let us endeavour to devise Workmen worthy of him.— Though Stone is far less employed than Brick for ordinary building purposes in this part of England, it will suit us to give it the precedence, following in this as well as in most of the subsequent departments, nearly the same order as that which I found convenient in organizing my Economic Museum, where Building Designs constituted the first, and Building Materials the Second Class.—I may here remark that in getting together my lithological Series, I soon was struck with the ignorance which prevailed among those who were constantly handling or dealing in these materials, as to the locality which supplied them, and as regards foreign Marbles, even as to the country whence they came. To dispel ignorance of this kind where it pre-

---

(1) Under this title are here included the ordinary Stone Mason, the Stone-carver, and the Marble Mason, whether employed for constructive or mortuary decoration, or for any other purposes not appertaining to high Sculptural Art.

vails, will of course be one of the aims of the proposed Institution, and its Museum replete with instructional series, fully labelled, will of itself do much, whilst it will help the technical Professors to do more.[1] As for the geological history of Petreous Materials, there is some use in knowing that certain Stones are of igneous, others of aqueous origin, and that among the latter many will for a simple reason, stand the weather best when the layers of deposit of which they are formed, are placed horizontally; but as for those more speculative portions of Geological Science which are apt to lead young men's minds beyond their depth, some discretion must be used in teaching them.

A comparatively small amount of Chemistry will be sufficient for enabling the Mason to understand an explanation of the chemical differences in the materials used by him, and also to take in the advice that may be given him as to Mortars and Cements. As regards Physics, he should pay particular attention to those applications of the Mechanical Powers, the intelligent use of which in moving such ponderous masses as those he has to deal with, may make to him a difference of life or limb.

An attainment which decidedly claims attention from the Mason is the measurement of Solids or Cubic measurement, simplified as much as possible in the Minor Studies, but rising into geometrical problems in the higher ones.

[1] One of the first results of instruction thus organized, will be a rational and fixed nomenclature. In the mean time I venture to use the plural "STONES" for designating collectively all kinds of Stone or "PETREOUS MATERIALS" used in the construction or decoration of Buildings.—As MARBLES are strictly speaking Carbonate of Lime, it would be convenient to have a separate word for designating Granite, Porphyry, Serpentine, and other Stones susceptible of a high polish, and used for ornamental purposes in lieu of Marble, with which they are too often confounded.

In this as in other Trades in which manual skill has a partly mechanical and partly artistic tendency, the Candidate for a Minor Certificate should be able, not only to work in the appropriate material a simple ornament from a Drawing, but also to copy that Drawing in free hand.

The Major Candidate will be expected to display both with chisel and pencil, much more advanced abilities; though artistic origination may still be dispensed with.—[1]As for a Diploma of Excellence, I do not at present see how it can be merited without stepping into the province of the Sculptor.

Summary.

Minor Examination.

GRADE 1 COURSE of G. S. F.
GRADE 2 COURSE. The portion of Physics relating to the Mechanical Powers.—The portion of Inorganic Chemistry bearing on Stones and Cements.—The Department of Applied Knowledge (Second Division of the Course) relating to Stones and Cements, unless it should be found preferable to include this Subject in the Introductory Portion of the Technical Knowledge.
MATHEMATICAL STUDIES. Grade 2 of Mensuration.
ART CLASS. Stage 1, or if possible Stage 2, of Drawing.
HANDBOOK. To embrace what a good Workman may fairly be expected to know concerning the ordinary

---

[1] Of course every encouragement will be given to promising young Students to try their best at Design and Modelling.

Appliances, Materials, and Processes. To include precautions for avoidance of Accident or Injury.

———

Manual Skill to be tested by some simple and appropriate piece of Carving.

———

Major Examination.

The G. S. F. and Preparatory Attainments to be advanced one Grade.

MANUAL. The Technical Knowledge to be advanced and expanded in the same proportion as the Preparatory Knowledge, and to include introductory notions of the Geology of the most useful Petreous Formations, and the leading particulars of QUARRYING, as well as a good knowledge of home and foreign Materials.

———

Sculptural Abilities to include all ordinary ornamental work.

THE BRICK AND TILE MAKER.

It has been repeatedly shown how essential it is for the success of the proposed national scheme of Technical Instruction, that it should embrace in its comprehensive grasp the whole Industrial Community, in order that every part of the educational structure may be co-ordinate with the rest, and that when located in its proper place it may support and be supported. At the same time it has been proved to be equally essential that the most cautious discrimination should be used in selecting for the beginning of actual operations, those among the multitude of Industrial Pursuits, and those only which urge undeniable claims to attention, and hold out unquestionable chances of success. Our selection will grow and ramify as our efforts become appreciated, our resources improve, and our steps, at first guided by forethought, acquire the aid of experience.

Such are the considerations which present themselves to the mind on arriving at a Trade like that of Brickmaking, in which we see a multitude of hands working with a minimum exercise of brains. Accordingly for the present it may be best not to recommend for the rank and file of the Brickfield, anything beyond an Examination and Certificate in Grade I of General Scientific Foundation, which being under a different name the same thing as the Science of Daily Life, is equally calculated to minister to their physical well-being, and to promote their technical advancement. On the one hand it professes to teach just what everybody has occasion to know, in a way that everybody is competent to understand, and may therefore without fear of wrong or ridicule be prescribed even for the most mechanically moving Labourer. On the other hand as Bricks happen to be things that every one in this Country ought to know something about, the rationale of their manufacture is clearly though simply explained in the Applied Division even of the Grade 1 Course. Any inclination to study the Grade 2 Course, or merely its Chemistry should be encouraged, but the only Trade Certificate which seems likely to be required, is one of the Major Degree for the range of attainments that should be possessed by a Foreman, or other person having the management of large Brick and Tile Works.

No apology is necessary for uniting Tiles with Bricks, but I am aware of the difficulty there is in clearly separating Tiles from Pottery, into the nature of which they gradually pass by refinement of material, glazing, and colouring. Generally speaking the Ceramic Arts are distinguished from Brick and Tile Making, by the use of what is known as the Potter's Wheel, instead of a Mould; but many exceptions occur, and we find ourselves in one of those numberless technical dilemmas, in which the mutual overlapping of kindred Industries, must be dealt with by making the Students of each, acquainted with all that stands not far beyond the border.

The following is a rough sketch of what might be comprised in the Curriculum in question :—

Major Examination.

GRADE 2 COURSE of G. S. F.

GRADE 3 of Inorganic Chemistry.

MANUAL, to comprise a sound knowledge of the various available kinds of Clay and Brick Earth, with their preparation and mixtures, as also of all usual machinery and appliances, processes and products, including a, as regards Bricks, not only all ordinary kinds, but also those peculiar in shape, colour, hardness, fireproof quality, etc. ; b, as regards Tiles, all usual kinds and qualities, with a few that might claim to be Earthenware.

THE BRICKLAYER, PLASTERER, &c.

From the prominence given to matters of Household Economy in the selection of the subjects composing the Second or Applied Division of the General Scientific

Foundation, results a certain advantage to Artisans whose occupation lies in that line. Thus the Bricklayer will find even in Grade 1 of that Foundation, which we may assume to have been initiated in his School years under the name of the Science of Daily Life, and to have been perfected during his Apprenticeship, simple yet sound notions concerning the rationale of Mortars and Cements, and nearly all other things on which his Trade has a Scientific bearing; and these notions will be very satisfactorily completed for practical purposes, if he acquires Grade 2. Accordingly as the technical manipulation of Bricks and Mortar is so much more a matter for Apprenticeship than for College Instruction, I would suggest offering no Trade Certificate, but that Bricklayers be recommended to try, as far as conveniently possible, for Certificates in the following Subjects.

JOURNEYMEN :—
    In Grade 1 of G. S. F.
    „   „   2 of Physics.
    „   „   „ Inorganic Chemistry.
    „   „   „ Mensuration.
MASTERS, to take each of these Subjects a Grade higher.[1]

It is obvious that the suggestions here made will more or less apply in the case of other Artisans whose pursuit may be benefited by Scientific or Mathematical Knowledge, but does not require a regular Course of Technical Studies.

[1] The Bricklayer's Labourer, who among other duties requiring no inconsiderable amount of intelligence, prepares the Mortar, raises the Scaffolding, &c., will constantly feel the benefit of a Grade 1 Foundation, and may not unfrequently be induced to strive for more.

WOOD-WORKERS IN GENERAL.

Workers in Wood, will no less than Workers in Bricks and Mortar, derive constant advantage from a sound and early acquired acquaintance with the Science of Daily Life. Those whose simple labour is more a matter of practice than of study, such as Timber-fellers, Sawyers, and the like, will find in a Grade 1 of G. S. F. just the sort of Elementary and Applied Knowledge they require, whilst for the Young Carpenter, Joiner, or Cabinet Maker, that Knowledge will admirably serve as a stepping-stone to higher things. After learning at School the rudiments of Physics, and making good use of the Mechanical Powers during their Apprenticeship, they will readily take in any further instalment of Physical Science, and as regards Chemistry, Glue, Stains, Varnishes and Preservatives from Dry Rot will be quite within their compass. Even as regards Botany they will already have some notion of the difference between Endogens and Exogens, and of the leading phenomena in the growth of the latter connected with the nature of the Wood produced; and they will be competent to feel an intelligent curiosity about all that most concerns them in the secrets of Forestry, or the incidents of the Timber Trade.[1]

There is much to be learnt in common by the various denominations of Workers in Wood, but much also which they have to learn divergently. The House Carpenter deals with less sorts and larger masses of Wood than most of the others; his calculations of

---

(1) The advantage of assuring to every one a good acquaintance with the origin and history of the Materials with which he has to deal, has been duly recognized in organizing the Studies at that excellent establishment the Engineering College at Cooper's Hill.

quantities require a good arithmetical foundation, and there is scope for scientific thought as well as practical training in the art of providing support for every pressure, and resistance for every thrust, and of effecting a maximum of strength with a minimum of materials.

As we pass from Building Carpentry to Joinery, Cabinet-making and Upholstery, we find a ponderous and powerful construction replaced by accuracy of joint and elegance of design, and quantity, by variety of Woods employed; to say nothing of the introduction of other materials which creep in till Wood is in some departments, as in Bedsteads and the like, almost superseded. Some of the wood-using Trades are contracted into a narrow specialty; as for instance, that of the Stair and Handrail Maker, or of the Chair Maker, or again that of the Cooper, whose all-absorbing thought is to keep liquids in. That of the Boat-builder is on the contrary to keep liquids out, and this specialty is far from being a narrow one. It is part of a vast range of important Industries appertaining to NAVAL ARCHITECTURE, and concerning which it may be best to say nothing till we see what development is given to the Nautical College so admirably located at Greenwich.

Considerable discrimination must be exercised in selecting the Industrial Pursuits for which Trade Examinations and Certificates are to be appointed at the beginning, and as regards various Handicrafts of which the uncertain outlines melt as it were into each other, it may be best to wait before inscribing them in Category B, till we ascertain in what proportions the respective Candidates are likely to make their appearance, and how they may best be grouped for economy of Teaching Power.—There can be no doubt as to the eligibility of such Trades as those of the House Carpenter, the Joiner, and the Cabinet-maker, but it will not be so easy to determine to what extent they should be united or separated in the Technical Teaching, and consequently in the Handbooks and Manuals. Under these circumstances the following suggestive Summary of Studies prepared with a view to them, is necessarily of a somewhat vague character. No Examination for a Diploma of Excellence is indicated, because the rising Carpenter would probably aspire to become a Builder, and the rising Cabinet-maker to become an Upholsterer.

Summary.

Minor Examination.

GRADE 1 COURSE of G. S. F.

GRADE 2 COURSE. Mechanical Physics, as far as and including the Subject of Wheel Carriages (see p. 174). — The part of Botany relating to the growth of Exogens.—The Department of the Second Division of the Course relating to the different sorts of Wood and their uses. It might be well further to include in the case of the House Carpenter, all the departments relating to Dwellings, and in the case of the Cabinet Maker, other appropriate selections.

MATHEMATICAL STUDIES. Grade 2 of Mensuration.

ART CLASS. Stage 2 of Drawing.

HANDBOOK. Junior Technology of the particular branch of Woodworking in view. In each case to include a scientific review of the appropriate Workshop, with its appliances and tools, and of all ordinary materials and operations; the item of common accidents, with means for avoidance or relief, not being forgotten.

Manual ability to be appropriately tested.

Major Examination.

GRADE 2 COURSE of G. S. F.[1]

MATHEMATICAL STUDIES. Grade 3 of Mensuration. In some branches Geometry should be specially attended to.

ART CLASS. Stage 3 of Mechanical and Freehand Drawing.

MANUAL. Practical considerations will determine to what extent the several Wood-working Industries should in their advanced Technology be united or taken separately. At all events the Scientific Review of Materials will not only enter fully into the origin, properties, and uses of the Woods ordinarily used in each, but extend to less known ones, whilst all other Departments will be treated with equal thoroughness.

Manual ability to be tested by some superior feat of Workmanship.

[1] By differences in the Examination Syllabus, the portions of the Elementary and Applied Divisions of this Course which have been pushed forward a Grade beyond the rest in the Minor Curriculum, may now receive a nearly equivalent promotion without being raised to Grade 3, of which the scientific phraseology might be considered too arduous by the class of Artisans likely to become Candidates. In due time the rise in the general standard of Education may allow of prescribing Grade 3 more freely.

THE PLUMBER, THE GLAZIER, THE PAINTER, AND ALLIED TRADES.

Continuing the construction of a Dwelling from the point to which it has been brought by the previous Trades, we find ourselves confronted with a new difficulty resulting from the frequent practice of uniting several more or less distinct Trades in the hands of one Tradesman. The selection of them is mainly regulated by considerations of practical convenience, and is of course more comprehensive in country places than in large Towns. Sometimes one and the same person is Slater, Zinc Worker, Plumber and Sanitary Mechanician, Glazier, House Painter, White Washer and Paperhanger. Elsewhere one sees a few of these Handicrafts omitted at the beginning of the list, and others tacked on to the end, such as Gilding and other branches of decoration. The best way of overcoming such difficulties will probably be as follows; though of course here, as in every other branch of the proposed operations, the plan to be adopted will to some extent depend on the actual and relative number of the Students in the various departments.

— Each regularly organized Trade or Handicraft, which by its importance, the amount of knowledge involved, and the distinct nature of that knowledge, deserves to have a separate autonomy, will have its own special Minor and Major Curriculum, Examination, and Trade Certificate.

— Where on the contrary two or more Handicrafts, frequently and conveniently carried on by the same Man, involve almost identical preparatory knowledge, and abilities of closely kindred nature, they will be published together in order to avoid unnecessary repetition of the information they require in common. Thus for instance there will be but one "Handbook" and one "Manual" for the House Painter, the Grainer, and the Writer, whereby with a little arrangement and a few remarks, the same review of pigments and other materials and appliances, may be made to serve for them all.

— The foregoing will not prevent the Instruction, Examinations, and Certificates in these branches of Industry, from being distributed in any manner found desirable, thanks to the plan of clearly defining the Curriculum of every Student, by means of the explanatory Syllabus published in his Examination Progamme.

— Where conveniently possible, a central or chief Trade will be selected, round which affiliated branches of Industry, may naturally group themselves, and Students who have obtained a Minor or Major Certificate in the one, will be encouraged to try for Supplementary Certificates in the others. Many Industries of small compass, but not undeserving of special notice, may be either taken up thus as SUPPLEMENTARY PURSUITS, or studied by themselves on a suitable foundation.

— Every encouragement will be given to Artisans to extend their knowledge to pursuits allied to, or anywise bearing on their own, and indeed some Curricula of high degree, as for instance the Superior Curriculum of the Decorator, will compass obligatorily a mixture of many Trades.

— Candidates for additional Certificates will not, as a rule, be re-examined in branches of knowledge in which their present Certificates prove that they have already passed a satisfactory examination.

A glance at a few of the Occupations appertaining to the present group, may tend to elucidate the foregoing suggestions.

a.—ROOFING. Tiling is now seldom an independent trade, being mainly in the province of the Bricklayer. Slating is sometimes in the same hands, or in those of the Plumber, but it is frequently also, and not undeservedly, taken as a specialty. Should it become customary to unite in the special occupation of a ROOFER (*Couvreur, Dachdecker,*) the covering of Buildings, especially with non-metallic materials, this might not inappropriately be considered as a Standard Trade, and educationally dealt with accordingly; not only because the variety of available materials is considerable, but also because the process of using them is often attended with danger, and the means for rendering it less so deserve special study.

b.—The title of PLUMBER and HYDRAULIST might conveniently unite among other things:—firstly, all kinds of metallic Roofing, Guttering, and the like; secondly, Appliances connected with Domestic Hydraulics, *i.e.* with Domestic Water supply and Sanitary conveniences.—A certain improvement on the usual knowledge of these matters is a general desideratum, and the Summary will be found to be framed with this view. Moreover suggestions for a more special and thoroughly scientific acquaintance with them under the title of " Domestic Engineering," will be found in the next sub-heading.

c.—HYDRAULIC, SANITARY, and DOMESTIC ENGINEERING. As Hygiene becomes better appreciated, and the attempts to promote its public and domestic applications are regularized and extended, we may hope not only to see its importance fully recognized in the Curricula of the Medical Faculties, but also to see Hygienic Professions and Trades thrive and prosper on the good they do through the evil they prevent.—HYDRAULIC ENGINEERING, or the Science and Art of Water supply, and SANITARY ENGINEERING, or the Science and Art of Drainage and Sewerage, are Professions that may be expected to be attended to at the Civil Engineering Colleges, but with margin enough for our co-operation, especially as far as they are connected with the physical well-being of the Working Classes. As for DOMESTIC ENGINEERING, it is a much more homely affair, and quite within our compass. It simply unites in a course of distinct studies, for the purpose of better casting on them the light of Science, those domestic applications of Hydraulics and Mechanics to purposes of Health and Comfort, which are too often undertaken without being properly understood. The Master Plumber who is led by natural affinities to include them in his business, will be induced to go into them with much more scientific thoroughness, by being offered the opportunity of a special Course of Studies, success in which will enable him to add to his Trade-title that of "DOMESTIC ENGINEER." Should his genius or his ambition prompt him to covet a professional standing in this useful line, he must strive to earn a Diploma of Excellence by a Course of Studies widely compassing the domain of Domestic Hygiene. They will embrace, besides domestic sewerage, sanitary appliances, water supply, and other necessities of life, many of its comforts and conveniences, such as Baths of various kinds, Bells and their substitutes, Alarms, Fire-escapes, &c.

d.—GLAZING. If under the title of Glazier were included the knowledge of all kinds and descriptions of Window Glass, plain, worked, or coloured, and the art of using the latter for decorative purposes, we should pause and consider; but Glazing as carried on in the ordinary way by the Painter and Plumber, is decidedly not above being taken as a Supplementary Pursuit.

e.—The HOUSE PAINTER. This important Trade is one on the scientific development of which no pains should be spared. It should have a Minor and a Major Certificate. The Candidate for either of these should be recommended to take up in a similar degree the Supplementary Pursuits of Distempering and Paperhanging, or those of Writing and Graining, in preference to including the Trade of the Plumber, &c. The Painter aspiring to higher honours should go in for them as Decorator. (See sub-heading *h*.)

f.—DISTEMPERING. This may be either *plain*, including Whitewashing and other simple coatings for neatness or preservation,[1] or *ornamental*, including stencilling or other Decorative Work. The kindred nature of the knowledge and abilities involved, suggest at once that Distempering should be included among the occupations taken up by the Painter; though it should be so far treated as a distinct Subject, as to allow of its being taken up by persons prevented by their health or otherwise from pursuing Oil-painting.

g.—PAPERHANGING is often taken up by the House-

---

(1) A rough kind of Whitewashing, as well as Limewashing, and applying Tar or the like, are mostly performed by Men acting as Labourers under Bricklayers or Carpenters.

painter, and not inappropriately, but the two Subjects of Distempering and Paperhanging, supported by a Second Grade of G. S. F. would form by themselves a very eligible means of livelihood.

h.—The DECORATOR. Whilst it is exceedingly desirable to promote the development of chastened and refined architectural taste in the designing of Dwellings, it is no less desirable to carry the same principles of true refinement into all the details of their internal decoration. Hence it is important that the Technical University should offer every facility and encouragement for the proper training : 1stly, of the Working Tradesman who adding a Major Certificate of proficiency in Decorative Work to that which marks his cleverness at painting in Oils and Distemper, will take the title of PAINTER AND DECORATOR ; 2ndly, of the PROFESSIONAL DECORATOR, whose ability to devise and superintend all kinds of Decorative Work, will be attested by a Diploma of Excellence. Like the Builder who, without being necessarily a good hand at all the Trades engaged in the construction of a Building, knows enough of them to be an intelligent supervisor of the whole, he may render greater service by his fertile origination and adaptation of Designs, his judicious selection of patterns and colours, his patronage of freehand work in preference to Stencilling, and his careful surveillance of all that is done, than by any manual participation therein. Mastering the history and vicissitudes of Decoration in various Countries, and soaring through the Æsthetics of Economic Art to a standpoint whence he can control the tendencies of Fashion, he may be able to exercise a powerful influence on the manifestations of wealth in the dwellings and furniture of the upper ten thousand, and cause genuine artistic taste to permeate downwards to nearly every social level.

i.—GILDING. Notwithstanding the importance of this operation, and the scientific interest attached to the extraordinary malleability of Gold, it is a matter of so much more practice than knowledge, and it is associated with so many different Handicrafts, that for the convenience of these, it had better be treated as a Detached Subject. This need not in the least prevent its being made an obligatory part of the Curriculum of the Trade described in the next Paragraph.

j.—The CARVER AND GILDER. The Trade thus named for convenience sake, and in conformity to the common practice, is intended to include the making and mounting of Frames of all kinds, for Looking-glasses, as well as for Pictures, Prints, &c. It must however be borne in mind that the Carving of Wood for Gilded Frames, has been to a great extent superseded by the moulding of Stuccos and similar compounds, and that the working of Wood Mouldings is largely done by machinery. In fact a Carver who could execute Artistic ornaments in Wood, would deserve a special Certificate as Art Workman. This Trade further includes, or might appropriately include, all ornamental work in Wood, Stucco, Carton-pierre, Papier-mâché, and other materials wholly or partially gilt, employed in the decoration of Dwellings.—It is evident that this Trade is one of those that should come under the cognizance of the professional Decorator, whose control over the selection and arrangement of Furniture, might also extend to the following Subject.

k.—PICTURE-HANGING. This Occupation involves more special knowledge and cleverness than is generally supposed, both as regards symmetry and artistic rules, and also as regards permanent safety from accident; and if treated as a Supplementary Pursuit it will form a convenient adjunct to the business either of the Carver and Gilder, or of the Upholsterer.

By the foregoing plan of separating educationally a number of OPTIONAL SUBJECTS from the various Trades with which they are commonly associated in practice, the confusion which met our view on first accosting this group, is considerably smoothed away, and the number of Standard Trades for which Curricula should be provided, is reduced to a few, of which the following are examples.

THE PLUMBER AND HYDRAULIST.

Minor Examination.

GRADE 1 COURSE of G. S. F.

GRADE 2 COURSE. Parts of Physics relating to the Mechanical Powers and to Hydrostatics. — The Chemistry of the Familiar Metals.—Department of the Applied Division of the Course relating to the extraction and elaboration of the Metals; also a selection from the Departments relating to the supply and qualities of Water.

MATHEMATICAL STUDIES. Grade 1, or if conveniently possible, Grade 2 of Mensuration.

ART CLASS. Stage 1 or 2 of Drawing.

HANDBOOK. The Junior Technology of the Trade, duly supported with Introductory Matter, but not extending beyond what every intelligent workman may be expected to perform and to understand.

Major Examination.

GRADE 2 COURSE of G. S. F., supported with any suitable Works on Mechanism and Hydraulics.
MATHEMATICAL STUDIES, Grade 2 or 3 of Mensuration.
ART STUDIES. Stage 2 or 3 of Drawing.
MANUAL. Advanced Technology of the Trade, with abundant introductory matter, bringing Science into clear connection with its Technical Applications.

DOMESTIC ENGINEER.

(There will be no Minor Examination.)

Major Examination.

GRADE 2 COURSE of G. S. F.
GRADE 3. Inorganic Chemistry.—Also Statics, Dynamics, Mechanism, and Hydraulics from appropriate Sources.
MATHEMATICAL STUDIES. Grade 2 or 3 of General Mathematics.
ART STUDIES. Stage 3 of Drawing.
MANUAL. Special Technology, of the character indicated in the Analysis.

Superior Examination.

GRADE 3 of G. S. F.

The other Preparatory Attainments need not be raised much above the level indicated for the Major Degree. A far greater expansion will occur in the knowledge to be derived from a full use of the Manual supported with abundant references to special Works.

HOUSE PAINTER.(1)

Minor Examination.

GRADE 1 COURSE of G. S. F.

GRADE 2 COURSE. The parts of Chemistry most essential for understanding the nature and properties of the Pigments, Vehicles, Varnishes, &c., commonly used. The selection to be as usual indicated by the Examination Syllabus.

MATHEMATICAL STUDIES. Grade 1 of Plane Mensuration.

ART CLASS. Stage 2 of Drawing.

HANDBOOK. The Junior Technology of the Trade. The Introductory Portion to include leading notions of Chromatics. — Also the construction of suitable scaffolding.

Major Examination.

GRADE 2 COURSE of G. S. F.

MATHEMATICAL STUDIES. Grade 2 or 3 of Mensuration.

ART CLASS. Stage 3 of Drawing.

MANUAL. The Advanced Technology of the Trade. The Introductory Subjects, including Chromatics, to be effectively treated, and Hygienic Advice to be freely given.

---

(1) The Preparatory Knowledge here indicated will suffice for several of the occupations which have been mentioned as suited for being annexed to the present Trade. Should additional attainments be involved in any of them, they would be indicated in the respective Syllabuses.

THE DECORATOR.

(No Minor Examination.)

Major Examination
For a Certificate as WORKING DECORATOR.

To pass this Examination the House Painter holding a Major Certificate must in addition to the attainments therein attested show the following :—A Grade 3 of the part of Physics relating to Light.—Considerable expertness at the calculation of intricate surfaces.—Stage 5 of Drawing in appropriate Styles.—A good knowledge of the laws of Taste in form, colour, and subject.—Some acquaintance with Standard Works on Decorative Art.

A satisfactory Manual Test will be *de rigueur*.

Superior Examination
For a Diploma as PROFESSED DECORATOR.

GRADE 3 of G. S. F.

The other Preparatory Attainments will not require to be on a much higher level than in the preceding curriculum, and Manual Skill will scarcely be so necessary; but the Technical Knowledge must include a thorough acquaintance with the laws of Artistic Taste, with the history of Decorative Art, its varied Styles and its manifold resources, and with the appropriate works, to which numerous references will be made in the Manual. For further particulars see the Analysis.

The Practical Test to bear chiefly on Origination.

METAL WORKERS IN GENERAL.(1)

It may seem at first very discouraging to have to meet the educational requirements of a host of Trades and Handicrafts in which so many Metals are applied in so many ways, to so many purposes; but on consideration we shall find the task greatly lightened by the conclusions which we have already arrived at in cases more or less analogous to those which now present themselves.

The first step towards Metal-working is to obtain the Metal that is to be worked, and here an easy distinction may at once be made between the uppermost, and the remainder of the Brains and Hands engaged. For reasons already explained, the advanced Studies required for holding superior positions in connection with Geology, Mining and the like, may with advantage remain entrusted to the Government School of Mines, being likely to be pursued chiefly by high-class Students capable of enjoying to its full extent the refined scientific culture centred at South Kensington. But below these Students set aside as belonging to Category A, are οἱ πολλοί of the Mining Population, having neither the means nor the previous mental training required for availing themselves of that high privilege, and who will be only too thankful if under the proposed system of National Technical Instruction, Science of equally sound principles but more condescending disposition, stoops to meet their wants and their comprehension. Further, we find that this

---

(1) Plumbers and Zincworkers have been already adverted to.— Machinery, Watchmaking, and the like, will be treated of in a subsequent group.—Many of the higher branches of Metal-working, such as those of the Gold- or Silver-smith, escape from our present notice into the domain of Art Workmanship.

circumstance equally applies to a crowd of Workers engaged in the manifold occupations which purify, mix, fashion, and turn out as useful or ornamental articles, the crude products of the Mines.—Let us take first those which belong to what it has been proposed to call Category C, engaged mainly in manual labour, with a certain amount of skill, but without a sufficient extent or variety of Technical Knowledge to deserve special Curricula and Trade Certificates.

Category C.

These occupations will all be benefited by that Common Sense which even Grade 1 of a General Scientific Foundation, purposely devised as the Science of Daily Life, cannot fail to develop. When it is remembered that the Second Division of the Course, especially of Grade 2, touches on the Technology of nearly all the Trades and Manufactures which minister to our daily wants, and particularly on the extraction and elaboration of the familiar metals, it becomes obvious that a Metal Worker who merely possesses such a foundation, will be a *Savant* as compared with a fellow worker who has never looked into Science at all.[1]

Category B.

It is unnecessary to repeat what has been said of the convenience of not only using the General Scientific Foundation as a stepping-stone to further attainments, but also in many instances of borrowing from a higher Grade of the same, certain portions of advanced in-

---

[1] For further indications see "Studies for Subject Certificates" at the beginning of this Chapter.

formation. This expedient will supply in most of the Metal-working Trades, nearly all the Preparatory Knowledge required. There are however some Metal-working Industries in which a Supplementary Knowledge of Metallurgy is demanded, and a few in which Mechanism must be understood. As for Artistic Training, the main point is as usual, that the respective Workmen should be able to do their appointed work correctly from drawn patterns, and the respective Masters to draw the Patterns required.

The present Group affords us an opportunity of illustrating a principle which should be constantly kept in view in devising Curricula and preparing Text Books for the various Industries; viz., that while the elevation and commercial success of each Industry, and the health and well-being of those engaged in it, should be primary considerations, nothing should be neglected which can confer a benefit on the community at large. There is a branch of Technical Knowledge generally treated rather lightly as an incidental appendage of the Trade of the Furnishing Ironmonger, but the cultivation or neglect of which by himself and his Manufacturer, makes a very great difference in the domestic comfort of his customers; namely the knowledge, theoretical and practical, of all that appertains to the production and domestic applications of Heat. This highly important branch of instruction will I imagine be entrusted under the name of THERMOLOGY, or some equivalent Title, to a special Professor for being illustrated in the University Museum by a collection of Models, Appliances, and Specimens, more complete than any yet existing, and for being duly taught to the Students concerned, both through the Cyclopædias and *ex cathedra*. In the higher Grades, this thermological instruction should embrace a comprehensive range of hygienic as well as economic questions; including among others the following points:—Common causes of deterioration of the air of a room;—unseen and therefore too little studied play of atmospheric currents in a room under the complex influences of natural or artificial warmth, of moisture, of the admixture of Carbonic Acid, of draughts, &c.;—contrivances and appliances for the special purpose of Ventilation;—relative advantages of the various open Grates, close Stoves, and other heating appliances, as regards convenience, economy, ventilation, and action on the Air when overheated;—rationale and cure of smoky chimneys;—comparative heating power of the various Fuels, including Gas, taken in connection with their price and relative convenience;—best construction of Grates, Stoves, and Ranges, for Cooking Purposes;—contrivances for a supply of Hot Water, either connected with the foregoing or devised for Laundry Purposes, and the like.[1]

Few points in Domestic Hygiene are more perplexing than the vexed question, "open Grates versus close Stoves," complicated as it is on either side by a multitude of ingenious devices, and rendered more perplexing by the conflicting experience of different countries, whilst to make matters worse, many clever Philosophers have found it more convenient to argue in favour of whatever they had a personal preference for,

---

[1] Among the Manuscripts alluded to elsewhere as being available for inspection at the Twickenham Economic Museum, are Papers relating to the leading types of Grates and Stoves, and to the modes of testing them. We may hope shortly to have the benefit of important additions to this branch of knowledge, in the form of an official Report by the Committee charged with the testing of Grates and Stoves at the International Exhibition in connection with the Society of Arts.

than to take the trouble of ascertaining whether this preference was founded on habit or on reason.—Notwithstanding the labours of the influential Committee appointed at the suggestion of Mr. Chadwick in 1866, and followed up in the ensuing year by a series of practical investigations at the Paris Universal Exhibition, much remains to be done which it will be peculiarly within the province of the National Technical University to undertake. Endowed with a spacious Museum, constituted on the principle of not merely exhibiting to the best advantage things voluntarily brought to it, but also of actively getting together all that may be required to complete each instructional Series, it will possess rare advantages for clearing up matters like these by practical investigations on a comprehensive scale; for turning at once to the best educational account the information thus elaborated, and for demonstrating that Industry guided by Science, may in serving its own interests, confer vast benefits on the community at large. It may safely be hoped that the progress made in the industrial line at the National University, will be duly responded to in the physiological line at the Medical Faculties;[1] for the study of Hygiene is beginning at last to find in this country that appreciation which has for some time been accorded to it on the Continent, and would at once make rapid strides, if admitted to the obligatory Curriculum of University Examinations.

Returning to the more direct purpose of our Analysis, and considering the subject of Thermology in relation to those who will have to work it into practical results, we take the example of a London Furnishing Iron-

---

(1) The appointment of Dr. Corfield to a special Chair of Hygiene at University College, is in this respect an encouraging omen.

monger in a large way of business, one of those who are at the same time rather inappropriately called Working Ironmongers because they undertake to do or get done all Work connected with Ironmongery. We find that he employs among other Workmen, one or two who are sent out whenever a Grate or Stove is to be fixed or altered, or anything else to be done in the heating line. These Men ought certainly to have a few sound notions of what they should do and avoid, such as would be afforded by even a moderate acquaintance with Thermology.—Then above them we see perhaps a Foreman who like the French *Fumiste* is especially competent to conquer a smoky chimney, and who is equally ready to act as consulting Engineer for a kitchen range or any other heating appliance. What the above Workmen ought to know something of in a *minor* degree, this special Foreman ought to possess in a *major* degree, and each will accordingly strive to add an appropriate Supplementary Certificate to the one marking his proficiency as Worker in Ironmongery.

The Diploma of Excellence with the title of PROFESSED THERMOLOGIST, will only be conferred on the well trained Candidate who masters an exhaustive Course of Studies.

The Professor of THERMOLOGY at the proposed University will probably be also Professor of PHOTOLOGY; that is to say of the knowledge or subject of LIGHT economically considered. This subject, which was included with that of the applications of Heat, in the labours of the South Kensington Committee, will often be taken up by the same Students, but arrangements will be made to allow of its being taken up separately by Lampists and Chandlers, not specially interested in

Thermology. The Superior Degree conferring the title of Professed Photologist will like that of the sister Course of Studies, be very expansive, including:—a knowledge of the best means for ascertaining the intensity and quality of any given Light;—a knowledge of all the most notable materials and appliances available for the production of Light, with their comparative merits as to brilliancy, quality, steadiness, safety, convenience, and economy;—a knowledge of the use of Reflectors, including those for transmitting Day-light to the interior of Buildings in close-built Towns;[1]— a knowledge of the effects of Light and Darkness on the Human Frame in general, and of the amount and quality of Light most congenial to the Visual Organs, according to personal temperament and occupation.

In the same way that some of the metal-working Trades plunge rather deeply into the subjects of Heat or Light, others go off at a tangent in other directions involving requirements which must equally be provided for in the Preparatory and Technical portions of their respective Curricula. Thus for instance there is much special chemical and photological Knowledge, some amount of Decorative Art, and a good deal of Technical Contrivance involved in the Gasfitting Trade.

A question arises in this Group which will frequently occur in future ones, namely as to the manner of dealing with Commercial Trades. In theory a Tradesman ought to know the whole history of every article he

---

[1] This is one of the many resources which the spread of intelligence through a scientific education is likely to enhance, and as an example of the results may be mentioned that it has been shown by experiment to present a remedy for the internal darkness which is reckoned the chief objection to the mode of construction of Urban Dwellings for the Working Classes known as the "Internal Corridor System."

sells, but in practice the number and diversity of Articles sold in a single shop, are such that all we can expect of the Dealer is that he should sufficiently possess the rationale of each, to be able to judge whether it be good and genuine of its kind, and to some extent whether it be deserving of preference over another article of a different make. Thus taking for example the Furnishing Ironmonger, it is impossible that he should know thoroughly the details of manufacture of the whole multitude of Articles stored for sale in his warehouse ; but he ought to be taught to recognize the quality of the materials of which they are made, trained to discuss intelligently the principle and contrivance of any new patent Lock, improved Bottle Jack, or Hydrostatic Coffee Machine, and instructed to recommend things with a substantial "reason why," and not simply because they are the "last out."

Another question is to what extent Trades should be amalgamated or kept apart, and what minor Occupations should be taught as Supplementary or Detached Subjects, so that it may be optional with the Students to associate them with Standard Trades, or to learn them as distinct means of livelihood. As examples of such Occupations may be named :—Nail-making, Wire Work, Bell-hanging, Copper-beating, Tinning, Electroplating, Lacquering and Bronzing, etc.—Nothing positive need be decided till the actual working of the proposed Institution brings out certain circumstances which it would be difficult to divine beforehand, but which it will be easy to conform to when become practically apparent. In the mean time appearances are in favour of the rule that many small Occupations which may conveniently be taken separately by Minor Students, should be united by Major Students.

Outlines for Summaries.[1]

Minor Examination.

GRADE 1 COURSE of G. S. F.

GRADE 2 COURSE. Inorganic Chemistry, and any other portion of the Elementary Division that may be specially required.—The Department of the Applied Division relating to the extraction and elaboration of Metals, and any other Departments that may have a bearing on the particular Handicraft in view.

MATHEMATICAL STUDIES. Grade 1 of Mensuration if required.

ART CLASS. Stage 1 of Drawing.

HANDBOOK. To include in its Introductory portion any Preparatory Knowledge of a Technical Character that may be required by the Trade in view, as for instance the description of the Horse's Foot in the case of the Journeyman Farrier.

Major Examination.

GRADE 2 COURSE of G. S. F.

GRADE 3. Inorganic Chemistry, supplemented with some suitable Work on Metallurgy. The several Syllabuses to indicate likewise any other branches of Preparatory Scientific Knowledge required, and the best Sources.[2]

MATHEMATICAL STUDIES. Mensuration if required.

---

(1) See also the analogous Suggestions at the end of the Article on "Miscellaneous Chemical Occupations." In respect of Students belonging to Category C, see the indications at the beginning of the present Article.

(2) As an example of such supplementary Studies, may be mentioned that the Master Scale and Weight Maker should be well up in the Decimal System.

ART CLASS. Stage 2 or 3 of Drawing.

MANUAL. Advanced Technology of the Trade in question, duly supported with Introductory Matter, and as usual not only of a higher educational level, but also of a much more comprehensive character than the Technology of the Handbook.

Thermology.

Supplementary Studies for a *Minor Certificate* to be obtained by a Journeyman House Smith, or Worker in Ironmongery, possessing a Minor Certificate in that capacity.

GRADE 2 COURSE. Part of Physics relating to Heat, and portions of Organic Chemistry required for understanding the formation, properties and products of Coal and other Fuels.—Departments of the Applied Division of the Course relating to Warming, Cooking, Ventilating, &c., unless superseded by Introductory Information on these matters inserted in the Handbook.

HANDBOOK. To include all that a Journeyman may fairly be expected to know concerning Thermology as described in the Analysis.

Supplementary Studies for a *Major Certificate* to be obtained by a Master House Smith, or the Foreman of a Working Ironmonger, possessing a Major Certificate in such capacity.

GRADE 3 of the parts of Physics and Organic Chemistry mentioned in the foregoing.

MANUAL. Advanced Course of Technology, well supported with Introductory Information of the nature above described, and affording a sound knowledge of Thermology as described in the Analysis.

Studies for a *Diploma of Excellence* as PROFESSED THERMOLOGIST.
With the exception of raising the whole range of G. S. F. to Grade 3 there will not be any essential difference from the foregoing as to Preparatory Knowledge. It is as Technical Knowledge, or introduction thereto, that will be treated the important range of subjects in view, and the Manual, supplemented by its Notes, Appendices, and numerous references, will form a very complete guide book of Thermology, theoretical and practical.

Photology.

Supplementary Studies for a Minor Certificate, a Major Certificate, and a Diploma of Excellence in this Subject, will embrace in a gradation analogous to that shown for Thermology, the questions indicated in the Analysis as belonging to the Sister Science, as well as the appropriate Preparatory Knowledge to be borrowed from Physics and Chemistry.

The union of the two branches of Practical Science will be encouraged, promoting also, in the higher degrees, an infusion of Artistic Taste for combining beauty with utility in the respective appliances.

THE UPHOLSTERER.

Properly speaking, the Compound Trade here in view, is more correctly expressed by the frequently used compound title of CABINET-MAKER and UPHOLSTERER. According to Worcester's Dictionary a "Cabinet Maker is one who makes articles of wooden furniture which require nice workmanship," whilst "An Upholsterer is one who furnishes houses with beds and furniture." As however the supplying of Furniture includes Cabinet-work much better than the latter name includes all the Joinery that is done in the Workshop of the Upholsterer, the latter Title will for brevity be used alone.

This comprehensive Trade will be found to be one of those which demand a Major Certificate and a Diploma of Excellence, but are not likely to require a Minor Certificate. In fact the Master in a regular business generally employs special Journeymen for the several departments, but should himself possess a good knowledge of them all. Besides the Joinery and Cabinet-work executed in his Workshop, he should understand the quality of other articles in Wood, Cane, and other materials, which it may better suit his purpose to procure from a special Artisan, say the Chair-maker. Though he does not as a rule manufacture his Iron and Brass goods, he must know something about them. Then the subject of Stuffing Materials is much more vast and interesting than is generally supposed, and as for Woven Fabrics, besides those which form the indispensable skin of Stuffed Goods, a great many more, including Curtains, Carpets and the like, should be thoroughly gone into by the Tradesman "who furnishes houses with beds and Furniture." If to all this we add the Artistic Element, that is to say Taste in Form, Colour, Symmetry, and Effect, and a moderate proficiency in Drawing, we have five chief branches of technical attainment, namely Wood-work, Metal-work, Stuffing Materials, Fabrics, and Artistic Talent, to say nothing of accessory attributes.[1]

Before proceeding to a suggestive Summary, it may be well to remark how much power this Trade possesses when directed by intelligent and conscientious minds, for introducing in the domestic habits of the People, not only improved Comfort, but also improved Taste. Nor is it too much to say as regards many of the dwellings of the poorer classes in particular, that between the shabby-fine, impure old furniture that is there, and the unpretentious, but neat and cleanly furniture that might be placed there at equal cost, under the guidance of scientific knowledge and hygienic principle, often lies the difference between Disease and Health.

Summary.

(No Minor Examination.)

Major Examination.

GRADE 2 COURSE of G. S. F. A large portion of the 2nd Division of this Course is so peculiarly suited for supplying the necessary Preparatory Scientific Knowledge, that probably nothing further will be required.

---

[1] In the Dictionary above quoted, we read that the term UPHOLSTERER is derived from UPHOLDER, which originally meant the bearing of Coffins, as well as the *undertaking* of Funeral rites in general. This office is still to a great extent performed by the Upholsterer, but need not be included in his Curriculum.

MATHEMATICAL STUDIES. Grade 2 of Mensuration. Advanced Arithmetic.

ART CLASS. Grade 3 of the appropriate Styles of Design, including Origination.

MANUAL. To include, besides a good practical acquaintance with the Materials used and the Operations performed in the various branches of the Trade, a sound knowledge of the laws of Artistic Taste applied to Articles of Domestic Furniture, with a view to uniting neatness, and where desired elegance, with the requirements of Health and Comfort.

Superior Examination.

The whole of the Preparatory Attainments to be raised one Grade, particularly including in the Artistic Training cleverness at originating and preparing Designs adapted to any required purpose, taking into account available space and means.

TECHNICAL TRAINING to include :—A complete knowledge of all materials and contrivances available for the appropriate Furnishing of Dwellings of various levels of pretension ; special attention being paid to the means for turning limited incomes to the best account.—A comprehensive knowledge of the various styles of Furniture and Decoration which prevail in various Countries, or have prevailed in former times.

Upholsterers who have obtained Diplomas of Excellence, will be encouraged to try also for Diplomas as Professed Decorators, which will show their ability to contract for all that relates to the embellishment as well as the furnishing of first-class Mansions.

SECTION 4.—TRADES CONNECTED WITH FABRICS AND CLOTHING.

This Group of Industrial Pursuits is intended to embrace the whole range of Class IV. of the late Twickenham Economic Museum (p. 129), including not only Textile Materials and Fabrics in the various stages up to actual use, but also Waterproof Articles and the substances used in their manufacture ; also Leather, Furs, and various accessories of Dress, useful or ornamental. It is another added to the many groups which have already proved that Technical Instruction cannot be dealt out to the various Industries in uniform portions, and that any scheme for dispensing it fairly, must as a first condition be based on a detailed review of the whole Technical Kingdom, and secondly be made as pliable in its various adaptations, as it must be firm in its main principles.

Reviewing the vast field of operations now before us, from the growth of Textile Materials to the ultimate use thereof, and from the coarsest Matting to the most delicate Tissue, we find comparatively few of those prominent and well-defined types of Working Trades, whose intellectual needs can conveniently be met by a regular Curriculum. Among these few, two of a chemical character, those of the DYER and the TANNER, were selected as the first examples in our analytical review. At the other extremity of this Group, where FABRICS are converted into APPAREL, we shall see the

Tailor, Hatter, Shoemaker, and a few others, working collectively at that conversion, in a manner to deserve our special attention, though it will not be easy to find a collective title for their Occupations.[1] The greater part of the Group however consists of Pursuits less easy to compass with outlines of special instruction.

First come what may be styled AGRICULTURAL INDUSTRIES, such as the growing of Hemp, Flax or Cotton, and the producing of Wool or Silk, to which may be added the supplying of Hides and Skins in whatever way obtained. All these will in due time claim, directly or indirectly, their share in the benefits of the proposed system of Technical Instruction; but it would be unwise to attempt to include them in its first operations.

Then come the Processes by which these Materials become Fabrics. They are usually conducted as large MANUFACTURING INDUSTRIES, on the plan of subdivided labour, and many hands directed by few minds, and afford opportunities for applying most of the expedients which have been recommended for similar cases. The legitimate ambition of the directing minds, will find a suitable recompense in Diplomas of Excellence granted for comprehensive knowledge, and commanding abilities. The Foreman of a Department will in most cases only aspire to a Major Certificate, but his knowledge should include the essential particulars of all the other Departments.—Descending to what may be called the *minor* level of factory employees in the line before us, we are reminded that assuming they can only be classed as belonging to Category C, they may be greatly benefited by adding to a Grade 1 of the whole G. S. F., a

---

[1] The French word "*confection*" might be very convenient, but would be liable to misinterpretation in English.

Grade 2 of Chemistry and also of that Department of "Applied Knowledge" which relates to Fabrics and Clothing.

This applies also to many branches of Manufacture carried on at home by more or less independent Artisans, as Stocking Weavers, Silk Weavers, and the like. The more advanced or more intelligent of these, might take Grade 2 for the whole of their G. S. F., supplementing it with some suitable book on Mechanism. They would also do well to learn Drawing.—Perhaps after a time it may be found worth while to include in the Advanced Cyclopædia, certain Manuals that may prove valuable sources of information for whole groups of Artisans; as for instance a Manual of Weaving in general, from which any Weaver in particular may be directed to omit the few portions which do not concern him, but to learn all those which do.[1] This knowledge will be supplemented from other sources. Thus the various Silk Weavers will borrow from the Manual of the Silk Mercer such information as the following :—

SILK ; its History, Production, and Trade.

BLEACHING AND DYEING applied to Silks.

SILK FABRICS most deserving of notice, reviewed under the united considerations of utility, beauty, and price.

It is obvious that a Workman engaged in some other department of the manipulation of Silk, may be substituted for a Weaver, or that a Weaver of some other Material may be substituted for a Weaver of Silk. In each case the principles of selection of the most appropriate SUBJECTS will be the same, though different the application.

(1) Hygienic hints for the special benefit of Weavers will of course be included; especially concerning eyesight.

After the Manufacturing Industries, a number of COMMERCIAL TRADES throw across our path fresh difficulties to be considered and overcome. Some of them are indeed the mere creatures of circumstance. Their assortments of Goods are so optional, and vary so much according to localities, whether populous or secluded, that it would be futile to attempt to fetter them with strict formularies of technical instruction. We have moreover to encounter the extreme laxity and confusion which prevail in Commercial Nomenclature in general, and in the nomenclature of Textile Fabrics in particular. If, by way of illustration, we submit to a searching review the articles included in the warehouse of the ordinary Woollen Draper, we find that a feeble attempt at regular classification is almost lost in a confusion of identical or nearly identical Fabrics bearing different names, and of identical Names given at different times or in different places to dissimilar Fabrics. In this as in many other branches of Technology, the Museum of the proposed Technical University, will if elaborated with proper resources under the direction of competent Professors, each responsible for his speciality, form a Standard of Reference of inestimable importance. Its enlightening influence on the *consuming* as well as on the *producing* portion of the community, will not only bring order where ignorance and confusion reign, but may usefully control and guide the prolific manufacturing ingenuity which now runs wild for the simple reason that *Novelty* attracts purchasers irrespective of *Improvement*.

What however chiefly concerns our present argument, is the usefulness of the Museum as a means of Technical Training. Being specially contrived for imparting the right kind of knowledge in the most convenient form, its several Departments will equally correspond to the instruction given in the Technical Cyclopædias, and to that delivered *ex cathedra*, and the present is one of those Subjects in which a Commercial Student inspecting the Illustrations with his Text Book in hand, may prepare himself for an Examination, with very little help from a Teacher.

It must be remembered that, like the Furnishing Ironmonger whose case was described in the previous Section, the Tradesman who unites in his Shop the products of many distinct Industries more or less connected with Apparel, cannot be expected to go deeply into the origin and manufacture of every article; the essential point being that he should know enough concerning them to be able to judge of their quality and value, so as to keep himself, and consequently his customers, from being taken in. Thus the GENERAL OUTFITTER, especially in the Country, may possibly find after considering the many resources and devices which the proposed educational system places before him, that his best plan is to earn a Major Certificate in Woollen Drapery, Linen Drapery, or Haberdashery, and then only to try for a Minor Certificate in each of the Supplementary Trades represented in his Shop.[1]

Passing now to the class of Tradesmen engaged as a rule in protecting and adorning the human frame, and by way of exception in pinching and disfiguring it, we are struck by the prominent and representative position occupied by the TAILOR, the HATTER, and the SHOEMAKER. Looking closer we notice that many other Trades hook themselves on to these by links that must

---

[1] In passing Supplementary Examinations of this description, a Candidate need not be examined again in Attainments in which his Major Certificate shows that he has already satisfactorily passed.

not be ignored. To the Gentlemen's Tailor, who himself presents several varieties, must be associated the Habit-maker, and Ladies' Tailor, the last mentioned doing to a certain extent identical work with the Dressmaker; and thus step by step we might go through a number of occupations entrusted to male or female hands. Something similar occurs in the Trade of the Hatter, with its feminine offsets in the form of Bonnets, Caps, and other Head-dresses.

The making of Shoes, like Glove-making, has to a certain extent become Factory Work in certain localities; cutting or clicking, closing or joining, welting and soleing, are often done by different hands, and binding is mostly done by females; yet the ordinary Country BOOT and SHOE MAKER who includes Ladies' Shoes, affords a very instructive example of a substantial and well-defined Manual Trade well deserving of an instructional lift, and it will suit our convenience to take him first, making his requirements a point of comparison for others.

Science in general can introduce to the Shoemaker new resources, and teach him to make an improved use of old ones. Hygienic Science in particular will show him on the one hand, how to avoid the injurious effects to which some parts of his occupation occasionally subject a Workman constantly employed at them, and on the other hand, it will teach him how he may benefit very materially his customers, especially those of the feminine gender, by carefully studying the anatomy and physiology of the Human Foot, together with the effects of pressure in producing Corns and other Ailments, and by effecting accordingly the best possible compromise between real and conventional beauty, Nature and Fashion. There is scarcely so comprehensive a range of subjects as to afford scope for a Diploma of Excellence, but the stages of Knowledge by which the Apprentice should become Journeyman, and the latter become Master, are sufficiently clear to warrant the following outlines of a Suggestive Summary :—

Minor Examination.

GRADE 1 COURSE of G. S. F.
GRADE 2 COURSE. Portions of Chemistry bearing on the Tanning and Currying processes. — Departments of the 2nd Division of the Course relating to the production and uses of Leather, as also of India-rubber and Gutta Percha.
ART CLASS. Stage 1 of Drawing.
HANDBOOK. The Introductory Portion to include leading notions of the structure and functions of the Human Foot. The Technology to conclude with brief advice on the Hygiene of the Trade.

The test of Manual Efficiency may be resorted to more easily than in most of the Trades hitherto reviewed.

Major Examination.

GRADE 2 COURSE of G. S. F. This Preparation will probably suffice; the chief difference from the foregoing Curriculum, being in the Technical Studies.
ART CLASS. Stage 2 of Drawing.
MANUAL. The Introduction to include a thorough insight into the Anatomy and Physiology of the Foot, and the normal and abnormal growth of the Skin. The Technology to include the His-

torical and Ethnological Aspects of the Trade, or in other words, a Review of the various Foot Protectors or Contractors used at various times, and in various Countries; attention being particularly directed to the latest and best contrivances devised in this country or abroad for supplying the Poorer Classes with durable waterproof Boots and Shoes, uniting cheapness with other essential conditions. The Hygiene of the Trade to be fully gone into as regards the health and comfort both of the Workman and of the Customer.

As a man may be a good Master Shoemaker without being a first-rate hand at fine stitching or the like, the standard for these more manual parts of the business need not be higher for him than for the Journeyman.

Taking the foregoing indications as a starting point, there will be no real difficulty in preparing for other Clothing Trades, analogous schemes of study; provided care be taken, whilst making every necessary allowance for the requirements of each, to cut away everything that may fairly be dispensed with. Thus for instance as regards the HATTER, there will be no necessity for him to have as thorough an insight into the contents of the Head, as the Shoemaker should have into the internal structure of the Foot. Fortunately the human Cranium cannot, like the more tractable nether extremities, be squeezed into conformity with any given Last which Fashion may prescribe, and consequently the transgressions of the Hatter against the laws of Personal Hygiene are few as compared with those of the Shoemaker. It is rather in the point of view of artistic taste that his influence is open to improvement. Besides striving to improve the formal stiffness of full-dress gentility by gradually relieving it of the incubus of the *cylinder*, he may exercise his taste more freely in fancy Hats. In these, and still more in the article of Caps for Men and Boys, he will do well to make himself acquainted with all that local and technical experience have at different times devised for special circumstances of climate, occupation, &c. Having thus arrived at the best contrivance he can devise under the influence of physical and hygienic knowledge, he will finish up by introducing to the utmost extent, the artistic element of TASTE in Outline and Colour. It is obvious however that these are matters for the Master, and not for the Journeyman, and that the knowledge of Materials should be much more extended in the case of the former than of the latter; whereas on the contrary in actual work of hand, there need scarcely be any difference between them.

The TAILORING TRADE presents a stronger case to the same effect. The Master, if he is in a large and rather stylish way of business, must not be expected by any means to be so clever at the use of needle and thimble as some of his Journeymen; for his superiority lies not in the fingers, but in the mind and eye.[1] His knowledge of the various kinds of Materials that pass through his hands must be widely extended, unless it be found preferable to make a distinction between the simple Master Tailor, and the Tailor and Draper who would be expected to earn his double Certificate by proportionate knowledge.—Whether or not the tailor-

[1] Cases of this description naturally suggest the idea that the established Master who wishes to be certificated as deserving of his position need not pass a Minor Examination before his Major one. It would be however undesirable to make this a general rule.

ing Trade calls for a Superior Examination with its Diploma of Excellence, may remain for the present an open question.

We have seen all along the subject of Hygiene, as applied to Technical Industries, dividing itself into two questions: the Health of those employed, and the Health of the Customer. As regards the former, the Tailor's Workshop suggests some little advice, which however to be useful, must not be exaggerated. As regards the Health of the Customer,—first of all we must relieve the Tailor of much responsibility respecting Military outfits, and all manner of conventional costumes and liveries ordered by those who do not wear them, and worn by those who have no voice in the contract. Then as regards Gentlemen's Tailors in general, it must be confessed that anything their good sense might accomplish in the furtherance of rational or well-shaped garments, sinks into insignificance by the side of what might be done by the Sister Trade for promoting healthful comfort and artistic grace amongst its fair customers.

FEMININE OCCUPATIONS CONNECTED WITH DRESSMAKING, MILLINERY, &c.

Here we find ourselves for the first time called upon to consider a Trade, or rather a cluster of Trades or Occupations, that are or ought to be in female hands, and before attempting to assign to them their appropriate share in the educational benefits of the proposed Institution, we must scan the special difficulties that are to be overcome.

It has been stated in a previous Chapter, that a large proportion of those Industrial Pursuits which are most frequently assigned to the softer sex, involve a limited range of ideas, but that there are interesting exceptions to this rule. The Trades now specially before us, that is to say those which make up Garments and Adornments of various descriptions for female use, frequently exhibit the rule and the exception under one roof. As in the case of the analogous male Trades, we see large establishments where brain-work is reserved for the few, whilst hand-work more or less split into distinct departments, often becomes a mere manual performance, measured out in uniform stitches by a practised but unreasoning eye. We see little necessity for imparting any comprehensive Technical Knowledge to this needle-armed rank and file, but all the greater is the responsibility, and the need of intellectual Training, in the leaders of this patiently and submissively industrious host. In fact the number and diversity of the Articles called into requisition as Materials or Accessories in the endless work of female clothing and adornment, so vastly exceed even those which perplex the male Outfitter, that a Diploma of Excellence would be well deserved by any Lady Superintendent possessing anything like a knowledge of the whole range of producing Industries, and manufacturing processes involved. At the same time it is obvious that nothing beyond a carefully reduced and simplified Compendium of this information, can be proposed as a suitable educational standard for the Mistresses of the ordinary dressmaking and millinery Establishments or Shops in Town or Country. This Standard or Degree approximately coincides with that to which we have hitherto assigned a Major Certificate; whilst the Minor Degree suggests itself as about the right thing for the Young Women employed in a business of this description, in a capacity analogous to that of Artisans or Journeymen.⁽¹⁾ In the smaller establishments, and especially in the country ones, the various branches and stages of handicraft, instead of tending to split and become mechanical as in those on a large scale, show a tendency to coalesce; the social as well as mental interval between directing and working is diminished, and at last we find the two capacities merged in one person. There is nothing new in this. Many of the male occupations previously reviewed present similar gradations from the large establishment to the individual; but this does not interfere with the feasibility of appointing a Minor and a Major Certificate at fixed levels, each to become a goal for those below it, and a recompense for those who reach it.—In drawing however a comparison between the resources of the Females now before us, and those of the Males previously reviewed, we are led to the following considerations :—

— Let us assume that in due time some such system of early training in the Rudiments of Practical Science as the one suggested in a previous Chapter, will have been generally adopted in the Primary Schools of the People, and that the instruction of this kind required by Females, though leading in some respects to different applications, will have been recognized to be fundamentally the same as that required by Males. On this assumption we might feel entitled to expect that almost every Girl intended to be apprenticed to Dressmaking and Millinery, or to any other feminine Handicraft, should come from School with a good preliminary insight into that range of Knowledge which, considered

(1) If the word Artisan were always pronounced as it should be, with the accent on the first syllable, I would propose ARTISANNE for the female equivalent; but the ordinary practice of accentuating the last syllable of Artisan, might engender confusion.

in its direct and homely bearings, is the SCIENCE OF DAILY LIFE, and which considered in relation to Technical Studies, has been denominated GENERAL SCIENTIFIC FOUNDATION. Yet I fear that this expectation, however correct in theory, will be found practically to demand considerable allowance.

—Girls are useful at home when Boys would be in the way. From 9 years of age, and sometimes earlier, they can in a variety of ways, and especially by minding the babies, afford their Mothers valuable aid, and enable them to do remunerative work. The consequence is that Parents who, especially in towns, would be glad that their Boys should have a regular schooling up to at least 12 years of age, are inclined to stint and curtail that of the Girls.—For this and other reasons, we must consider the latter as likely to enter into their Apprenticeship with a very scanty scientific foundation; especially as the substitution of machine for hand in many departments of needle-work, lowers the age at which they can begin their technical career.

—The hindrances during Apprenticeship are not less serious than those which preceded it. It has been proposed as regards male Apprentices, that every possible facility should be afforded them for attending Science and Art Classes in the evening. For them this would be in every respect desirable; but a young Apprentice to Dressmaking would on the one hand be likely to find plenty of employment for her needle in evenings at home, and on the other hand the propriety of her going out at that time to join the Science or Art Classes, would in many cases be very questionable.

From the foregoing considerations it results that considerable allowance must be made for the difficulties which Females have to contend with, especially as regards the studies for Minor Certificates. It need not be so great as regards the studies for Major Certificates; nevertheless in presence of the overwhelming variety of the Materials used, care must be taken to reduce to a brief and simple form the account of their origin and manufacture. Here again we shall find that up to the Major Degree the battle of technical study is half won by mastering with the aid of the first, the second Division of the General Scientific Foundation; embracing as each Grade does in proportion to its power of grasp, all that is most essential or most interesting in the Technology of Daily Life.

Reverting to the Ladies SUPERINTENDENTS or DIRECTRESSES of large Dressmaking, Millinery, and Ladies' General Outfitting Establishments, it is evident that the management of a business of this description is, as before hinted, a task involving for its proper accomplishment such a depth and variety of Technological Knowledge, and so much Artistic Refinement, to say nothing of Hygiene, as to claim a Diploma of Excellence as professed LADIES' OUTFITTER, or EUCLAMYDIST. Positions of this kind are now often reserved for Gentlemen Directors; but without denying to these the opportunity of acquiring and getting recognized by a Diploma the scientific, artistic, and technical attainments required, every encouragement should be afforded to female competitors. By encouragement I mean a convenient arrangement of the Classes and Examinations, and an explicit understanding that Female Students are expected, and not any lowering of the high scientific character which should invariably attach to Diplomas of Excellence; for though we have seen that considerable allowance must be made for the difficulties which young Girls of the Working Class have to encounter in the pursuit of Knowledge, yet whenever Female Students have been in a position favouring the full development of their natural abilities, they have abundantly proved that they could cope with an amount of earnest study not only equal to anything required for the intelligent management of a large concern, but such as to make them the worthy depositories of that improving influence in the Councils of Fashion, the want of which has often been so lamentably apparent.[1]

Among the thousands of female hands at work for clothing and adorning their own sex, or for providing appropriate portions of the male wardrobe, a very large percentage are still employed under conditions alike adverse to the healthy development of body and of mind. They have been purposely omitted from the foregoing review, because the regeneration so praiseworthily begun by various philanthropic agencies, must be further advanced before the educational influence of the Central Technical University can be satisfactorily exercised in this direction. Even as regards the members of the needle-working community to whom the present article applies, our operations must for a time be regulated by extreme caution, and it is in a tentative sense that must be accepted the following suggestions.

Summary.

Minor Examination.

GRADE 1 COURSE of G. S. F.
ART CLASS. Stage 1 of Drawing.

---

[1] Should the movement find favour, there would be an excellent opening for the appointment of female Professors.

SECT. 4.] *DRESSMAKING AND MILLINERY.* 361

HANDBOOK. The Technology to include a review of the Appliances and Materials, sufficient for recognizing their quality, without going much into their origin and manufacture. More attention to be paid to the practical work of cutting out, making up, and adorning articles of apparel or head-dress, with the aid of appropriate patterns. The Hygiene of the Trade not to be overlooked.

Practical Tests here come in as a regular part of the Examination, and should include, besides plain needlework of all kinds, with marking and the like, the actual carrying out of the practical work described above. The time occupied to be taken into account.

Major Examination.

GRADE 1 COURSE of G. S. F.
GRADE 2 COURSE.—Department of the Applied Division relating to Fabrics and Clothing.
ART CLASS. Stage 2 of Drawing.
MANUAL. The Introductory Portion to give a good insight into the Laws of Taste applied to Dress. The Technical Knowledge to include a sound acquaintance with the nature and mode of production of the Appliances employed, such as Needles and Thread, and of the materials used, such as Fabrics, Trimmings, Fastenings, and Ornaments; so that the Student may be enabled to judge *en connaissance de cause* of their quality and relative advantages.—Also an acquaintance with

the chief types of Sewing Machines.—The Hygiene of the Trade to be fully gone into.

The Practical Tests, which will form an essential feature of the Examination, should show ability to take measure, and to devise and carry out any ordinary articles of Apparel or Head-dress, according to instructions, without patterns, without waste of materials, and with becoming "*chic.*" Also cleverness at various kinds of fancy as well as plain needlework.

Superior Examination.

GRADE 2 COURSE of G. S. F.
ART CLASS. Stage 3 or 4 of Drawing, chiefly applied to costume, and including origination.
MANUAL; to be so supplemented by Notes, Appendices, and References to, or Analyses of Special Works, as to give a considerable expansion, and at the same time a much more scientific tone to the Curriculum of the Major Degree, especially as regards the following purposes:—To establish on sound Hygienic, Economic, and Artistic principles, permanent rules of guidance in matters of Dress and Adornment, and also a correct appreciation of the respective qualities and capabilities of the various Materials, taken singly or appropriately assorted.—To make from the advantageous standpoint thus obtained, a critical review of the costumes of various times and countries.— To deduce from these data, full of suggestive

interest, a lively as well as rational guidance to rising talent, developing a spirit of origination fraught with expedients for all circumstances, and capable of adapting itself at will to the gorgeous, the fanciful, the retiring, and the plainly useful.— To combine a thorough knowledge of Hygienic requirements with the other qualifications necessary for the proper management of a large commercial establishment.

Actual skill at needlework to be less a consideration than the ability to appreciate and utilize the skill of others.

SECTION 5.—TRADES AND OCCUPATIONS CONNECTED WITH FOOD.

There is in the educational aspect of this group as compared with that of the foregoing, a considerable diminution of the Mechanical Element, which we have seen at work in the spinning and weaving of Fabrics, and the making up of Garments, and a vast increase of the Chemical Element, which there was almost confined to Dyeing, but which here assumes paramount importance, not only as the chief agent in numberless processes of production or preservation, but as supplying the analytical gauge by which the comparative value of various Food Articles is ascertained. Botany and Zoology come also into more extensive request, but not, as a rule, beyond the limits to which they are proposed to be included in the respective Grades of General Scientific Foundation.

A. *Industries yielding Raw Products.*

Much as we might delight in bestowing our attention on pursuits so replete with the brilliant applications of Technical Science as are many Agricultural Operations, they must, for reasons explained elsewhere, only be taken cognizance of so far as can be done without interfering with the beneficial working of existing Institutions. The same may be said of Pisciculture, which indeed, together with Fishing, Shooting, and other means of procuring Food from wild sources, should be included under the term Agriculture, or rather under some more comprehensive term devised for the purpose. A subject deeply deserving of the attention of the Council of the proposed University, is the abundant accession of Supplies derivable from a methodical and earnest study of Colonial Resources; but this is a subject which less concerns the general Body of Students, than the select Body of Young Professors to whom the proposed Institution, after freely imparting to them the most practical as well as the most advanced lessons of Technical Science, will entrust the double mission of conveying this Knowledge to the Colonies, and of sending to the Mother Country in return, new materials for the employment of successful industry, and for the diffusion of improved comforts among the People.

B. *Preparing and Preserving Industries.*

These are intermediary stages between the raw production of certain Food Articles, and the ultimate Culinary Process which fits them for pleasing the Palate, as well as for agreeing with the Stomach. A few of these Industries are almost purely mechanical as for instance the Trades of the Miller who grinds Corn into the convenient consistency, and sifts it into the desired quality, and of the Butcher who by mechanical means effects the vastly more than mechanical difference between a living Animal, and so many joints of Meat. It is true that he is often obliged to have recourse to chemical aid for keeping fresh what cannot be disposed of at once.—Other Industries have for their express object to keep good for a future time, and perhaps for a distant land, Food subject to decomposition. Some of these call in the aid of culinary knowledge, which will be discussed bye and bye, but as for the simple and non-culinary modes of preservation which consist merely in the abstraction of Water, or in the use of common Antiseptics, they are mostly carried on by a class of persons whom we could scarcely expect to present themselves as Students. If they did, we could do nothing better than advise them to master a Grade 1 Course of G. S. F. with a Grade 2 of Chemistry, and also of the Department of the Course relating to Food, where all the chief points concerning it are intended to be dealt with hygienically as well as chemically.

Recurring to the two distinguished preparers of Vegetable and Animal Food mentioned above, namely the MILLER and the BUTCHER, we find on looking first into the ordinary Flour Mill, that whether it be turned by Wind, Water, or Steam, the manual work done there requires little mental culture, except on the part of the Master, or if the establishment be a large one, on the part of the Foreman or Foremen, whereas the duties of these afford some scope for the utilization of scientific knowledge. From the study of Mechanical Physics and Mechanism they will derive ability to understand the machinery built up for them by the Millwright, and to judge whether it be on a good principle, and of good materials and workmanship. Through a study of the various kinds of grain commonly ground, and of their various modes and causes of deterioration, they will become better capable of obviating sources of mischief or loss. In short there will probably be scope for a special Curriculum somewhat to the following effect, without a corresponding Minor Degree; a case of not unfrequent occurrence.

Summary.

Major Examination.

GRADE 1 COURSE of G. S. F.

GRADE 2 COURSE. Part of Physics relating to the Mechanical Powers. (To be followed up by an appropriate work on Mechanism.)—The whole of Chemistry, or select parts to be indicated by the Examination Syllabus.—The Department of the Applied Division relating to Food.

MANUAL. To include :—The rationale of Water,—Wind,—and Steam Power, and the Machinery for applying each to Mill purposes, especially for the grinding of Corn, and the sifting of the results. —A knowledge of all kinds of Grist, and chief varieties of products, including the most notable foreign ones.—Causes of deterioration of the Grain, and means for obviating ill results.—The Hygiene of the Mill; prevention of Accidents, &c.

The BUTCHER'S TRADE presents a somewhat analogous case, inasmuch as it would probably be a failure to constitute special Studies for Journeymen Butchers, whom, like the Journeymen Millers, it will be best to consider as belonging to Category C; whereas on the contrary, there are few Tradesmen whom a good scientific training could more effectively help to raise their own position, and at the same time to benefit the public, than the Master Butcher. Among the various branches of knowledge indicated in the following Suggestive Summary, he should apply himself specially to some parts of Comparative Anatomy, and above all to those departments of veterinary art which bear upon the recognition during life or *post mortem*, of all diseases or imperfections by which the flesh of domestic animals is rendered unwholesome. Sound practical knowledge in this direction attested by a Major Certificate, and of course united with a conscientious determination to do honour to it, would find a sure reward in public estimation.

Summary.

Major Examination.

GRADE 1 COURSE of G. S. F.

GRADE 2 COURSE. Chemistry.—The part of Zoology comprising the Mammalia.—The Department of the Applied Division relating to Food.

MANUAL. To include :—An appropriate sketch of the comparative Anatomy of the Ruminants, and the portions of Veterinary Art bearing on the Diagnosis of their Diseases.—The best and most humane modes of killing.—The Hygiene of the Slaughter House and of the Shop, including the use of ice.—Curing and other means of preserving Meat applicable regularly or occasionally.—Utilization of by-products.—Review of the laws, regulations, and practices affecting the Meat-market which prevail in this country or abroad, and especially of the system of Public Slaughter Houses or *Abattoirs*.

Were it not that great care must be taken by the University Authorities not to lay themselves open to the reproach of patronizing "good things," I might suggest that the *Charcutier*, an improved Continental edition of the Pork Butcher, should be encouraged to acquire a full development in this Country; very great care being taken at the same time to ensure those sanitary measures which are more necessary in dealing with the flesh of Swine, than with any other.

The Fishmonger has some points in common with the Butcher, especially as regards the resources which Science supplies for retarding decomposition, and from the studies sketched out for the latter there would be no great difficulty in arriving at those suited for the former, substituting Fishes, Mollusca, and Crustacea, for Mammals,—a similar transition to Domestic and Wild Fowl, Venison, Game, &c., would transfer our advice from the Fishmonger to the Poulterer.

C. *Manufacturing Industries.*

It might be contended that several of the Occupations included in the foregoing Sub-groups, as for instance the preparing of Desiccated Milk, or the curing of Fish or Meat, are manufacturing Industries, but in their case the object is to obtain a preserved article approximating as nearly as possible to the fresh one, whereas the title of Manufacturing Industries, best applies to those which produce either a new article altogether, or an essentially improved one.—In most of them, Chemistry performs a prominent part; so much so that several, such as Sugar Refining, Brewing, Distilling, Vinegar Making, and the like, have been included at the beginning of these Analyses under the title "Miscellaneous Chemical Industries." From the indications there given, it will not be difficult to deduce suggestive Curricula for certain somewhat less Chemical branches of Food Manufacture, as for instance that of Cheese or Chocolate.

D. *Commercial Industries.*

Under this title are comprised those Industries less occupied with the preparation of Food Articles, than with their purveyance by wholesale or retail Trade.— Here we see before us the same difficulties that met us in the Fabric and Clothing Department, as to the endless variety of Goods, the fallacious irregularity of their nomenclature, and the dissimilarity of contents exhibited by Shops of the same title. Sometimes indeed these are confined to noted specialties, whilst at other times they embrace for the convenience of a rural district, the most heterogeneous assemblage of articles; or again we may see Grocers selling Wine and Spirits, and Publicans in revenge selling Tea.

Still however we may hope to make satisfactory progress by means of our usual expedients. It has been seen that Domestic Requirements in general are to be methodically canvassed in the Second or Applied Division of each Standard Course, bringing to bear on them in succession the Elementary Science acquired in the First Division, and that the subject of Food in particular, will be the object of an attention proportionate to its surpassing interest. Even in the Grade 1 Course, the rationale of Nutrition, and the varied nature of Food Resources, will be explained as far as it is possible to make such things plain and easy to beginners. In the Grade 2 Course the treatment of these matters will be earnest and comprehensive, and for Grade 3, till a special Course can be provided, such Books will be selected as may best give to Food Economy a thoroughly scientific development. An obvious means therefore suggests itself for the common benefit of all Trades dealing in Food Articles, and for that of the Grocer in particular, whose assortment embraces Nutrition in almost every form. Conformably to the plan indicated in previous Summaries, let the Grocer's Assistant build up on a Grade 1 of G. S. F. a Grade 2 of Chemistry, and then take from the Applied Division of the same Grade 2 Course, the whole of the Department relating to Food, or let the Master Grocer raise the whole of these Studies one Grade higher, and they will respectively have achieved much practical and even technical progress before diving into the technicalities of their Handbook and Manual. At the same time it may be noted that as the Preparation will be the same for a number of Food Trades, the desideratum of economizing Teaching Power by making the Students travel together in their Studies as far as they can, will be satisfactorily accomplished.—As regards the Technical Knowledge on the contrary, we must resort to division of Trades, in order to reduce to compact proportions the Studies to be assigned to each. Thus on the broad groundwork of a common foundation will rise distinctly side by side, Grocery proper relieved of every branch that can conveniently be detached, Chandlery including Soap, &c., Cheesemongery with its concomitant Bacon and Eggs, and a few other autonomies more or less susceptible of definition, concluding with the rather ambitious one of the Wine Merchant. The variety and convenience of the combinations that can be thus effected, is as obvious as the advantage accruing to each Trade in particular. An expansive Grocer for instance, following the plan suggested for a General Outfitter in a previous Section (p. 350), might add to his Major Certificate in Grocery proper, Minor Certificates in each of the collateral branches he might wish to embrace. Should he be capable, he would take so many Major Certificates; or gifted with a mind and means for still higher things, he would try for a Diploma of Excellence as GENERAL PURVEYOR, and unite the functions of the Wine and Spirit Merchant, the Grocer and Tea Dealer, the Oil and Italian Warehouseman, the Cheesemonger, the Wax, Tallow and Soap Chandler, &c. &c.[1]

There are certain purposes which should be carefully studied and never forgotten by persons engaged in any of the Food-purveying Trades, and which should be constantly kept in mind by the class of Superior Candidates now in view; viz. to avail themselves of the light shed by Chemistry and Physiology, on the rationale and resources of nutrition,—firstly, for assigning to each aliment or condiment its real action on the human system, 2ndly, for fostering cheap improvement in the Dietaries of the People, and 3rdly, for recognizing and peremptorily discarding every injurious adulteration and unworthy fraud.

[1] Some establishments are confined to a single article, or a small group of articles, with a view to acquiring within these narrow limits, increased variety and perfection. The measures by which judicious exertions of this kind might be encouraged and recompensed, must be left for future consideration.

E. *Pistrinary and Culinary Industries.*

The former will mainly consist of the Trades of the BAKER, PASTRY-COOK, and CONFECTIONER, whilst the latter will embrace COMMERCIAL and DOMESTIC COOKERY; all of them being employed in bringing Food Articles to that condition in which they are welcome to the palate, and supposed to be acceptable to the stomach.

The Trade of the Baker, whilst it decidedly is one of those regular Working Trades for whose benefit the proposed instructional system is more particularly designed, and whilst it is second to none in importance in popular estimation, is nevertheless not comprehensive in its range of necessary knowledge, and would become still less so if we could venture to strike out the subject of Adulterations.—Most Bakers are inclined to wander more or less over the vague frontier which separates their own domain from that of the Pastry-cook, but for purposes of study the distinction should be maintained. The same applies to the distinction between the Pastry-cook and the Confectioner.—The Summary for the Baker might be somewhat to the following effect.

Summary.

Minor Examination.

GRADE 1 COURSE of G. S. F.

GRADE 2 COURSE. A selection of the most necessary parts of Chemistry. The portion of the Applied Division relating to the Cereals.

HANDBOOK. To include a review of the most notable Breadstuffs, Yeast and its substitutes, the rationale of Panification, the construction and use of Ovens, the chief varieties of common and fancy Bread and Biscuits, the baking of Meat, &c., with a few moderate suggestions respecting the Hygiene of the Trade.

Major Examination.

GRADE 1 COURSE of G. S. F.[1]

GRADE 2 COURSE. The Portion of Chemical Physics relating to Heat.—The whole of Chemistry.—The entire Department of the Applied Division relating to Food.

MANUAL; to embrace much the same subjects as the Handbook, but more comprehensively and scientifically; adding a careful review of foreign materials, appliances and products, also an impartial consideration of existing English and foreign laws, regulations, and practices concerning Bread Supply, and compassing all reasonable suggestions for an improvement of the Hygiene of the Trade.[2]

There is scarcely occasion for a Diploma of Excellence, and if a Baker should wish to go beyond a Major Trade Certificate, in which desire he should meet with every encouragement, he might try for a Subject Certificate in Grade 3 of Chemistry, whereby he would greatly improve his insight into the rationale of his operations. Perhaps still more useful to him however

---

[1] Should an entire Grade 2 Course be found practicable, the Curriculum would be altered accordingly.

[2] I have heard of a Baker's Shop of which the shelves were painted with the well known poisonous bright green (Arsenite of Copper). It adhered to the hot loaves, and the consequences were very serious.

in a practical way, would be to add to his Baker's Certificate a supplementary one as Pastry-cook; in fact it may be said, that the most satisfactory way of settling the boundary question between two provinces like these, is to make oneself master of both.[1]

The Baker's Trade offers ample scope for honesty, but little for genius. Not so those of the Pastry-cook and of the Confectioner, who to all that is best calculated to gratify the palate, are at full liberty to add what may fascinate the eye; provided their designs are in good taste, and their colours are not poisons. Accordingly in altering the Baker's Summary to make it suit either of his confederates, the element of variety, borrowed or original, should be made prominent, and Drawing, with the rules of Artistic Taste, be made obligatory, especially in the studies of the Major Degree.

In speaking above of COMMERCIAL and DOMESTIC COOKERY, the distinction intended to be conveyed, and the educational motive thereof, were as follows:— There are Factory Kitchens where for instance canistered or potted Meats are cooked and seasoned wholesale, where Liebig's process reduces each pound of Beef to a representative spoonful, or where if we may listen to scientific gossip, things not pleasant to look at before they go into the cauldron, come out as exquisite sauces of a mysteriously dark colour. Now establishments of this description should be under the superintendence of a well-trained culinary Mind, to say nothing of that scientifically inventive genius which blessed the Prussian Army with its *Erbs-Wurst*. On the other hand the numerous *Marmitons* at work under the direction of this *Chef*, may individually be so trained to mono-

---

[1] See the article on the Painter, Plumber, &c., in Sect. 3.

tonous routine, as to be the ignorant instruments of knowledge, and the unthinking conductors of thought. Here then Cookery on a MANUFACTURING and COMMERCIAL footing, only invokes our scientific instruction for a limited number of Male Experts, for whom the general level would correspond to the Major Degree, and who might be taught together with Domestic Cooks of the same calibre.[1]

To define what is meant by DOMESTIC COOKERY, would be superfluous. As for the question whether any and what place might be assigned to it in the proposed system, this is a matter concerning which I feel that it is safest to offer at present only a few suggestive remarks.—Considering on the one hand the *real* importance which attaches in a hygienic and economic point of view to the improvement of culinary knowledge among the Working Classes, and on the other hand the *artificial* importance attached to culinary matters by many in the higher social levels who can afford, or rather think they can afford to make them a prominent object of their lives, we are led to consider whether the willing expenditure of the wealthy in the form of liberal stipends for *Cordons bleus*, may not be made subservient to the special training of Culinary Instructors for the Daughters of the People; and thus we arrive at the following conclusions :—

a.—Cookery for the Homes of the People.—The training of the Daughters of the People in appropriate Home Cookery should be regularly included in their Education; efficient means being adopted for that pur-

[1] There is no reason why Australian Stews should not, as far as Culinary Art is concerned, bear comparison with the same dishes done over our own kitchen fire.

pose, as suggested of late by various eminent advocates of an improved system of Female Education.[1]

b.—Plain Cooks for Service.—Arrangements should be made in all parts of the Country, in order that Girls intended for Service as PLAIN COOKS might be enabled to learn at a minimum of expense, good plain Cooking on the best principles, intelligently adapted for moderate Families. The Kitchens of appropriate Institutions to be, as far as practicable, utilized for the purpose.

c.— Professed Cooks. — Advanced Schools for the training of PROFESSED COOKS, Male and Female, should be established where needed, turning to account any facilities that might be offered by Club Houses, Hotels, and other establishments of a high class.

d.—Instructors in Plain Cookery.—Thorough-good Plain Cooks who could satisfactorily prove their practical qualifications by reliable testimonials or otherwise, would be admitted to a Course of special studies at the Central College or one of its Provincial Affiliates, consisting say of Grade 2 of General Scientific Foundation, with such special insight into the Hygiene and Economics of Food, and such instruction in the Art of Tuition, as might fit them for becoming direct Teachers of the Daughters of the People, or for instructing Schoolmistresses, Female Students at Normal Schools, and others, in culinary rationale and resources; such fitness to be attested after due examination, by a Certificate equivalent to a Major Trade Certificate.

---

[1] See Mrs. Grey's Lecture on this subject, delivered at the Society of Arts, and other papers from the same pen; also the letter by Miss L. Twining in the 'School Board Chronicle,' of Nov. 11, 1871, p. 407.

It is to literary exertions like these that we owe in great measure the movement originated at Grosvenor House.

e.—Instructors in Refined Cookery.—First-rate Professed Cooks, Male as well as Female, with qualifications similarly proved, would be admitted to an analogous Course of a higher Grade and more aspiring character, embracing whatever might best fit them for teaching the most refined Gastronomy on Hygienic principles, in the advanced Schools of Cookery above alluded to. There they would introduce any new delicacies with which an expanding knowledge of the resources of the World might enrich the epicurean *menu,* and benefit Commerce, without transgressing the Laws of Health. Such abilities of a high order, would be attested by a Diploma of Excellence.

f.—Fees.—As Culinary Professors bearing such Diplomas might be expected to be eagerly engaged and highly remunerated, their College Fees might be fixed at exceptionally high rates, in order that a proportionate reduction might be made on those demanded for training Plain Cooks as Popular Instructors; whereby special encouragement would be given to an educational department so obviously useful that scarcely anything besides expense could be urged against it.

Details have been avoided in sketching these measures, because nothing more than a suggestive sketch, easily alterable, can be safely attempted till it becomes apparent what amount of success is likely to crown the efforts now being made to render more rational the education of the People in general, and female education in particular.

SECTION 6.—INDUSTRIES CONNECTED WITH WARMING, LIGHTING, AND CLEANLINESS.— INDUSTRIAL HYGIENE.

Coal constitutes as it were a pivot on which a number of Industries may be said to hinge. The operation of obtaining it from the bowels of the Earth, has like other MINING OPERATIONS its *head* in Jermyn Street, but its swarthy *hands* are below the level of the instruction which there prevails, and consequently open to new means of improvement.—The COMMERCIAL INDUSTRIES by which Coal is brought from the pit's mouth to the Grate or Furnace, similarly call for a popular adaptation of select *excerpta* from the note books of the Geologist and of the Mining Engineer. It is however difficult to say whether in carrying out the proposed Educational System, it may be found expedient to establish special Courses of Technical Instruction on Fuel in general, and Coal in particular, including accordingly these topics in the Cyclopædias, for the benefit of Coal Merchants, Coal Agents, and others specially concerned. But even without this, convenient sources of information will be at hand. The craving for knowledge on this important subject is so universal, that it will necessarily form a conspicuous department in the Applied Division of each of the Standard Courses. In the Grade 1 Course all essential points will be briefly and familiarly explained; in the Grade 2 Course they will be more fully though still popularly gone into, and in the Books of Grade 3,

WARMING, LIGHTING, AND CLEANLINESS.

they will be treated of in due scientific phraseology. As explained in the previous Chapter, the Grade 1 and Grade 2 Courses are to be prepared and printed in such a form as to facilitate their being either used for self-instruction, employed for Class-teaching, or delivered in the form of Popular Lectures on the Binary System. Now the subject of Fuel and Combustion affords an excellent example of the good to be effected by means like these among thousands of the Industrial Community who are out of reach of regular College Instruction. There are numbers of retail Dealers in Coal, Coke, Charcoal and Firewood, induced by circumstances to couple this mixture with Potatoes and Greengrocery, or with some other business in a small way, and who are as unlikely to attempt regular studies, as they are on the contrary likely to be both entertained and benefited, by attending at the nearest Industrial Institute, a Course of Lectures of the exact kind they want. Indeed, scarcely any subject is better adapted for being illustrated with striking Diagrams (p. 252), and relieved with brilliant though cheap experiments, or better calculated to make both sellers and buyers among the lower ranks of the community, cordially agree that there really is something in Science.

Next to the subject of FUEL, which should include MEANS FOR IGNITION, from the Faggot to the Lucifer Match, would naturally come the subject of Grates, Stoves and the like, were it not that it has already been considered in treating of the Metal-working Trades.[1]. The Manufacture of GAS has been included

---

[1] Furnaces for smelting, for the generation of Steam, and for numberless other industrial purposes, may be studied in two ways :—either in the aggregate as a branch of Civil Engineering, in which case they escape our consideration for the present, or in a retail manner in the review of the Plant and Machinery appertaining to each pursuit.

in the "Miscellaneous Chemical Manufactures," as also that of various denominations of Paraffin and Composite Candles, and that of Soap.

The last mentioned article reminds us of Laundry-work, which forms a Pursuit uniting several departments of considerable importance, in which a little Science should underlie a good deal of manual practice. Like everything else relating to CLEANLINESS, it will occupy a prominent place in the Standard Courses. First will come the physiological reasons for Cleanliness in the Elementary Division, and then in the Applied Division the substances and processes employed. For a common Washerwoman Grade 1 of G. S. F. would by itself be a great gain. The Superintendent of a large Laundry should add to it the portions most interesting to her of Grade 2, both Elementary and Applied, and might raise on this foundation such a structure of Technical Knowledge, as would not fail to bring her establishment to the foremost rank.

An easy step brings us from means of Cleanliness to means of Disinfection, and another to matters connected with Drainage and Sewerage, the correction of Nuisances, and the prevention or mitigation of Zymotic Diseases; in short to everything that medical minds can suggest, and administrative or engineering abilities can carry out, for the benefit of the Public Health. Many points might be mentioned where openings for useful intervention are likely to occur, but it will be best not to discuss this subject till it can be seen to what extent the ground may be occupied by the expanded operations of existing educational Institutions.—Perfectly distinct from Public, and even from Domestic Hygiene, is Industrial Hygiene, which is not only quite within the compass of the proposed University, but should be most sedulously promoted; every opportunity being turned to account for enlightening the Students on any dangers or unhealthy influences that may attach to their respective callings, for pointing out to them what science, prudence, and presence of mind can do to avert or remedy evil chances, and for impressing on them the paramount duty of every man to keep his bodily and mental powers in the best condition he can.[1]

SECTION 7.—INDUSTRIES CONNECTED WITH CART AND COACH BUILDING, SADDLERY, &c.

The occupation of the Wheel-wright, which in its rural attributes reaches little above the level of a Cart or Van, takes a totally different character when under the prouder title of COACH BUILDING, it ministers to the demands of wealth and fashion; filling a London Factory with an endless variety of materials, and employing a proportionate number of kinds and stages of Handicraft. This division and subdivision of labour, too often baffles the progress of the Apprentice in establishments which, though first rate as regards the article produced, are not so eligible as regards general instruction as many country ones less refined but also less split up, where moreover a more communicative spirit is likely to prevail. It is among the less sub-

---

[1] In the early part of 1854, I induced the Council of the Society of Arts to appoint a Committee for investigating under the collective title of "INDUSTRIAL PATHOLOGY," the Accidents, Injuries, and Diseases, which attach to various Industrial Occupations, and I should be happy to show to persons interested, the "Synopsis" then prepared for me by my kind friend the eminent sanitarian, Mr. John Simon.

divided Factories that may probably be found circumscriptions of work suited for being compassed by regular Curricula of Studies. Evidently however this is a field of operations which should be reserved, with many similar ones, till we can actually confront our available instructional resources, with the yet unformed ranks of the Candidates for Knowledge and Reward. One thing is certain. The Diploma of Excellence reserved for the master mind capable of directing and supervising the whole complex business of a large fashionable Coach-building Establishment, should embrace an amount and variety of obligatory proficiencies which few Trades need surpass. Not the least among them should be a lively appreciation of the tasteful and elegant, chastened by Artistic Studies, and rendered more useful by cleverness at the appropriate style of Design.

The connection which naturally exists between the Coach-builder and the SADDLER, would be a sufficient excuse for inserting here a few words concerning the latter, even if it were not convenient to do so for the purpose of showing that in a commercial connection may lie an educational contrast. Many obvious reasons tend to make the Coach-builder's business an extensive concern; but that of the Saddler and Harness-maker, excepting the manufacture of certain accessories, is almost everywhere to be seen carried on within the limits of an ordinary Shop and its dependencies, and whilst the one by a sort of natural cleavage resolves itself into departments, and is worked on the factory plan, the other may be cited as a very fair sample of a normal Working Trade, claiming something like the following outlines for its Summary of Educational Requirements.

Minor Examination.

GRADE 1 COURSE of G. S. F.

ART CLASS. Stage 1 of Drawing.

HANDBOOK; to include as far as required by way of Introduction to the Junior Technology of the Trade, the structure of the Horse, and especially of those parts to which Saddles, Harness, Bits, &c., are to be adapted. Also Hides and Tanning considered from the Saddler's point of view.

Practical cleverness at stitching, and generally at all work which a Journeyman may have to perform, to be appropriately tested.

Major Examination.

GRADE 2 COURSE of G. S. F.

ART CLASS. Stage 2 of Drawing.

MANUAL. The Advanced Technology of the Trade to be supported by an expansion of the matters mentioned above, and by any analogous ones which may strengthen its intellectual groundwork. It should moreover include in addition to all ordinary Saddlery, and besides a brief review of military and foreign trappings, all reliable means for the avoidance of danger to those riding or driving, by improved construction of Saddles, Stirrups, Bits, Safety Reins, Kicking Straps, and other portions of equine accoutrement.

Cleverness at stitching and the like to be less obligatory than with the Minor Candidate.

There will probably be no Diploma of Excellence.

SECTION 8. — INDUSTRIES CONNECTED WITH THE MANUFACTURING AND WORKING OF MACHINERY.(1)

This Group stands to the Mechanical Powers and Mechanism, in a relation somewhat similar to that of the "Miscellaneous Chemical Industries" to Chemistry. The considerations there discussed in reference to Breweries and other large establishments, may dispense with now discussing similar ones respecting ENGINE FACTORIES and the like, and we at once proceed to a few suggestive indications of attainments likely to be required, by the majority of the Men employed there.

Summary.

Minor or Workman's Examination.(2)

GRADE 1 COURSE of G. S. F.

GRADE 2 COURSE. The whole of Mechanical Physics relating to Heat.—The portion of Chemistry relating to the Common Metals, Coal and other Fuels, Combustion, &c. Also selections from the second Division of the Course bearing on the same Subjects.

---

(1) For Watch and Clock Makers, see Section 9.
(2) This only refers to Workmen who, though their actual employment may be limited to a single branch of factory operations, aspire to possess, as it is very desirable that they should, a general knowledge of the whole.

SUPPLEMENTARY STUDIES (on the Grade 2 level):—
Mathematics, Statics, Dynamics, and Mechanism.
ART CLASS. Stage 2 of Drawing.
HANDBOOK. The Junior Technology, consisting chiefly of the usual review of Appliances, Materials, Processes and Precautions, will include the chief kinds of Steam Engines commonly made, whether for Stationary purposes, for Marine use, or for Railway Locomotion, &c. It may however suffice for this Minor Examination that each Student be well acquainted with the construction and working of his own kind of Engine.

Major Examination.

(For Foremen or First Class Workmen.)

GRADE 2 COURSE of G. S. F.
GRADE 3. The Branch of Chemical Physics relating to Heat.—Inorganic Chemistry.
SUPPLEMENTARY STUDIES (on the Grade 3 level):—
Metallurgy.—Mathematics. Statics, Dynamics and Mechanism.
ART CLASS. Stage 3 of Drawing.[1]
MANUAL. The Introductory Portion to include all Preparatory Knowledge requiring to be developed with a special view to its intended applications, such as the subjects of Fuel and Combustion, the production of Steam, its power and the measurement thereof. The Technology to include the usual detailed Review of Ateliers, Appliances, Materials and Processes; embracing not only all

---

[1] This is not intended to interfere with the functions of the Engineering Draughtsman.

notable kinds of Engines and varieties of construction, but also the operations of transmission and establishment at destination. The consideration of Means for the avoidance of danger or injury to health as regards the persons employed, to be carefully gone into.

Examination for a Diploma of Excellence.

The character of the studies would differ as usual from those of the Major Degree, not only in being of a higher Grade wherever this might be of real advantage, but also in being much more comprehensive, and constituting in many respects an accomplished as well as a clever MANUFACTURING ENGINEER.

Equally deserving of our attention with the Men engaged in the construction of Steam Engines, are those employed in their use. They are mostly separated into three divisions, according as the Engines are, — 1stly, stationary; 2ndly, used for propelling vessels; or 3rdly, used for locomotion on rail or road.—As regards the 1st division, the rank and duties of the Men employed vary so much according to the construction and calibre of the Engine they work, as Drivers, or Stokers, or both in one, and also according to the purpose for which it is used, and of which, like the Engine Driver in the Brewery, they ought to know something, that for the present it may be best to defer any attempt at the forma-

(1) This Superior Degree, indicated with a view to Managers of Engine Factories, or Professors of the Science of Steam Power, would only be organized as far as this could be done without interfering prejudicially with existing arrangements for the training of Engineers.

tion of gradational Curricula. — As regards persons working or attending to Engines used for propelling Steamers, no better instruction could be offered them than that they would get at a special department in the Nautical School at Greenwich.—For Drivers, Stokers, and Cleaners of Railway Locomotives, it might be best that instruction should be provided at Establishments connected with and supported by Railway Companies, acting in concert for this purpose as they should for many others; but it would of course be desirable that the plan of these, as well as of all other technical studies, should be arranged in harmony with the system centred in the Technical University, and in that point of view the following suggestions may possibly prove convenient.

Even the Men employed at unskilled labour without prospect of promotion, should be offered the opportunity of acquiring thoroughly a Grade 1 of G. S. F., and of pushing forward if possible the most appropriate parts to Grade 2.

Cleaners of Locomotives and others of that level, whose position is often a stepping stone to that of Stoker, should be further encouraged to ground themselves as well as they can, in at least Mechanical Physics and Mechanism.

From these occupations which may be included in Category C, we rise to that of the Stoker, whom we will consider in some measure as a Lieutenant to the Engine Driver, since he should be able to replace him in an emergency. Accordingly we find that as regards Preparatory Knowledge we may nearly assign to him the Minor Curriculum sketched above for the Journeyman in an Engine Factory. He may require less Mathematics and Drawing, and rather less acquaintance with the Chemistry of the Metals employed, but more knowledge of the various Fuels and Lubricants. His Handbook will omit Factory-work, and only describe in detail the Engines used on Railroads, but it will touch upon all matters connected with their working. Thus, to the leading notions concerning the construction of ordinary Railways, their gradients, curves, &c., will be added an account of the management of an engine, and of the customary rules and signals.

The Major Curriculum contemplated for the Engine Driver, or as he is sometimes called, the Engineer, can by a similar process be deduced from the above Major Curriculum of a Foreman in an Engine Factory. The Preparatory Knowledge may be slightly curtailed, but with a more than proportionate increase of the Technical Knowledge. The general history of Steam power may be simply given in outline, but that of its applications to Railway Locomotion should be fully gone into, including a sketch of the Atmospheric and other special Systems, and a comparison of foreign practices with our own. In treating thoroughly everything that may qualify the Engine Driver for the efficient discharge of his duties, particular attention will be paid to the various causes of danger to which Railway Traffic is subjected, and to the resources available for counteracting them.

Here as in the various Summaries given for other Industries, the suggestive outlines supplied form, as has been explained at the outset, only very imperfect sketches which will have to be completed by detailed Curricula, or Syllabuses of Studies. These, it will be remembered, are to be inserted with appropriate Questionaries in the respective Examination Programmes, and it is thus that will become fully apparent the relative difficulty of the tasks assigned to the several Classes of Students. Nevertheless it has purposely been made apparent that a somewhat high standard of Knowledge is deemed desirable for the Locomotive Engine Driver. This is partly because the Railway Service, when organized as it should be, presents notable facilities for scientific and technical Instruction, but chiefly because the importance of the responsibility which rests on the Engine Driver of a Passenger Train, is such that nothing should be spared that may enable him to acquit himself cleverly of his task, and may induce him to perform it prudently and conscientiously.

Should it be deemed advisable to institute a Diploma of Excellence, it will probably be with a view to the superior and comprehensive abilities which should be possessed by general or local Superintendents of Locomotive Departments.

SECTION 9.—THE WATCH AND CLOCKMAKER.

In viewing the Trades or Occupations of this interesting denomination from the stand-point of our present purpose, we first notice a great manufacturing industry split up into a number of distinct occupations, each confided to a special set of Men, who mostly require much more dexterity, accuracy, and patience, than scientific knowledge; except the last set, who put together and finish up what the previous ones have prepared. This last class of talented Artisans will doubtless deserve a special Curriculum of scientific instruction, at which some of the others may do well to try their strength; but more particularly within compass of our system are those Working Tradesmen who, coupling on a compact scale a certain degree of production with every kind of repair, carry on nearly all the more intellectual branches of the business, and train Apprentices to do the same. To them is generally brought, in small Towns, everything in the way of minute metal-work and delicate contrivance. Few trades have done so much to deserve every encouragement and facility that can be afforded them in the pursuit of Technical Knowledge, and the Horological Institute is one of those meritorious undertakings which it will be a pleasing duty for the Technical University to foster and to multiply.

The study of the Materials employed in Watchmaking, will of course include Diamonds, Rubies, Enamel, and the like; but the Student who wishes to unite with this Trade that of the Jeweller, must go in for the special requirements of that profession, which will be indicated further on.

Minor Examination.

GRADE 1 COURSE of G. S. F.
GRADE 2 COURSE. Inorganic Chemistry.
SUPPLEMENTARY STUDIES. Mathematics. Statics, Dynamics and Mechanism.
ART CLASS. Stage 2 of Drawing.
HANDBOOK; to include in the usual order all essential details relating to ordinary Watches and Clocks.

Manual skill to be tried if a suitable test can be devised.

Major Examination.

The above Preparatory Knowledge to be advanced a Grade. The Technical Knowledge to include, with the exception perhaps of Marine Chronometers and a few other specialties of a high character, all horological devices and resources of any note, home or foreign; also a sufficiently refined acquaintance with the various modes of ornamentation, for judging of artistic merit as well as of good workmanship.—The Hygiene of the Trade not to be neglected.

A practical test of ability to be devised.

Superior Examination.

GRADE 3 of G. S. F. (or of the Elementary Division; the Applied Division remaining at Grade 2.)

SUPPLEMENTARY STUDIES. A thorough mathematical as well as practical knowledge of Dynamics and Mechanism, especially in relation to Horology.

ART STUDIES. Stage 3 of Drawing may suffice, but with a thorough knowledge of the Laws of Taste in Outline, Colour, and Subject, especially as applicable to ornamental Time-pieces.

THE MANUAL; fully supplemented, so as to embrace every department of the Profession, from its minutest details to its highest achievements, adding its History and its Literature to the various branches of Science it involves, and omitting nothing that may tend to uphold the character of this highly intellectual Industry.

SECTION 10.—THE OPTICIAN, THE PHILOSOPHICAL INSTRUMENT MAKER, &c., ELECTRO-TELEGRAPHY.

Few of the Groups hitherto reviewed equal this one in the high scientific character of the articles produced, and in the scientific knowledge which many of them demand; but the very reason that this demand is plain and indisputable, makes the task of meeting it with appropriate instruction more clear and easy.—We will take for our starting point the Trade of the Working Optician, which affords a very convenient and normal example of a Working Trade susceptible of deriving a fair share of benefit from the proposed system of Technical Instruction. The following are suggestive indications for a Minor and a Major Curriculum. No Diploma of Excellence is contemplated, as it would probably be merged in one embracing the general range of Philosophical Instruments.

Minor Examination.

GRADE 1 COURSE of G. S. F.
GRADE 2 COURSE. The branch of Chemical Physics relating to Light.—Inorganic Chemistry.
SUPPLEMENTARY STUDIES. Grade 2 of Mathematics.
ART CLASS. Stage 1 of Drawing.
HANDBOOK; to include all introductory information required concerning the structure, functions, and common failings of the Eye, especially concerning

Myopia and Presbyopia. In other respects to be conformable to the usual Minor Review of Appliances, Materials, Processes and Products.

Major Examination.

GRADE 2 COURSE of G. S. F. This will probably supply as much as is wanted of the Chemistry and Technology of Glass, and of the Metals commonly used.

GRADE 3. The portion of Physics relating to Light should be pushed forward to this Grade, without going into the higher problems of Mathematical Optics, or any theoretical controversies.

SUPPLEMENTARY STUDIES. Grade 2 or 3 of Mathematics.

ART CLASS. Stage 2 or 3 of Drawing.

MANUAL. The structural, physiological, and pathological knowledge concerning the Eye, should be such as to make the Master Optician a reliable adviser in case of need, not merely respecting the use of concave and convex glasses, but also as regards the use of coloured or neutral tinted Spectacles, and the various constructions of these, as well as of Shades, Reflectors, and other contrivances for protecting the Eyes from Glare or Dust, or for improving as to quality or convenience the adaptation of natural or artificial Light to various Industrial Occupations.—The more strictly Technical Knowledge will also include as usual many things that were omitted in the Minor Degree. Thus for instance the Master Optician, without being expected to manage the blow-pipe like a regular Barometer and Thermometer Maker, should know how to fill and graduate the chief varieties of these instruments, and indeed he should not only understand the rationale and quality of everything he sells, but should be competetent to undertake in a country place, the repairing of any article not too foreign to his line of business.

The making of Barometers and Thermometers affords an example of one among numerous branches of the manufacture of Philosophical Instruments, which it may be expedient to teach as so many distinct Pursuits, leaving optional the manner of grouping or associating them. Other examples are afforded by the manufacture of Mariners' Compasses, of Scales and Weights, of Foot Rules, &c. It will be time enough when the Technical University is established, to see what occupations or branches of manufacture associated by analogy or commercial convenience, may besides being taught separately, deserve to be bound together by a joint Curriculum and a common name.

It may be found that the broad autonomy of a general "Philosophical Instrument Maker," must in the Minor Degree be separated into departments rather according to the kind of work, such as Brass-turning, Joinery, &c., than according to the branch of Science which the respective Instruments are to serve; or rather there will, as in various cases which we have come across, be no Minor Degree. The Master Workman or Foreman of Department, who after putting together and adjusting the various parts of a Philosophical Apparatus, is competent to try its capabilities, will if his range of subjects can be satisfactorily determined, be a suitable candidate for a Major Certificate. —The directing mind at the head of a comprehensive establishment of this kind, will if worthy of the task, have no difficulty in winning a Diploma of Excellence, should it be found expedient to institute one.

ELECTRO-TELEGRAPHY has suddenly started up from the depths of Science, to be a Science in itself, and one of which the mere apparatus and paraphernalia employ in special establishments a vast number of skilled hands. Nothing presents itself however that may not be met by the means already suggested. As for the important studies connected with Telegraphic Engineering, the part to be performed by the Central University will depend on the provision made elsewhere, and the same may to a considerable extent be said as regards the training of the thousands of subordinate employees. I will only mention that however satisfactory may be the practical cleverness at transmitting and reading of despatches to which they attain at the schools established for the purpose, a Grade 2 Certificate in Static and Dynamic Electricity and Magnetism, and another in Inorganic Chemistry, would be very desirable.

SECTION 11.—THE STATIONER. ALLIED INDUSTRIES.

This line of business exhibits perhaps a more striking example than any preceding one, of the distinction which must be made between the kind of knowledge required by the Manufacturer who devotes himself to the production of perhaps a single thing, such as Paper, Parchment, Ink, Steel Pens, Black lead Pencils, or any similar article, and that required by the Tradesman who unites all these Articles, and ten times more if he is inclined, within a single Shop. He ought to know something about every Article he sells, and a good deal about all those of which the usefulness depends on quality, and of which the quality can be readily detected only by one who understands the origin of the respective materials, and the rationale of the processes respectively involved. But at the same time we must remember that our system of Technical Instruction is based on voluntary studies, encouraged by Certificates which would be a failure if their practical value as a recommendation, were outweighed by the time and trouble required for earning them. There must accordingly be a compromise between what a Dealer in manifold Goods like the Stationer ought to know, and what he can conveniently learn, or in other words we must to a certain extent sacrifice such knowledge as can best be spared, in order that he may be induced to apply himself earnestly to what most he wants.

Under the Master Stationer we see, according to the scale of his concern, one or more Assistants from whom we must expect much less than from himself, but who ought still to know enough about the many things they handle, to entitle them to a Minor Certificate bearing the usual proportion to their Master's Major Certificate.

If now from the Tradesman whose business compasses the little world called Stationery, we pass to some of the individual Manufacturers who supply him, we are met by difficulties of an opposite character. Let us take for example those engaged in the production of a single article, say Ink, Steel Pens, or Lead Pencils. Here we find in operation the same plan of many hands directed by few minds, which has been repeatedly discussed under the name of Factory Work. With the subordinates, a Grade 2 of General Scientific Foundation will mostly serve every purpose. The directing mind must possess some further amount of Preparatory Science, and add to thorough Technical Knowledge a good acquaintance with all available resources. Still there is comparatively little scope for a regular Curriculum, and in view of the vast Agenda before us, it will mostly be prudent to postpone attending to Industries of this description; the more so as they can in the meantime avail themselves of the resources in the way of appropriate Knowledge offered to Students of Category C. Besides this, as previously explained (Chapter IV. Section 6), all Industries of any note will be carefully epitomized in the Junior, and fully described in the Advanced Cyclopædia, and it is to be hoped that an earnest recourse to these standard sources of information, will be found attractive and self-remunerative by many to whom the incentive of a Certificate cannot be offered.

PAPER-MAKING presents a somewhat different case. This highly important manufacture, though very much a matter of mechanical routine with most of the subordinate Hands employed, undeniably claims a Major Trade Certificate for the Foreman or Manager. Though the application of his knowledge may probably be confined to the specialty of some particular Mill, his studies should embrace all branches of the manufacture of Paper, and of kindred Articles commonly produced in this Country, and they should be supplemented with a review of foreign doings. It may safely be said that few Industries offer at the present time such promising openings for remunerative improvement, as regards both the utilization of new materials, and the adaptation of recent scientific discoveries.

A Trade allied to that of the Stationer, and which is not unfrequently associated with it, is that of the BOOK-BINDER, concerning which a very few words will suffice. The compendious account of it in the Junior Cyclopædia, will be well worth the attention of subordinates engaged in any branch of the process. For the Master Book-binder, independently of technical skill in overcoming difficulties, and turning out a satisfactory article in every branch of the Trade, there is enough required in the way of knowledge of materials, and of artistic taste in the applications of embossing, gilding, &c., to make it probably worth while to institute a Major Certificate.

SECTION 12.—THE PRINTER.

In the course of our review, we have more than once come across Handicrafts in which a little scientific knowledge is wrapped up in so much technical routine of manipulation, that Apprenticeship has, so to say, everything to do, and College Teaching next to nothing; but one would scarcely have expected that one of the most striking instances of this kind would be presented by that Industrial Pursuit which more than any other is the vehicle of scientific progress. Yet such is the case. The Man who has charge of the Press, ought to understand its mechanism; but the Compositor in whose intellectual task lies the pith and marrow of Typography, needs for it less scientific than literary knowledge, and his Curriculum must in fact be somewhat *sui generis*, as shown by the following brief Summary.

Compositor's Minor Examination.

GRADE 1 COURSE of G. S. F.

SUPPLEMENTARY STUDIES. Orthography, Punctuation, and such other literary abilities as are necessary for setting up in correct type, ill-written and confused Manuscripts.

HANDBOOK. To include select information concerning the Types commonly used, and all that is done with them, as well as concerning Appliances, Machinery, Ink, Paper and other requisites.

Cleverness at composing, taking Proofs, and distributing, to be tested by the actual performance of appropriate tasks of moderate difficulty.

Major Examination.

(For Working Master Printers, Foremen and Superior Compositors, acting also as Correctors of the Press. "Readers" properly so called are not included).

GRADE 2 COURSE of G. S. F.

SUPPLEMENTARY STUDIES. Literary abilities of the kind described above, but much more fully developed.

MANUAL. The knowledge contained in the Handbook to be extended as well as raised in character, so as to include the latest improvements in Machinery, Appliances, and Materials of all kinds, the less common Types, the introduction of Woodcuts and other illustrations or ornamentations, tabulated matter, Stereotyping, and other processes commonly included in large Letterpress Establishments, and generally, all that may qualify a Printer for doing honour to his Trade.

Technical Cleverness to be practically tested as above. The ordeal to include severe, but not useless difficulties.[1]

---

[1] No Diploma of Excellence is contemplated, because a Director of a large Printing Establishment, whose attainments expand in a scholastic and bibliological direction, is likely, unless he should have risen from the ranks, to be deficient in those matters of clever practical performance which are the essence of the Trade.

SECTION 13.—PROFESSIONS AND TRADES CONNECTED WITH MUSIC.

It is unnecessary to dwell on the importance of MUSIC in dispensing sympathetic and humanizing influences throughout all classes of the Population, and in superseding undesirable ones, or to say anything of the positive and unquestionable enjoyment it affords, or of the thousands of honest livelihoods it maintains. If we are inclined to pause before inscribing it on our Agenda, it will be partly because we acknowledge the force of the French proverb, *qui trop embrasse mal étreint*, and partly because we are anxious to assist, rather than interfere with the action of other regularly constituted agencies. Few persons will be inclined to deny that Music ought, like Science and Art, to have the benefit of a distinct government Department, when the supreme direction of educational matters shall be in the hands of a responsible Minister. It will devolve on the person entrusted by him with that Department, to see that all is done with even hand to foster rising musical genius, to afford for the development of professional Music, whether theoretical, vocal, or instrumental, facilities and advantages equal if possible to those of the most favoured nations, and to promote by every reasonable means the diffusion of Music among the People, till it reaches and maintains that level which the example of many foreign localities, and of not a few of our own, has proved to be perfectly attainable.

It is from such a regular organization of efforts and resources, dovetailing with the other Departments of National Education, that could best be derived that amount of Musical Foundation, theoretical and practical, which ought to support and bind together all the occupations belonging to the Musical Craft, in the same manner that a standard Scientific Foundation will bind and support the various Industrial Pursuits hitherto reviewed. Accordingly, it may not be necessary that the Technical University should itself provide actual Musical Instruction, though its organizing and stimulating influence may have to be exerted in order that such Instruction may be everywhere efficiently provided.

But independently of the consideration that our University is not to usurp the functions of a *Conservatoire de Musique*, most of the Industrial Pursuits of which Music is the origin and end, will be found, on close examination, of a nature not to be easily brought within the compass of its Examinations and Certificates, though they may be aided in other ways. The most important among them may be roughly grouped as follows:—

a.—The publication of Music by Engraving, Lithographing, Printing, &c.

b.—The Manufacture of Organs, Harmoniums, Pianofortes, Harps and kindred Instruments of complex contrivance.

c.—The Manufacture of Stringed Instruments commonly so called. They mostly arrange themselves in two lots, referable to the Violin type, and the Guitar or Mandolin type.

d.—The Manufacture of Wind Instruments commonly so called, which may for convenience be divided into those chiefly made of Wood, and those made of Metal.

e.—The manufacture of Drums, and other miscellaneous Musical Appliances.

As regards Group *a*, the various processes by which Music is published, are rather matters of apprenticeship and practice than of didactic instruction. Passing to Group *b*, we are reminded of those Factories where the subdivision of labour, whilst it favours accuracy as well as celerity of workmanship, tends to make machines of men, and only calls for a broad intellectual development among the select few by whom those living machines are regulated. For supplying the kind of knowledge which these few require, a valuable resource will be presented by the Technological Museum of the Central University, taken in conjunction with their respective "Manuals," and with its own explanatory Catalogue. For instance it may display instructive examples or models of the many ingenious contrivances that have at various times been tried with more or less success in the manufacture of Pianos, Harmoniums, Organs, and other similar Instruments; and particularly of those having a special object, as for instance suitableness for tropical climates, where extremes of dry heat and warm moisture occur, and where many materials succumb to the ravages of Insects. Many of the smaller Industries included in Groups *c*, *d* and *e* may similarly derive considerable aid from the University Museum, on which, unless forestalled by exertions made elsewhere, will devolve not only the duty of exhibiting full chronological and ethnological illustrations of Musical Instruments of all kinds, as far as they may serve as guides or warnings for present and future Makers, but also the duty of showing to these in a technical and commercial spirit, the best resources at their disposal both at home and abroad.

SECTION 14.—ART INDUSTRIES. PHOTOGRAPHY AND THE POLYGRAPHIC ARTS.

Though Art in one form or another has met us in many of the Industries reviewed, occupying in some of them a conspicuous position, we have left as yet untouched most of the Pursuits which more particularly claim the title of Art Industries. It will in fact be found that with a few exceptions deserving of special attention, they lie less within the province of the proposed University than within that of the South Kensington Authorities, under whose management so much progress has been already made in improving the taste and execution of Art-workmanship in this Country. Independently of this fact, in Art Industries the necessity for carefully organized Tests and reliable Certificates of Competency, is less felt than in Industries of which Science is the foundation. The man whose technical abilities rest on a broad groundwork of Scientific Knowledge, cannot at all times supply ready proof of this on demand, whereas Artistic Skill can mostly be made manifest with ease, both in character and degree. Moreover by attaching a fixed standard of proficiency in a specified branch of Art to a given Certificate in an Artistic Industry, we might be rendering less service to a Candidate, than by encouraging him to enlarge his sphere of artistic attainments, at the same time pushing them as high as his natural turn and his opportunities might permit. It is true that many Employers of artistic labour, will scarcely appreciate our good intentions. The main qualities which they demand of their Journeymen are well-trained technical dexterity, a good eye and a quick hand, with patience for repeating any number of times a task ever so small and monotonous; but they should bear in mind that the dexterous hand and the knowing eye, though they may be only obtainable through the routine of a long apprenticeship, are best to be relied on when they have undergone the refining influence of regular attendance at Art Classes. The employer who wishes to uphold the character of his Craft through the abilities of his Men, should give them every reasonable encouragement and opportunity for perfecting themselves in appropriate branches of Art. —Where the duty of the National Technical University towards them will chiefly lie, will be;—*firstly*, in seeing that due facilities are afforded them in every populous locality, for acquiring these attainments in the manner most conducive to their practical interests, and to the satisfaction of their Employers;—*secondly*, in affording equal facilities for appropriate Scientific Studies to be recompensed by Subject Certificates;—and *thirdly*, in providing by means of the Junior and Advanced Cyclopædias, repertories of Technical Knowledge so obviously useful as to be consulted without the incentive of a Trade Certificate. — In order better to judge of the working of these measures, we will indulge in a cursory glance at some of the strata of which the *personnel* of Art Industry in general is composed.

Setting aside the few who earn fame in making a fortune, let us descend to the many who only earn weekly wages. We find that here as in many other industrial regions, there is a substratum consisting of hundreds of poor subordinates to whom an insight into the Science of Daily Life, as supplied by the Grade 1 Course, would be a boon without a drawback, but for whom it would scarcely be expedient to recommend anything beyond.[1]

Next to these let us consider the more or less skilled Artisans numerously engaged in the Artistic Departments of various Manufacturing Industries conducted on the usual plan of Factory Work. Here as in analogous Mechanical Departments, we find ourselves obliged by the subdivision of labour, and the consequent reduction of the range of knowledge required by the individual Hands, to class nearly the whole of these as belonging to Category C. They will however mostly be found well suited for a full enjoyment of the resources available for that category, and which have been described at the beginning of this Chapter under the title "Studies for Subject Certificates.[1]

Rising to the more talented members of the Art-industrial Community, we are surrounded in all directions by those skilled Workmen whose abilities have been so usefully brought into relief by the Society of Arts' Examinations. It is obvious that any measures for adding intellectual to artistic tests, should if possible be entertained in concert with that Society.

---

(1) A striking illustration of mechanical labour having a preparatory share in the performances, not merely of Art-industry, but of high Art, may be witnessed in the Studio of a Sculptor.

(2) As examples of the division of artistic labour may be mentioned the Japanned Tray, of which one painter does the fillets, another the border, and a third the centre; or Architectural Carving in which separate hands do the string-coursing, the foliage or enrichments, the heraldic ornaments, the drapery of a figure, and lastly the face and hands. Various stages of artistic industry may likewise be seen by following the production of a bronze ornament through its modelling, its casting in lead, its touching up, its recasting in bronze, its cleaning, and its ultimate chasing.

There are in the range of Art Industries a few types which it will be more particularly the duty of the Central University to select, for the purpose of fostering every branch of their efficiency. An interesting example is afforded by the WORKING JEWELLER. He is, or should be, an Artificer possessing scientific knowledge, artistic genius, and manual skill, who firstly, understands thoroughly those picked pages of Mineralogy where Gems occur, uniting for the purpose the required notions of Chemistry and Crystallography; secondly, can devise designs for displaying Gems singly or variously grouped to the best advantage, and thirdly, knows so well and can handle so cleverly the metallic materials involved, as to realize unassisted if it were necessary, the graceful jewel of his thoughts.—It must be remembered however, before attempting to frame a Curriculum of Studies in accordance with these indications, that in many respects the present organization of the Jeweller's Trade, as of other Industries, tends to a splitting and separation of attributes, and a mechanicalizing of processes, more and more at variance with the *ideal* of former times.

That influence would not be ill-spent on the part of the University, which could induce Pattern Designers never to neglect tempering their Art with Science. In fact the Designer should be as well acquainted with the operation or process in which his design or pattern is to be used, as the Musical Composer should be with the instrument on which his piece is to be performed. It is scarcely necessary to say that through neglect of this rule, the Glass-blower, the Thrower of Ceramic Ware, the Jacquard Weaver, and many others depending on Designers for their patterns, may lose their time and patience in producing under difficulties, what will fetch no higher price than an article which they could produce with ease. Similarly the Carpet Weaver may through the fault of his Designer, have to use worsteds of expensive hues, where a combination equally pleasing to the eye, and equally durable, could be produced with much cheaper ones.[1]

Whilst it is desirable that the Designer should show knowledge and foresight in adapting his origination to the convenience of the Performer, there are many cases where the latter would be vastly benefited by a gleam of original ingenuity; and indeed the fertility of resources resulting from a lively power of origination, has been supposed to be one of the chief merits for the sake of which many foreign Art-workmen are employed so willingly, and paid so well, in this country.[2]

PHOTOGRAPHY in which Nature supplies the design, and Science the execution, seemed at first a conspiracy against Art, but more and more we find that it affords scope for Artistic Taste, and that in its higher aspirations it opens the way to a considerable amount of real artistic employment. Something like the following Summary of desirable attainments suggests itself:—

Minor Examination.

GRADE 1 COURSE of G. S. F.

GRADE 2 COURSE. The part of Chemical Physics relating to Light.—The whole of Chemistry or at all events, of Inorganic Chemistry.

[1] Indications as to what Designers are, and what they should be, will be found in a Paper by Dr. Christopher Dresser on "Hindrances to Applied Art," in the 'Society of Arts' Journal' of April 12th, 1872.

[2] Respecting the unsuccessful attempts made by the Society of Arts to elicit proofs of origination among British Art-workmen, see the Report of the Society's Adjudicators quoted in Chapter II. Section 3 from the Society's Journal of the 5th of March, 1869.

HANDBOOK; to comprize only the common processes, including however all that is required for clearly understanding, not only the bearings of Optical and Chemical Science on those processes, and the means calculated to secure their success, but also the Laws of Taste applicable thereto.

A Practical Test to be resorted to, including plain touching up.

Major Examination.

GRADE 2 COURSE of G. S. F.
GRADE 3. Optics as far as the Science bears on Photography.—Chemistry, Inorganic and Organic.
MANUAL; to comprize the history and rationale of the various branches of the Art, with a thorough knowledge of every scientific device and artistic refinement, through which high-class results may be secured even under difficulties of climate, &c.

The Practical Test to include delicate finishing up.[1]

Through the facility with which any number of Proofs are taken, or to employ the usual term, *printed* from the Negatives, Photography may be considered as one of the POLYGRAPHIC ARTS, that is to say of the Arts producing multiplied impressions; but this title

---

[1] It may be left for the present an open question whether it be best to treat the colouring of photographic portraits as a distinct branch of Art, or to include this or any other artistic performance in the Practical Test, as a criterion of Candidates having attained the degree of artistic refinement required for being a first rate Photographist.

more properly belongs to such processes as the following:—[1]

  a. *Printing from raised surfaces.*
     Typography or Letterpress Printing.
     Nature Printing (as from Ferns and the like).
     Xylography or Wood engraving.
     Graphotype.
     Phototype.
  b. *Printing from surfaces rendered variously absorbent.*
     Lithography.
     Zincography.
  c. *Printing from hollowed surfaces.*
     Etching, Aquatinta, and the like.
     Copperplate, and Steel Engraving.
  d. *Moulding or embossing by means of hollowed surfaces, or matrices.*
     Seal engraving, Die-sinking, &c.

Reviewing these examples from the standpoint of Art Workmanship, we first set aside Typography, in which it has already been seen that the chief performance lies not in the original making of the Types, any more than in the printing from them, but in the intermediary task of the Compositor. We also set aside ordinary Nature Printing as being done mechanically, and the printing from surfaces prepared by photographic means, as coming more or less within the scope of the above remarks on Photography. The remaining Processes, unless removed from their original character by the prevailing tendency to reduce everything to

[1] This list is by no means intended to be exhaustive.

Factory Work, are decidedly *artistic* as regards the most difficult and important task in each, and what the Central University can appropriately attempt to do for them besides encouraging the development of Art Classes of every description will mainly consist,—1stly, in offering Instruction and Examinations in suitable "Subjects,"—and 2ndly, in displaying in the Museum a carefully made Selection of first-rate examples calculated to raise the standard of Excellence in each branch, and accompanied with the most recent and best materials and appliances.

SECTION 15.—COMMERCIAL PURSUITS.

We have hitherto left unnoticed the interval which in so many instances separates the production of Trade Articles from their display in the Retail Shop, and the Commercial System through the instrumentality of which the right thing is so conveniently brought from a distance to the right place. It will however be easy to deduce from the principles proposed for regulating the studies of the Retail Dealer, what should be those of the Import or Export Merchant, the Broker or any other Intermediary having such connection with the Goods in question as to involve a knowledge of their nature and quality. We have seen that the Shopkeeper should, for the benefit of his Customers and for his own, possess this knowledge to a notable extent; not indeed with every detail that should be familiar to the producer or manufacturer, but with that good practical judgment and that reasoning insight which are acquired by basing one's technical studies on a sound scientific foundation. This same plan must be carried out in the studies of the Commercial Intermediaries just referred to, and the indications given in this and the preceding Chapters have sufficiently shown how manifold will be the resources available for accomplishing this object according to the kind and grade of their requirements. For instance, as regards Preparatory Knowledge, they will have the Elementary portions of the Standard Courses from Grade 1 to Grade 3 ;—as regards Technical Knowledge, the Applied portions of the same, and two Cyclopædias purposely so arranged as to favour the selection of appropriate Subjects.—Then again, the Commercial Department of the University Museum will, in conjunction with its explanatory Catalogue, make it easy for them to acquire that familiarity with Goods of any given kind, which is, without Visual Instruction, so difficult to obtain. Some time may elapse before the University can conveniently institute regular Courses of Commercial Knowledge with their Composite Examinations and Certificates, but *en attendant* Commercial Students will be offered opportunities of striving for Subject Certificates, including a few topics special to the THEORY OF COMMERCE, such as Commercial Geography, International Exchange in Weights, Measures, and Money, and International Commercial Law. Independently however of any prospect of distinctions, it may be hoped their keen perception of profit will tell them that any leisure judiciously employed in availing themselves of the educational resources offered them, will be time well spent.

CHAPTER VI.

RECAPITULATION. CONCLUSIONS.

SECTION 1.—CONSIDERATIONS RESULTING FROM THE FOREGOING CHAPTER.

A. Manner and Place.

THE Analyses contained in the preceding Chapter, though purposely restricted to a certain number of representative types, have probably sufficed to indicate the general features of the instruction required by the various Industrial Pursuits, according to their more or less Chemical or Physical, Artistic or Commercial, simple or mixed character. Let us now briefly reconsider these requirements in special reference to the question,—*how* and *where* shall the Artisan obtain the Instruction he wants?

Grade 1 Studies.

The first thing which attracts our retrospective glance is that ubiquitous Grade 1 Course of Elementary and Applied knowledge, which besides underlying as a General Scientific Foundation the broad field of Technical Study, is intended to extend beyond its limits, affording to the whole of the Labouring Population the much needed Code of Common Sense which it has been agreed to name the Science of Daily Life. One might find more delight in fostering the advanced progress of a chosen few, and upholding them as proofs of our national capabilities, but more real importance attaches to a speedy and substantial organization of the Instruction required by the lower and broader strata of the Industrial Community. Accordingly the first charge on the energies of the proposed Movement, will be to carry into effect the various measures indicated for the general diffusion of the Rudiments of Practical Science and especially the following:—

a.—To prepare the required Grade 1 Course as described in Chapter IV. Section 1, to bring it out in print as recommended in Section 5 of the same Chapter, and to secure the production at cheap rates of the necessary Sets of Illustrations (pp. 113 and 244), arranging as far as may be found necessary for the circulation of such sets from place to place (p. 231, *l*).[1]

b.—To agree with the leading Training Colleges for supplying them, not only with the foregoing educational materials, but also with Professors specially trained for instructing the future School Teachers in an ingenious and impressive use of them (p. 79).

c.—To afford the present Teachers of the Children of the People opportunities for adding the Scientific Element to their present stock of knowledge (p. 80).

d.—To promote the extension of Schooling from 12 to 14 years of age for youths whose intended Pursuits involve something above mere mechanical labour, and where required, to organize special schools, or advanced

---

[1] This plan seems to have been first practised by the South Kensington Authorities with sets formed from the residue of the International Exhibition of 1851.

classes in ordinary schools, for consolidating in this Secondary Period, the groundwork of an industrial career (p. 14).

e.—To promote the establishment, judicious organization, and regular maintenance of INDUSTRIAL INSTITUTES of various calibres, according to the requirements and resources of the respective localities; it being understood that under this title are included, for convenience sake, all Institutions or Establishments offering or procuring to the youths or adults of the Industrial Community, the Class Lessons, Lectures, or other forms of Scientific, Mathematical, and Artistic Instruction which they require, (pp. 49-53).—A Grade 1 Course of G. S. F. with a corresponding Art Class, may be considered as the minimum of the capabilities of an Industrial Institute. Other branches of Instruction have already been indicated as coming more or less within their province, and will again occur as we proceed in collating the chief attainments noted in the Summaries of the preceding Chapter, with the chief sources calculated to supply them.

Grade 2 Studies.

The man who has found out his ability to support himself in water, looks at the latter with a sensation very different from his previous distrust, and feels that it now only depends on him to become a good swimmer. A similar change occurs in the mind of the illiterate Workman, who from viewing Science with the shyness of awe and prejudice, has passed to the happier condition of seeing in Science a pleasant means of support and progress. This change has indeed been throughout one of the most interesting and satisfactory features of the experience accruing from the delivery of my Popular Course to labouring-class audiences, and from the holding of examinations in connection therewith; and I feel confident that the generality of Artisan Students, when once set afloat by a well-conducted Grade 1 Training, will be cheerfully disposed to buffet with lusty mental sinews, any waves they may have to encounter in their further progress.

This particularly applies to Chemistry, of which the Grade 1 Course supplies the initiatory sketch, and Grade 2 the earnest and practical exposition for Working purposes. We have seen that independently of the Major Curricula, which mostly include the whole Grade 2 Course, its Chemistry, and more especially its Inorganic Chemistry, has been frequently prescribed as an item of Preparatory Knowledge for Minor Students, and indeed such is in every respect the importance of this science for technical purposes, that it is highly desirable that every well organized Industrial Institute should extend at least thus far its teaching capabilities. This remark is applicable, though in a less degree, to the parts of Mechanical Physics, which have been occasionally recommended as bearing on the mechanical Trades. Nearly the same may be said of certain Departments of the Second Division of this Grade 2 Course, which have not unfrequently been included in the preparatory portion of the Summaries. They are in fact particularly calculated to prepare Artisan Students for the Technology of their respective Trades, giving them that broad view of the surrounding ones, and of the manner in which one trade helps another, which their special Handbooks cannot be expected to afford. It would indeed be a pity not to include in the Curricula of certain pursuits connected either with the construction or Decoration of dwellings, or with Fabrics and Clothing, or with Food, the portions of the Grade 2 Course which respectively embrace the applications of Science to those Subjects, making plain to the Artisan Student that most important part of his educational duties, which consists in thoughtfully bringing the knowledge he has acquired, to a practical bearing on his Trade.

The frequent occurrence of Mathematical Studies in the Summaries, reminds us that Mathematical Classes are among the desiderata of Industrial Institutes, though more important still are the Art Classes. If we add Reading and Writing Classes for those who labour under educational deficiencies, and Arithmetic of any Grade that may be required, and if we introduce in the more favoured localities, Classes for French or German, and Music, we shall have without going farther, a tolerably full programme for the Industrial Institutes of Country Towns. Yet considering how much the contemplated advance in the general standard of knowledge and efficiency among our Artisans, will depend on placing what they are to learn within their convenient reach, there is no question, that it will be exceedingly desirable to include all this and even more, in the attributes of these local Institutions, at the same time multiplying them throughout the length and breadth of the land, with a profusion far exceeding anything yet attempted. This may seem hopeless to those who have anxiously watched the many feeble, desultory, and consequently futile and transient efforts made at different times in this direction, but I trust that an unprejudiced consideration of the signs of the times, will convince them that the horizon is clearing up in the wind's eye.

Industrial Institutes, and equivalent Organizations for Industrial and Popular Instruction.

There is among our Industrial Classes a growing stir of progress, a legitimate ambition of improvement supported on the one hand by leaders chosen with a different aim, and on the other hand by time-honoured institutions of guardianship awakening to the sense of former usefulness; but even if all this were not the case, our Artisans could not long remain passive spectators of the efforts made on the Continent to raise still higher a system of Popular Education already above our own. It is impossible for Englishmen to read without a restive feeling of emulation, the account given by Dr. F. Leibing in his excellent periodical " Der Bildungs-Verein," of the exertions made in Germany by the Improvement Societies bearing that name, to develop and consolidate the advanced Instruction afforded to Apprentices and others by the so called Fort-Bildungs-Schulen. (Schools for further Improvement.)[1] It is almost equally exciting to read the details of analogous efforts in Belgium, given in the " Bulletins " of the " Ligue de l'Enseignement " by the enlightened Secretary-General of that powerful association, M. Charles Buls, or to study the School regulations prevailing among the Swiss, well known as the active pioneers of educational progress, or to scan the proceedings of the Ecole Centrale of France and its provincial

(1) Schools of this description exist in Prussia, Hanover, Saxony, Nassau, Würtemberg and other parts of Germany. In the Grand Duchy of Baden, attendance at these Schools used to be compulsory, and the voluntary system tried for the last six years having proved unsatisfactory, a compulsory law is again under consideration, which finds advocates in many parts of the German Empire. See the Publications on this and kindred subjects indicated in List 1 at the beginning of the present Work.

affiliates, or to peruse the annals of the Conservatoire des Arts et Métiers, where Parisian Workmen belie their reputation of fickleness by regularly attending Courses of not less than 40 Lectures.[1] However, the most earnest aspirations, whether of home growth or prompted by foreign competition, might continue to wear themselves out in ineffectual efforts, were it not that through the proposed system, many of the material obstacles which have hitherto stood in the way of Industrial Instruction, will be quietly but effectually removed. Then, not only in large Towns but in every small centre of Industrial activity, let a few earnest men come forward with energy to undertake a working part, whilst a few public spirited members of the wealthier class subscribe a moderate amount for unavoidable expenditure, and a few influential friends take on themselves the duties of Trustees,[2] and let communication be entered into with the Central Technical University. Advice and assistance will at once be forthcoming. Text Books will be supplied at an almost nominal price, the use of which on the Binary System will be explained and organized by Delegates sent down for the purpose, and thus the necessity for expensive Professors will be obviated in several important branches of instruction. Devices for the production of home-made Illustrations as far as practicable, will be freely taught by the organizing Agents,

(1) For a description of the system of Free Popular Instruction for Workmen, &c. given by the Conservatoire des Arts et Métiers see Society of Arts Journal, January 2nd, 1874, p. 120, and the Fourth Report of H. M. Commissioners on Scientific Instruction, &c., p. 21, with Appendix III. thereto. Also the "Enquête sur l'Enseignement Professionnel," List 1, p. xxii.

(2) Respecting the co-operation of the Clergy in matters of Popular Instruction see page 140, and respecting that of Schoolmasters see page 237.

whilst any necessary complement will be forwarded at the lowest possible rates from the Supply Department of the Central Museum (see p. 113). Any required steps will at the same time be taken to obtain for Provincial Institutes, the full benefit of the liberal assistance offered by the South Kensington Authorities towards reducing the cost of Scientific and Artistic requisites, no less than towards the erection of appropriate Buildings. (See Notes to pages 50 and 51.)

As already explained, the term " Industrial Institute " embraces any institution or organization for supplying Apprentices and Workmen with the instruction they require, whether in special Lecture Halls and Classrooms, or at the Local Schoolrooms,[1] or at the Mission room, the Temperance Hall, the Vestry Hall, the Town Hall, &c. It is however essential that there should be convenience for the safe custody of the educational materials, and very desirable that there should be sufficient space to allow of exhibiting permanently as a Museum of Domestic Sanitary and Industrial Economy, such of the Illustrations as are suited for Instructive Display (see p. 94).

It is by means like these, supplemented if necessary by the circulation of the more expensive sets of Illustrations from the nearest centres, that the various kinds of preparatory Knowledge indicated above, may be brought within the compass of Minor Students.

Mechanism, which has been included in the Preparatory Knowledge suggested for certain Pursuits, is a Subject requiring Models and Diagrams not every-

---

(1) At the Meeting of the London School - Board on Wednesday, March 11th, 1874, a measure was adopted whereby, subject to the sanction of the Board, the School Management Committee was empowered to let Board Schools on Week-day evenings for appropriate purposes.

where to be had; but it will be mostly wanted in large Towns where Technical Colleges may be expected to exist, or at the leading Railway Depots where, as stated before, the educational arrangements connected with the subject of Locomotive Machinery ought to be particularly good (p. 387).

Self-Instruction.

Generally speaking, self-instruction in scientific and technical subjects, however suitable the Student's Book may be, is very uphill work if he has never tried himself at such matters before; but the case is very different when after attending a regular Grade 1 Course of G. S. F., he endeavours further to impress it on his memory by home study. A sentence once made plain to him in oral discourse, will be plain in a book, and a woodcut is nearly as good as a Lecture diagram to him after he has seen and understood the latter. Having thus overcome a difficulty partly arising from mere diffidence of his own powers of comprehension, the young Artisan will be able to make good headway by himself in many branches of knowledge involving no experiments, as for instance in any department that may be prescribed for him in the Applied Division of the Grade 2 Course. Hence the importance attached above to the permanent display of the respective Illustrations in the form of a Museum and in the order of the Text.[1]

[1] It is with peculiar satisfaction that I find H. M. Commissioners on Scientific Instruction proposing in their Fourth Report measures long since advocated by me for rendering Museums more available for *Self-Instruction.* "By the skilful selection and arrangement of the specimens exhibited to the public; by providing descriptive Labels instead of the meagre indication of names at present adopted; and, above all, by sup-

Minor Technical Studies.

Passing to the Technical Studies prescribed for Minor Candidates in the Summaries, we find in most cases self-instruction equally feasible. We must remember that the student finds in his Handbook not only the exact knowledge he wants, carefully expressed in a style suited to the level of his culture, but also such friendly advice as may best smooth his path and make his exertions profitable. Further we may assume that he is serving his time as Apprentice, or working as Journeyman in the trade of which he now aspires to master the rationale, and consequently if his Master's establishment is anything like what it should be, he enjoys excellent opportunities for coupling visual with printed instruction. These facilities are the more likely to be sufficient, because aspiring only to a Minor Certificate, he will be only expected to make himself acquainted with the ordinary resources and processes of his Trade, leaving the introduction of new improvements to his Master.

Taking into account the various foregoing considerations, we shall feel justified in reckoning that if the development of Science Classes at Industrial Institutes, or in any other way, progresses satisfactorily, a Candidate for a Minor Certificate who earnestly does his best, may ere long hope to acquire in his own locality, if not the whole, at all events by far the greater part of the attainments required for passing his Examination. Any necessary complement will have to be acquired at the

plying Explanatory Catalogues suited to the wants of unscientific people, the public might be enabled to teach themselves with as much efficiency as they would be taught by the majority of Lecturers." (Art. 135.)

See the "Brief Account" of my Food Collection exhibited at the S. K. Museum, 1857, p. 11, and "Science for the People," 1870, p. 98.

Technical College where that Examination is to take place. This supposes of course that Minor Examinations will be held at Technical Colleges by Delegates from the Technical University of the Province, or from the Central one in the Metropolis.[1]

Studies for Major Certificates.

It may have been noticed in comparing the Summaries of the Minor with those of the Major degree, that as regards the PREPARATORY KNOWLEDGE, the chief difference has generally consisted in raising the respective attainments one Grade.[2] It is more particularly in the TECHNICAL KNOWLEDGE that we see the advanced Student's task grow and ramify. We have seen that in a Trade having many branches, it may be expedient to allow the Minor Candidate to confine his studies to one, whereas the Major Candidate will mostly be called upon to take up several if not all. But independently of this, we observe that whereas the Minor Student need only include in his Scientific Review of Appliances, Materials and Processes, those in ordinary use, learning to make the best of them and be content, the Major Student must cultivate an ambitious acquaintance with matters beyond the ordinary routine of manufacturing and commercial Industry, with materials and processes both home and foreign, not generally known, or not sufficiently appreciated, and with the

[1] Cases may occur, especially when popular culture is more advanced, in which on the one hand the Candidate can learn all he requires without leaving the place of his abode, and on the other hand the personal presence of the Examiner is not required as he can conduct the Examination from a distance in the manner previously described at p. 170.

[2] The retaining of Grade 1 of G. S. F. in some cases, has been prompted by special reasons. Thus for instance Dressmaking is a feminine occupation, and the trade of the Baker when conducted in the ordinary way requires less learning than honesty and hard work.

last touches given to Industrial Art by the improving hand of Science. Now in many Industries, a thorough acquaintance with these things cannot be acquired without visual illustration, nor can they be properly illustrated without expensive sets of scientific paraphernalia, such as only a first-rate educational establishment could afford. Besides this, as regards the more important Industries, there is in the very atmosphere of a University conducted on sound and liberal principles, something wonderfully conducive to that mental expansion which best entitles the Master to direct the work of his men, always uniting the idea of their welfare with that of his own success.

On the whole a retrospective glance at the Analyses leads us to the conclusion, that much in the same way as the Industrial Institutes will probably be called upon to supply the instruction required by Minor Candidates, often leaving however a certain complement determined by the nature of the Trade and other circumstances to be supplied by a Technical College, where the Examination is to be passed, so likewise may we, generally speaking, look to the Technical Colleges for supplying nearly all the Preparatory Knowledge, and a varying proportion of the Technical Knowledge required by Major Candidates, who will similarly afterwards repair to a Provincial or to the Central Technicial University, for completing and polishing their Knowledge and taking their degrees. There are indeed Trades that can be mastered almost anywhere, with the aid of a Text-book so well got up as will be each of the Manuals taken from the Advanced Cyclopædia, and with the help of the working Illustrations which the Student may have before him in the very establishment in which he is employed. But exceptions like these cannot exempt us from making due provision for meeting less favourable conditions by the multiplication of Technical Colleges, and by giving every encouragement to the formation of Scholarship Funds and Endowments.

Many Major Students if they must leave home at all, will prefer going at once to the University where they are to pass their Examination, and where, besides finding annexed a model Technical College for preparatory Studies (p. 94) they may if clever and industrious, obtain the following advantages :—

—Employment at the instruction of Junior Students, especially in the Ateliers (note to p. 60).

—Employment at Museum work (p. 112).

—Employment in their respective Handicrafts on the plan adopted at the Cornell University (p. 60).[1]

Candidates for Superior Certificates or Diplomas of Excellence.

Though not every pursuit analysed in the foregoing Chapter has called for a Curriculum of this level, sufficient examples have been given to show that the studies will bear to those of the Major Degree, much the same relation as the studies of the latter to those of the Minor Degree, except that the interval will mostly be greater. The Candidates may as a rule be expected to become matriculated Students at one of the

[1] The following particulars concerning this interesting establishment are from a speech by Professor Goldwin Smith at the opening of the present Session of the Working Men's College in Great Ormond Street.

Israel Cornell, a working man in the employment of a Telegraph Company, having become possessed of great wealth, founded the University bearing his name at Ithaca, Wisconsin, with the aid of a State subvention. It numbers at present 500 students, a portion of whom forming a "labour corps" unite the labour of their hands with that of their brain.

Technical Universities, but no glut or overcrowding need be feared even at the Central one, for at this high level we find that Scientific Instruction has been much more attended to than in the lower industrial regions, and the consequence is that many existing Institutions will in some measure be our competitors. Independently however of a Museum unsurpassed in Industrial Illustrations, we may safely reckon on being able on several other points to offer attractive facilities to men of good abilities but limited means. Whilst they will be required to devote themselves with thorough earnestness to the practical applications of well-established theories and standard facts, they will be kept as much as possible clear of speculative and controversial questions. Thus the time necessary for completing the Technical Training, say of an Advanced Chemical Student, will be greatly diminished by concentrating his attention on proximate and tangible results, whilst to others more favoured by circumstances, will be left the privilege of studying Chemistry for its own sake, and of opening by original research new careers of thought which may lead to something tangible in the future. Principles like this cannot fail to be appreciated by those who necessarily and avowedly are less anxious to serve the cause of Science, than that Science should serve their purpose, and with whom moreover the question between one year's study, or two, may make the difference between the possible and the impossible. This is the more worthy of consideration as it will be one of the objects of the proposed University to encourage the rise of genius from the ranks.[1]

---

[1] For indications respecting the studies of Candidates for Professorships or other positions connected with Science Teaching see Chapter II. Section 4.

Feminine Industries.

The manner in which the Industrial Instruction of Females should be provided for, can scarcely be discussed with advantage till the movement for improving the Education of Women has had time to mature its plans (p. 87). Should these be found capable of supplying the kind and quantity of scientific, artistic, and technical knowledge required, it will be our policy to favour their acting separately from, though in harmony with the Institutes and Colleges for Males, and to encourage them to complete as far as possible the required attainments. Thus perhaps Female Candidates need only come to the Universities for Examination; except as regards those branches of information which involve comprehensive collections or other expensive paraphernalia.

B. Time.

Our task would be a comparatively easy one if we only had to determine the Instruction best calculated in the abstract to benefit our Artisans; but at every step we are reminded that the proposed System is essentially *voluntary*, and that instead of simply stating what *we* think our industrial Friends ought to know, we must consider what we can make *them* think, in spite of hindrances and temptations, that it is worth their while to learn. This consideration which in the previous Section has pervaded the question of PLACE, must now equally pervade the question of TIME. We must consult as much as possible the convenience and inclinations of the recipients of Instruction in selecting the Hours, Days, and Seasons to be allotted to it, and we must reduce to a minimum, the time for which Candidates preparing for Examination may be called away from home; for we aspire to deal with the bulk of the Industrial Population, and not merely with the favoured few for whom Scholarships or other exceptional advantages can be provided.

The studious Season at an Industrial Institute in the country, may be said to last from the middle of September, or the beginning of October, to about the beginning of May. During this time, and in some instances beyond these limits, a part of the Saturday afternoons might be spent in Drawing or Modelling, Music, Drill, or the like. At all events week-day evenings may be considered available from 6 or 7 to 10, that is to say for the holding of at least three consecutive Lessons in each Class Room. Thus generally speaking as far as Time is concerned, Artisans will have no difficulty in preparing themselves at the Industrial Institute, during their Apprenticeship, for a Minor Examination to be passed after its close; or should facilities be denied them during that period of their career, it is to be hoped that they may be able to make up for the lost time in the first year or two of their Journeymanship.—The time that may have to be spent at a Technical College for simply passing the said Minor Examination, will be a mere trifle, and any amount of previous stay there that may be necessary for completing their Preparatory Studies, ought not to exceed a few weeks.

The Technical College has been shown to occupy a half-way position between the Industrial Institute and the Technical University (p. 54), uniting certain features of both. Its preparatory knowledge for Major Students will, except in being a Grade higher, resemble that offered to Minor Students at the Minor Institute.

On the other hand, it will approximate to the University, in introducing Technical Instruction supported with Laboratory and Workshop Practice.

As regards the question of Time, we must separate the Students into two lots :—firstly, those living at hand and devoting to study the *leisure time* of their weekday evenings and Saturday afternoons, and secondly those prepared to devote for a short period their *whole time* to study. As regards the *leisure time* Students, the College will only take matters a little more in earnest than the Institute, packing somewhat closely with Lectures and Lessons, the long Winter evenings,(1) and opening Laboratories and Ateliers on Saturday afternoons during, and even beyond a regular Session from Michaelmas to Midsummer, but all along carefully avoiding to interfere with the day's occupations. It will, on the contrary, be in the interest of the *whole time* Students, that everything should be devised for shortening the period of their Studentship, by making the best and most compact use of it; and this will naturally produce an approximation to University arrangements.(2)

There is however one feature in which the College will as a rule differ essentially from the University. For reasons already explained, its operations will in some directions cover a broader area; for the means of culture which it will offer to Artisans will extend beyond the limits of their technical necessities, and the

(1) The example of the City of London College, well known for its judicious and highly successful organization, proves that as many as 4 classes for different batches of Students may be held consecutively in the same room between 6 and 10.

(2) Whether or not Students of this description should pass an Examination before admission to the benefits of the College, and be then called Matriculated Students, may for the present remain an open question.

Public at large will be admitted to its benefits (p. 96). In short, it will in most localities be the all in all of Science and Art, Practical Knowledge, and elevating or invigorating Recreation. Even the Summer will to some extent, and especially on Saturday afternoons, be a season of progress in Laboratory Practice, Art, Music, Drill, Gymnastics, Swimming, &c. Of course the relative proportions of the technical and popular, didactic and recreative elements in a Technical College, will vary according to local circumstances, and there would be no harm if in certain instances other titles were preferred, such as Athenæum, College of Arts, Philomathic Institution, and the like. I attach little importance to uniformity of names, provided everywhere the same aims be pursued by concerted and concordant means.

In reference to the studies to be pursued at the Central Technical University, we may assume that the scholastic year, beginning at Michaelmas, will have three terms, ending respectively at Christmas, Ladyday, and Midsummer. As for the time required by Candidates for completing their respective Studies, this will vary so much according to the nature of their Pursuits, and the amount of instruction previously obtained elsewhere, that the loose computation in the following Schedule must only be considered as a starting point for discussion. Only Students belonging to Category B are referred to. Candidates for Minor Certificates who have to complete their studies at a Technical College, will probably give the preference to one attached to a University, if within their reach.

[Sect. 1.] PLACE AND TIME. 431

N.B.—I. I. stands for Industrial Institute; T. C. for Technical College; and T. U. for any Technical University. C. T. U. marks cases in which the Central University should be preferred.

Candidates for	Studies begun at	Studies completed and Examinations passed at	Duration of Stay at the same.
Minor Certificate	I. I. or elsewhere	T. C.	3 weeks to 1 Term.
Major do	T. C. do.	T. U.	1 to 2 Terms.
Superior do	T. U. do.	T. U.	Mostly 3 Terms.
Technological Professorships[1]	T. U. or elsewhere	C. T. U.	1 to 3 years.
Professors for Training Colleges	C. T. U.	C. T. U.	Mostly 3 Terms.
School Teachers	T. U.	T. U.	1 to 2 Terms.

[1] The Technological Professorships here referred to, are those directly connected with the proposed System of Technical Training, and its propagation in the Mother Country and the Colonies.

SECTION 2.—PROPOSED ORGANIZATION OF A CENTRAL TECHNICAL UNIVERSITY.

It is well known that for many years an eminent benefactor of the Industrial Population of the Eastern Districts of the Metropolis, Sir Antonio Brady,[1] was engaged with indefatigable energy in preparing the formation of an East London Museum of Science and Art, designed to be at once attractive and instructive. Through the liberal assistance of his numerous friends, and the willing concurrence of Parliament, he was enabled to effect the purchase of an exceedingly eligible site in Bethnal Green, and its transfer to the hands of Government on conditions that seemed calculated to ensure the People's benefit in a thoroughly educational sense. That such an institution thus situated within easy reach of nearly a Million of Working Men, might become the nucleus of a National University for diffusing Technical knowledge to every part of the Empire, was an idea so feasible, that it naturally occurred to me in writing my "Science for the People."[2] It was equally natural that it should be adopted in 1871 by the Com-

---

(1) Knighted for long services rendered to the Country in an important department of the Admiralty.

(2) See page 114, where a passage relating to this subject concludes as follows:—"I need not dwell on the instructive lessons of Manufacturing and Commercial Industry which that end of London would afford, or on the cheap resources which it would present to a concourse of intelligent and thrifty Working-class Students, or on the satisfaction with which their advent would be hailed."

mittee of gentlemen then associated as already stated (see Introduction) for promoting the establishment of a National Technical University; the more so as there then was good reason to believe that the Government would agree to the scheme. I accordingly prepared a Draft of a constitution for the proposed University, based on this supposition, and though the East London Museum has since been otherwise appropriated, I give it almost textually, as it may serve to point out a kind of arrangement that may be found to answer at some future time either in that or some other similar locality.[1] It must be understood that my suggestions suppose Government to favour and possibly to subsidize the Scheme, but to refuse undertaking its responsible management. Should the latter point be conceded, the whole affair would assume a different, and of course far more promising aspect.

PROPOSED OUTLINES OF A CONSTITUTION
FOR A
CENTRAL TECHNICAL UNIVERSITY.

The Lands, Buildings, &c., to be vested in the names of seven Trustees, representing—one the Committee of Privy Council on Education,[2] one the Board of Trade, one the Council of the Society of Arts, and four the Council of the Central Technical University itself.[3]

[1] A site near Battersea Park was at one time thought of, and that of Horsemonger Lane Gaol in the Borough Road has also been suggested.
[2] Or any other supreme Educational authority existing at the time.
[3] The Permanent Commission on State Scientific Questions proposed by Col. Strange, might also be entrusted with certain powers of appointment, and it would be easy to enlist similarly by addition or substitution, the assistance of the Corporation of the City of London.

Such Trustees to be at the first appointed by the Deed of Incorporation, and afterwards at each successive demise, by the parties respectively represented.

The Permanent trust thus constituted, to be entitled the Central Technical University Trust,[1] and to form a Body Politic and Corporate, with perpetual succession, a common seal, and the various powers, privileges, and immunities usually granted to such Bodies.

The purposes of the said trust to be as follows:—

To institute, develop, and maintain in constant activity and efficiency, a Central University for Technical Training, which may be more briefly designated as the Central Technical University, and which shall be organized for advancing Technical and Commercial Industry in all parts of the Empire,—

> By offering to properly qualified persons at a moderate cost, first-rate facilities for perfecting themselves in the scientific, artistic, and manual attainments proper to the various industrial, technical, and commercial pursuits, so as to afford practical examples of the most improved resources, methods, and management of Technical Instruction.
>
> By training Professors and Teachers for diffusing in all parts of the Empire, among the Working Classes themselves, or among those whose business it is to instruct or direct them, the most appropriate and practical knowledge in the various grades and departments of Elementary and Applied Science and Art; thus raising everywhere the standard of industrial well-being

---

(1) For the motives explained at p. 56, I have altered to "Central" the title "National" which had been adopted in my original draft.

and efficiency, without interfering in political or religious questions.

By testing through the means of Examinations, and recognising through the means of appropriate Degrees, Certificates, Diplomas, Prizes, and other Rewards, the attainments of persons either themselves engaged in technical pursuits, or aspiring to be the instructors of others, in Preparatory or Technical Knowledge.

By so organizing, developing, and displaying the various collections of Specimens, Models, Apparatus, Diagrams and other Illustrations required for purposes of study and useful intellectual culture, as to constitute a popular Museum of an attractive, instructive, and elevating character.

By promoting actively throughout the United Kingdom and the Colonies, the organization and healthy development of Institutions of various grades for instruction in Science and Art applied to Industrial Purposes.

By inducing as far as possible all Educational Institutions of an industrial, technical, or commercial character, to place themselves in relations of voluntary affiliation or association with the said University, for mutual interchange of information, assistance and support, and especially for ensuring uniformity of Principle in Industrial Instruction, and uniformity of Standard in Awards.

By entering into and maintaining correspondence with Societies, Institutions, and Eminent Men in this and other Countries, with a view to obtaining and supplying early and authentic information on all matters having a bearing on industrial and educational progress.

By conferring distinctions on persons eminent for practical knowledge or for services rendered to the Industrial Community.

And generally by adopting and carrying out, or aiding others to carry out, all judicious and practical measures for propagating among the Industrial Populations, knowledge calculated to enhance their technical efficiency, or to ameliorate their physical condition, through affording them useful guidance in the concerns of Daily Life.

The Board of Trust to elect for life a Chancellor or Honorary President of the University, and not less than three, nor more than five Vice-Presidents.

The general Administrative Authority to be exercised by a COUNCIL of twelve, who will be at first appointed by the deed of Incorporation, and afterwards will be yearly elected or re-elected, six by the Board of Trust, and six by a Body of Superior Professors of which the constitution will be presently explained.

The Council will have comprehensive powers for making Bye-laws and Regulations, for appointing the Administrative, Financial, and Professorial Staff, together with any Boards or Committees required, and for defining the duties and advantages attached to such appointments;[1] for fixing the nature and duration of the studies, as well as the Categories of Students to be admitted, the terms of Admission, and the Ex-

[1] Particular care and forethought will be required in constituting the "Board of Instruction," to which allusion has been repeatedly made.

aminations, Degrees, and Rewards: in short for carrying on the whole government of the University.

It will also be incumbent on the Council to see that provision is made for rendering the University Museum or Museums conveniently accessible as well as attractive to the local Public, and that through a correspondence actively carried on with all parts of the United Kingdom and of the Colonies, the important duties of a scientific, artistic, and industrial propaganda are conscientiously discharged.

The Professorships will be of two kinds :—SUPERIOR and ORDINARY. The SUPERIOR PROFESSORS will constitute a recognised Body with certain rights, including as stated above, that of electing one-half of the members of the Council.

The duties of Members of the Board of Trust and of the Council, will be *honorary*, and every possible measure will be adopted for ensuring in every department a judicious union of efficiency and economy.

SECTION 3.—PROSPECTS. PROPOSALS.

Many circumstances concur at the present juncture in favour of a national effort to place Industrial Instruction on a footing equal to that which it has obtained in other Countries. The most important is doubtless the earnest manner in which public attention is now directed towards the subject of Popular Education in general, with a genuine desire to recognize deficiencies, and as far as practicable to remedy them. Thus a new order of things is rising into existence, already impressed with features of methodical improvement, yet still in that transitional and plastic condition best suited for receiving the germs of scientific knowledge which we are so anxious to introduce; germs which properly developed, will unquestionably be a manifold blessing to future generations. But it is equally evident that whilst we prepare the harvests of the future, we must not neglect the urgent demands of the present. Steps must be taken, and that without delay, for placing Science and Art within reach of the Adult Working Classes throughout the Country, in an acceptable form carefully adapted to their wants and to the present state of their comprehension. To create a Central Institution worthy of being the head-quarters of this propaganda, requires no ordinary amount of energy and resources; but happily the patriotic effervescence moves, in favour of Educational Improvement, the wealth as well as the intelligence of the Country,— witness the munificence of a Josiah Mason, a Joseph Whitworth, and other noble-minded pioneers of Industrial Progress. At the same time, as before stated, a praiseworthy devotion to the weal of the Industrial Community is being manifested by many of our ancient Guilds and Corporations, who mindful of the patriotic exertions of their forefathers, feel that the time is come for a general effort to give to the measures originated in various quarters, a substantial, well connected, and progressive form.[1] Nor is the powerful Society specially constituted for the encouragement of Arts, Manufactures, and Commerce, slumbering on its well-earned laurels. Not a year elapses without some new proof of its willingness to assist in rendering effective that

---

(1) Particularly suggestive are the offers of Prizes by the Paperstainers' and Goldsmiths' Companies recently adverted to in the Journal of the Society of Arts.

public demand for Industrial Instruction which it has done so much to originate.

Perhaps the most promising sign of the times is the frankness with which our Working Men themselves acknowledge the deficiencies of their education, and the necessity for being provided with scientific and artistic advantages similar to those enjoyed by their foreign competitors; advantages without which it would be impossible to maintain the advanced position they have achieved in the industrial race. This clear-sighted appreciation of the real state of things may be said to have first dawned on their minds, when Delegates selected from their own ranks visited the Universal Exhibition at Paris in 1867, under the auspices of the Society of Arts, and of the Club and Institute Union, and reported so intelligently on the educational progress of their continental Brethren. It has since been manifested by many tokens of legitimate ambition, but by none perhaps more strikingly than by the formation of a special "Guild of Learning" directed by representative Men alike ready to accept with genuine thankfulness the co-operation of disinterested friends, and determined to put effectively their own shoulder to the wheel.[1]

Now there are two ways in which a Scheme of National Technical Training like the one described in the preceding Chapters, may be carried out. The one is to establish first of all the Central Institution, which like the heart in the human frame, is to transmit a

[1] See the account of the Meeting of the "Guild" held at the rooms of the Club and Institute Union, 150 Strand, and of the proposals liberally made by the Syndicate of the Cambridge University.—I may mention that at a late conference of Miners assembled at Manchester, a resolution was passed declaring "that the time had arrived for providing technical education for the mining portion of the population, and requesting the Executive Committee to draw up a plan for carrying out this purpose."

vivifying influence to all parts of the system. The other is to create by degrees the *réseau* of subordinate Institutions, and so to induce the establishment of the Central one, by making it felt to be indispensable.

If the advice offered some years since by the Society of Arts, and borne out by the opinion of many of our best educationalists, had been adopted, and a responsible Ministry of Public Instruction now existed in this, as in most other Countries, strengthened by the respect and confidence of the Nation, and well supported by departmental Secretaries each eminent in his specialty, we might hope soon to see a Central Technical University occupying the eligible space by the East London Museum which would become its Department of Visual Instruction.[1] Springing at once like Minerva into the fulness of an adult organization, this Central Guardian of Instruction, endowed with the commanding power of National resources, would soon provide the Staff, Literature, and *Matériel* of Technical Tuition, and in concert with the many valuable existing Institutions, it would rapidly cover the Country with ramified Conductors of Knowledge.—Next best to this would be a Central Technical University instituted with the concurrence of Government, by a combination of municipal, corporate, and individual resources like that contemplated in 1871 by the Committee referred to in the Introduction. Unfortunately the opportunity which then seemed so promising, failed through extrinsic causes. The creation of a Central focus now appears more likely to be the *result* than the *origin* of a number of minor creations. I will accordingly proceed to sketch a few of these, and as they must necessarily be in ascending order, I find myself compelled to begin

[1] See Chapter II., Section 8.

with my own humble efforts. I sincerely hope that they may encourage others favoured with better health and opportunities, to attempt what is beyond my reach.

For the reasons explained at length in Chapter III., think myself justified in assuming that the select Rudiments of Useful Knowledge embodied in my Popular Course of 9 Lectures represent somewhat approximately the Grade 1 of the Elementary Division of a General Scientific Foundation, such as is required by the generality of Working Men, and as considerable importance attaches to imbuing them with a notion of the Domestic as well as the Technical advantages which such a Foundation is calculated to afford, I intend, as soon as released from my present work, to proceed with preparing that Course for the press. Everything will be done to secure extensive circulation, profit being left out of the question, and the various devices will be carefully carried out which have been described as likely to enhance the educational value of those future official Text Books of which my Course can only aspire to be the humble forerunner. A bold type and abundant woodcuts will favour home study, whilst special facilities will be contrived for its delivery on the Binary System, either by professional or amateur Readers and Demonstrators; arrangements being contemplated for enabling some London Firm or Society to produce commercially the required Series of Illustrations on an extensive, and consequently cheap scale. Should this step succeed, something will already have been done to make Local Institutes alive to the progress they can effect through their own resources, with some slight assistance from an educational centre.—If my health should allow, the Second or Applied Division of this Grade

Course could be rapidly prepared from the materials in hand, and I would take care so to organize the production of its serial Illustrations of Hygiene, Domestic Economy, and Technology, as to facilitate the introduction of these subjects in Primary Schools, whilst Industrial Institutes would be supplied with Materials equally available for Lecture purposes, and for initiating small Economic Museums. Everything would be done to induce Working Class Students to prepare as far as possible among themselves in their leisure hours, the Illustrations of their Studies, and to open the way for the improvements anticipated from the development of the new system. Adding to this the Course of Industrial Chemistry mentioned in a previous Chapter (p. 198) as intended to represent the level of Grade 2 Studies, and taking further into account the " open-handed Examinations" described in Chapter III. Section 5, and Chapter IV. Section 7, we have partly in action, and partly susceptible of being put in action within a reasonable time, a set of actual examples, which without aspiring to be adopted in the machinery of the future Educational System, may certainly give some notion of the way in which the definitive wheels may be expected to work. Such notions are not without their value; for whilst on the one hand it would be imprudent to adopt on a large scale any untried contrivance, it has on the other hand become only too evident that several bold innovations are indispensable in order to make scientific instruction permeate steadily and systematically through the lower industrial levels. To be convinced of this, one need only sift and carefully scrutinize the complaints constantly echoed through our educational atmosphere — of Science Classes that, thanks to a speech from the Organizing Agent from South Kensington have been initiated with the most promising impetus of vitality, but which have died away into apathy and neglect in a few weeks time,—of small satisfaction given by Class Teachers who from the fact of their having been certificated in Science, have been supposed to be necessarily acquainted with the art of imparting it,—of sublime confusion which has pervaded the minds of Working Men listening to the high-flown Science of first-rate Professors ,—and a score of other forms of dissatisfaction uttered not by the detractors, but by the best advocates of the diffusion of Science among the People.—Now most of the innovations explained in detail in the preceding Chapters, are specially calculated to strike at the root of complaints of this description, and I therefore earnestly hope that they may be not only received with favour, but taken up and pushed forward by persons better qualified for the task than myself. In this hope I am further encouraged by noticing that a similar good fortune has already been vouchsafed to several measures which at various times I have suggested. This may in some instances have been due to my having published my ideas, but I attribute it mainly to the fact that conclusions suggested to one thinker by prevailing circumstances, are likely to arise in the same way in the minds of others. I may mention by way of example two principles for which I tried in vain many years ago to gain recognition in the Society of Arts System of Examinations :—Firstly, that Industrial Candidates should be induced to take up in preference, subjects specially bearing on their pursuits ; secondly, that not any one Science distinct and pure, but special combinations of scientific knowledge carefully devised and brought home to their practical applications, are what our Artisans severally want. I am happy to say that these principles are now becoming recognized in various directions, and have been adopted by Major Donnelly in his Technical Examinations, which only want a few further innovations to become as popular as they are in many respects judiciously organized.

Another example of an analogous kind is afforded by the pains which are at last beginning to be taken to adapt Scientific Examinations, and especially the means of preparation for them, to the convenience of Candidates.[1] It is true there is still at South Kensington a certain disinclination to undertake the responsibility of publishing, for the use of Scientific Candidates, official Text Books imparting the precise knowledge they are desired to acquire. Nor does any detailed Syllabus of Reference of the kind described at p. 204 point out for each subject, the particular Books or parts thereof to be selected from among the many indicated in the Science Directory as more or less available. On the other hand however, I gladly acknowledge that the Examination Syllabuses have now, thanks to the pains bestowed on them by Professors of first-rate eminence, become something infinitely superior in quantity and quality to the vague and scanty ones of former times. With such guidance, a judicious friend may point out to the Student, in the official lists, the names of publications far more to the purpose, as well as far more compact and cheaper, than those he would have had to purchase some years ago. This class of Literature has indeed of late made rapid

---

(1) I have much pleasure in again drawing attention to the progressive spirit manifested by the Indian Civil Engineering College at Cooper's Hill, whose Programmes of Studies are remarkable for method and clearness, and form of themselves excellent Examination Syllabuses.

strides, owing to the laudable emulation which has prevailed among the educational publishers. I could name several cheap, compact, methodical, and yet agreeably written Compendiums, which give intelligent and industrious Candidates a very fair chance of obtaining the coveted South Kensington Certificates. Very different was the case of the Student who some years since, taking up elsewhere for the first time even the vast subject of Chemistry, had a ponderous work thrown open before him, with nothing but a Syllabus of two paragraphs to tell him what he should select, so that in fact he had to trust to chance for making his studies coincide with the views of those who instituted the Examination.

So altered is the policy observed towards Candidates, that the only one of my suggestions in their favour now liable to be considered as startling, is that which I have described as the " open-handed system of Examinations," and even as regards this I am happy to find in a book like Page's " Geological Examiner," so evident an approximation to my QUESTIONARY, that I indulge in a hope of seeing my plan tried ere long by those whose approbation, far more than the practical success already achieved, would stamp it as a feasible and natural way of proceeding.[1]

The preliminaries of a scheme which embraces the intellectual organization of the whole Industrial com-

[1] Amongst the little books prepared purposely for Candidates for the South Kensington Examinations, some have under the name of "Home Exercises," sets of Questions for each division of the work, and this device when carefully carried out, is perhaps more conducive to conscientious study than the one adopted in some instances, of appending to the several subjects, Questions that have been asked concerning them at various

munity, can only be satisfactorily dealt with by *union of Principle in division of Labour*. Whilst I reckon on some of my Educational Friends for assistance in pursuing the tentative development of the above various means for regularizing the diffusion of knowledge, and for testing the results obtained, I trust that others will take up the complex subject of Apprenticeship, as one on the satisfactory solution of which will much depend the future ability of young Artisans to avail themselves of the proposed means of Instruction. It is not necessary that I should go again over the whole of the ground surveyed in the third Section of Chapter I., nor will I on the other hand launch into a review of the various forms of organization by which in other countries the legitimate interests of the Apprentice are placed under the care of a competent and authoritative Guardianship. The devising of such a Guardianship may perhaps be most usefully considered in connection with that important desideratum, the creation of a regular system of impartial and conciliatory Arbitration between Employers and Employed. I may however mention a few points to which the present friends and future guardians of our young Artisans may at all times usefully devote their attention, and which in case the "Guild of Learning" or any analogous association should become sufficiently powerful to have affiliations in all parts of the country, might afford appropriate scope for the exertions of each Local Branch :—

— To note without favour or prejudice, the Masters

Official Examinations. But the more one looks into these devices, the more one feels that something is still wanting to establish a straightforward and mutually satisfactory connection between the mind of the Examiner and the mind of the Examinee. The object of my plan is to supply that missing link.

who have the position and qualifications required for taking Apprentices.

— To afford, when sought, advice to the lads and their parents.

— To facilitate arrangements for payment of premiums by instalments (see note to p. 16).

— To see that all agreements entered into are equitable on both sides, and are equitably carried out.

— To see that proper facilities are afforded the Apprentices for attending appropriate evening classes.

— In localities where the Apprenticeship system proves unsatisfactory or insufficent, to promote the instituting of "Technical Schools," or such other establishments as may supply the best complement or substitute.[1]

— To favour by all available means the intellectual and technical culture of Workmen of all degrees, especially by the establishing and maintaining of Industrial Institutes and the like, on the best principles of efficiency.

In endeavouring to turn to account existing Institutes of various calibres for the purposes of the proposed movement, it will be advisable to look well into their character and resources. In some of them social difficulties would stand in the way, in others the mixture of recreation with study, however desirable in theory, has through various causes led to the ascendency of the

[1] Some light was thrown on the present condition of this question at a Meeting of the Endowed School Commissioners on Feb. 20th, 1874. After a few words on the part of the Rev. H. Solly as to the inadequacy of the present System of Apprenticeship, and the advantages derivable from a suitable appropriation of Apprenticeship Endowments, Lord Lyttleton acknowledged that this came within the Educational scope of the Commission, and stated that Technical Schools had been thus provided at Wakefield, Halifax, &c.

lighter, and the sinking of the weightier element, or on the contrary, didactic prejudices prevail that would prevent Science from assuming that practical and utilitarian character which connects in the mind of the Student the principle with its application.

It would be of infinite value in order to fix the ideas of the public and to keep our own from swerving from the prescribed path, to establish with the least possible delay in some convenient part of the Metropolis, a Technical College with an Industrial Institute attached, respectively designed to serve as standard models of their kind. In this part of our operations I trust we may look for support to the City Corporations, Trades Companies, and other enlightened City Friends, with a confident hope that their zeal for the furtherance of Industrial Improvement will induce them to use the splendid facilities they possess, in a manner worthy of London's industrial and commercial fame.

It may be well also to wake up the spirit of association which, in various parts of the metropolis, has at times been seen on the eve of producing great results in the way of Educational Museums and the like, but which, has mostly spent its energy in the discussion of divergent interests and opinions. A movement of this kind, guided by disinterested and conciliatory principles towards the attainment of a well-defined purpose of unquestionable benefit to the whole Industrial and Commercial Community, will present the rallying ground that was so much wanted, and show the way to make local aspirations thrive in the general tide of intellectual progress.[1]

---

[1] I can lend for perusal to persons interested, a Manuscript Programme of a People's Museum and Institute on educational principles, designed for a part of London yet unprovided.

Having found or established an Institution capable of serving the purpose of a model Technical College, experiments might at once be made with a few Minor and even Major Technical examinations; but we must be cautious in judging by these feeble attempts, of the results obtainable by a Central University favoured with the services of men of eminence, acting in concert with the most important Technical Institutions of the Country, and whose Certificates of Competency would at once command universal confidence. This is one of those instances in which social, like mechanical masses, are not to be moved by any amount of small hammering, but require that their inertia be overcome by applying a ponderous and dignified momentum. For giving our National Industries a rate of progression equal to that of the age, no less motive power will suffice than that of an Institution on a national scale, effectively supported, or still better, actually constituted by the Government, which as Professor Williamson has rightly remarked is the "embodiment of the national will."[1]

---

[1] In his late Inaugural Address at the opening of the Session of the British Association at Bradford, Professor Williamson expressed himself to the following effect:—"Science will never take its proper place among the chief elements of national greatness and advancement, until it is acknowledged as such by that embodiment of the national will which we call the Government."—Among the earliest appeals in favour of a Ministry of Public Instruction, as constituting the best means for securing a legitimate exercise of Government influence on the intellectual development of the People, may be mentioned that of John Scott Russell in the conclusion of his work on Systematic Technical Education. Among the latest may be reckoned that of the National Union of Elementary Teachers, expressed at their meeting of April 1874. Among the most influential I am happy to note the persistent exertions of the Society of Arts, and the Recommendations of H. M. Commissioners for the Advancement of Science.

SECTION 4.—CONCLUSION.

I trust it has been apparent from the whole tenor of this Volume, that the system it advocates is intended to assist and complete, not to supersede, or prejudicially to interfere with, our present educational Institutions. It will benefit the Schools of the People by making Instruction in the Rudiments of natural Science, easy to the Masters, and attractive to the Children. Apprenticeship will be rescued from an ignominious collapse, and restored to more than its former usefulness. Working Men's Clubs and similar meeting places of various descriptions, will as Industrial Institutes, have fresh resources, and a definite purpose to work for, and the same applies to Institutions capable of undertaking the duties of Technical Colleges. Establishments organized for the advancement of any special branch of Technical Knowledge, will receive the most friendly support, and Training Colleges in particular will be the object of devoted attention. The Government System of Art Instruction will receive active and disinterested furtherance, and the arrangements also centred at South Kensington for the advancement and recompense of Scientific Studies, (perhaps the only important educational machinery with which the proposed system might at first sight seem likely to clash) will be found on closer consideration to gain by the proposed innovations.—It is one of the essential features of the proposed System, that it is to give Working Men the benefit of a sound practical acquaintance with just as much Science as they require, without any avoidable technicalities, or any time unnecessarily bestowed on those ever-changing theories and forms of nomenclature which confuse the most practised memories. But as explained in Chapter IV., Section 2, everything will be done at the same time to prepare the Student for encountering those difficulties. For instance the Grade 2 of Chemistry will conclude with information and advice offering him every inducement to undertake more advanced studies with Text Books like those indicated for elementary instruction in the South Kensington Directory. Knowledge of the right sort gives an aptitude and a thirst for more, and there can be no doubt that thousands of Artisans initiated in the enjoyment of practical attainments, who have passed through the ordeal of an Examination, and are probably elated with success, will leave no opportunity unimproved for trying their strength at competitions of a higher standard. Moreover generally speaking the Technical University will only invite to its Examinations, Students whose purpose is best served by an aggregate or *Complesso* of Knowledge formed of selections from different departments, and carefully arranged to constitute a connected whole.[1] The South Kensington Examinations on the contrary, may be said to deal with distinct branches of Knowledge irrespectively of the purpose to which they are to be applied. Whilst then the mission of the one Institution will be necessarily and avowedly utilitarian, let the other unfold the Standard of "Science per se" and invite to its advanced Instruction all who seek the pleasurable exercise of their intellectual faculties, or the honourable privilege of extending the domain of knowledge. It is in this latter sense that I hope to see South

---

[1] Even in the case of "Detached Subjects," so called from not being taken as parts of a connected series, no Candidate will be examined who does not possess at least a Grade 1 of General Scientific Foundation. (See Chapter V., Section 1.)

Kensington acquire an importance worthy of the master mind that saw in its future development the dawn of a new era for England's Scientific renown. Whilst instruction in the Fine Arts will be dispensed, from the first rudiments to the highest finish, and Studios provided on the most attractive terms for the use of Artists of acknowledged eminence, whose influence thus concentrated will tell with vastly increased power on the artistic world, let the several Sciences in the pride of their distinct autonomies, attain to the highest degree of refinement, in appropriate laboratories rivalling the most perfect foreign types in the good taste of their construction, and the completeness of their apparatus. Here we may imagine Men of first-rate eminence, conducting in a luxury of convenience worthy of their fame, the most advanced and delicate investigations; or we may fancy them concocting new theories and nomenclatures, or settling old controversies and starting new ones, in reading and discussion rooms fitted up with every rational comfort, and replete with every intellectual resource. In adjoining Lecture Halls, *Scientia excelsior* sublimely theoretical and aspiring, will be dealt forth from Academic chairs to disciples duly prepared by high mental culture, and will inspire them with the glorious ambition of spending their lives in original research, enriching the realms of Knowledge through the conquest and annexation of the unknown.

All these things will come naturally within the province of South Kensington, and will thrive there all the better for being freed as much as possible from sublunary considerations, and relieved of the burdensome responsibility of conducting, from the lowest Grades upwards, the utilitarian culture of the great masses of the Industrial Community. In taking charge of this responsibility as its legitimate share in the EDUCATIONAL SYSTEM OF THE FUTURE, the Central Technical University will deserve the good wishes of every pioneer of Industrial Progress. Above all, it will by its disinterested and pains-taking energy, deserve the devoted co-operation of all cognate and affiliated Institutions; for it will seek no influence over them which may not be the means of propagating useful knowledge, and securing general prosperity, nor will it claim any power or importance which cannot be made to shed a new lustre on England's industrial honours. It is thus that an appeal can confidently be made to Government, for a patronage which every patriot will applaud, to the Society of Arts, for its support in that career of progress for which its indefatigable exertions have so admirably opened the way, and to Industrial Institutions throughout the Empire, for assistance in sowing broadcast the good seed of which they will abundantly share the produce at harvest time.

APPENDIX.

CONSISTING OF

ADDENDA FOR THE LISTS GIVEN IN CHAPTER IV.[1]

For Lists 3, 4, and 5 (pp. 205, 210 and 215).[2]

Introduction to Natural Philosophy, by Charles Tomlinson. 1/6 (Lockwood.)

Course of Natural Philosophy, by R. Wormell. 12mo. 4/- (Groombridge.)

Physics, by A. Garrot. 7/6 (Longmans.)

An Introduction to Experimental Physics, Theoretical and Practical, by Adolf Weinhold, translated, etc., by B. Loewy. (Longmans.)

Notes on Natural Philosophy, by G. F. Rodwell. Fcap. 8vo. 5/- (Churchills.)

Rudimentary Mechanics, by C. Tomlinson. 1/6 (Lockwood).

First Lessons in Theoretical Mechanics, by J. F. Twisden. (Longmans.)

Principles of Mechanics, by T. M. Goodeve. Text Book of Science. 3/6 (Longmans.)

Elementary Statics, by C. J. Ellicott. 8vo. 4/6 (Deighton.)

Elementary Hydrostatics, by J. Thurlow. Cr. 8vo. 2/6 (Whittaker.)

Elementary Course of Hydrostatics and Sound, by R. Wormell. 3/- (Groombridge.)

Electricity, by Sir W. S. Harris. Weale's Series. 12mo. 1/6 (Lockwood.)

---

[1] As explained at p. 205, this Appendix is intended to comprise works of insufficiently determined level, or of which information has been obtained too late for insertion in the main lists.

[2] Several of the works here given come more or less within the scope of List 2 (see p. 201).

Student's Text Book of Electricity, by H. M. Noad. Post
8vo. 12/6 (Lockwood.)
Elementary Treatise on Magnetism and Electricity, by E.
Walker. 8vo. 15/- (Deighton.)
Frictional Electricity, by Sir W. S. Harris, edited by Tomlinson.
8vo. 14/- (Strahan.)
Galvanism, Animal and Voltaic Electricity, by Sir W. S. Harris.
(Weale's Series.) 12mo. 1/6 (Lockwood.)

Familiar Lectures on Chemistry, by Liebig. Small 8vo. 7/6
(Walton.)
An Easy Introduction to Inorganic Chemistry, by Rev. A. Rigg.
Cr. 8vo. 3/6 (Rivington.)
Introduction to Organic Chemistry, by W. G. Valentin. 8vo.
6/6 (Churchills.)
The Owens College Junior Course of Practical Chemistry, by
Francis Jones, supervised by H. E. Roscoe. (Macmillan.)
Handbook of Chemical Analysis, by J. W. Slater. Post 8vo.
6/- (Mackenzie.)
Qualitative Chemical Analysis, by Thorpe and Muir. 3/6
Text Book of Science. (Longmans.)
Quantitative Chemical Analysis, by T. E. Thorpe. Text Book
of Science. 4/6 (Longmans.)
Systematic Course of Qualitative Analysis, by George Jarmain.
8vo. 1/6 (Longmans.)
Analytical Tables for Students of Practical Chemistry, by J.
Brown. 8vo. 2/6 (Churchill.)

Glossary of Mineralogy, by Bristow. Cr. 8vo. 8/- (Ward
& Lock.)
Treatise on Crystallography, by W. H. Miller. 8vo. 7/6
(Deighton.)
General Biology, by T. C. Macginley. 1/6 (Collins' Elementary Science Series.)
Botany for Class Teaching, by F. E. Kitchener. (Rivington.)
Outlines of Botany, by Dr. Silver. 5/6 (Renshaw.)
Earth, Plants, and Man, by Schow, edited by Prof. Henfrey.
12mo. 5/- (Bohn.) (Bell & Daldy.)
History of the Vegetable Kingdom and its Appliances to the

use of Man and Animals, by Rhind. Royal 8vo. 31/6 (Blackie.)
Elements of Zoology, by M. Harrison. 1/– (Blackie.)
First Steps in Zoology, by R. Patterson. 2/6 (Simpkin.)
The Student's Guide to Zoology, by A. Wilson. 6/6 (Churchills.)
A Manual of the Anatomy of the Vertebrated Animals, by T. H. Huxley. Fcap. 8vo. 12/– (Churchills.)
Manual of Animal Physiology, by J. Shea. 12mo. 5/6 (Churchill.)
Animal Physiology, by Jno. Cleland. 2/6 (Collins' Advanced Science Series.)
Handy Book of Anatomical Plates, by E. Bellamy. 12mo. 21/– (Trübner.)
A Manual of Chemical Physiology, by J. L. W. Thudichum. 8vo. 7/6 (Longmans.)
Elements of Health, by J. Thurnam. Post 8vo. 6/– (Bohn.)
Sanitary Arrangements for Dwellings, &c., by Wm. Eassie. Post 8vo. 5/6 (Smith & Elder.)
The House Owner's Estimator, by J. D. Simon. Post 8vo. 3/6 (Lockwood.)
Food and Dietetics, by F. W. Pavy. 8vo. 16/– (Churchills.)
Milk Analysis, by J. A. Wanklyn. Crown 8vo. 5/– (Trübner.)
Tea, Coffee, and Cocoa Analysis, by J. A. Wanklyn. Crown 8vo. (Trübner.)

For List 6 (p. 224).

Elementary Course of Theoretical and Applied Mechanics, by R. Wormell. 12mo. 4/– (Groombridge.)
Practical Hydraulics, by Thos. Box. Post 8vo. 5/– (Spon.)
A Manual of Metallurgy, by W. H. Greenwood. Vol. 1. 2/6 (Collins' Advanced Science Series.)

For List 9 (p. 245).

The Museum of Science and Art, edited by Dr. Lardner. 6 double vols. 21/– (Lockwood.)

Science and Commerce, by P. L. Simmonds. 8vo. 6/- (Hardwicke.)

A Manual of Industrial Chemistry, edited by J. D. Barry. (Longmans.)

Handbook of Chemical Technology, by Rudolf Wagner, translated, etc., by W. Crookes. 8vo. 25/- (Longmans.)

Workshop Appliances, by P. B. Shelley. Text Book of Science. 3/6 (Longmans.)

Practical Cotton Spinner and Manufacturer, by R. Scott. 12/- (Simpkin.)

On Cutting Gentlemen's Dress, by F. Pickles. 4to. 5/- (Simpkin.)

First Lessons in the Principles of Cooking, by Lady Barker. 1/- (Macmillan.)

Particulars of the following sources of information in Applied Science and Technology, have been supplied by Messrs. Hachette & Co., French Publishers and Booksellers, 18 King William Street, Strand, W.C.

Bulletin du Musée de l'Industrie de Belgique, edited by the Director Eugéne Gauthy. 16 francs yearly.

Dictionnaire des Arts et Manufactures et de l'Agriculture, par Ch. Laboulaye. In 44 parts, 2 francs each.

Fontaine. Description des machines les plus remarquables et les plus nouvelles de l'Exposition de Vienne en 1873. Enriched with 60 folio plates. 35 francs.

Les Applications de la Physique aux Sciences, à l'Industrie et aux Arts, par Amédée Guillemin. Illustrated with plates and maps. 20 francs.

Jacquemart's Illustrated History of the Ceramic Art. 25 francs.

Jacquemart's Wonders of the Ceramic Art, illustrated in 3 parts.

Particularly deserving of attention is the classified and descriptive Catalogue of the "Bibliothèque Lacroix," comprising a comprehensive series of French works, published or in preparation, on the various Industrial and Agricultural Occupations.—May be also had of the above Firm the "Rapilly" Catalogue of Works on Pure and Applied Art.

LONDON:
PRINTED BY WILLIAM CLOWES AND SONS,
STAMFORD STREET AND CHARING CROSS.

www.ingramcontent.com/pod-product-compliance
Lightning Source LLC
Chambersburg PA
CBHW051847300426
44117CB00006B/291